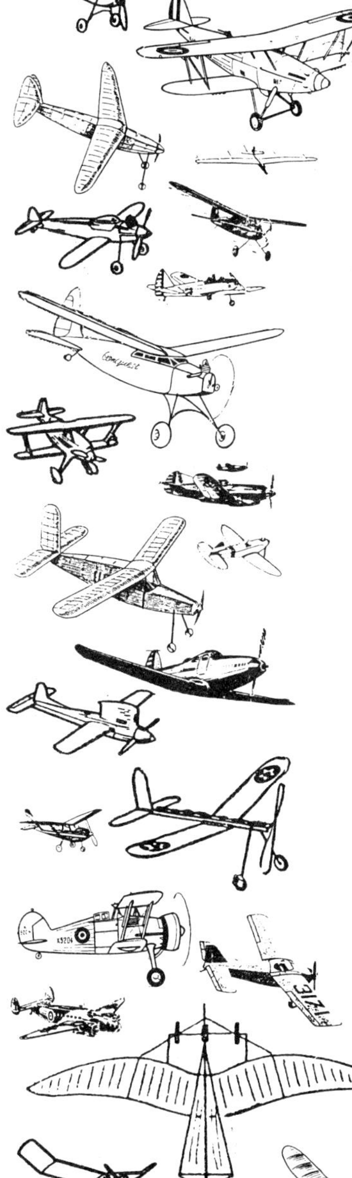

Model Flying – the First Fifty Years

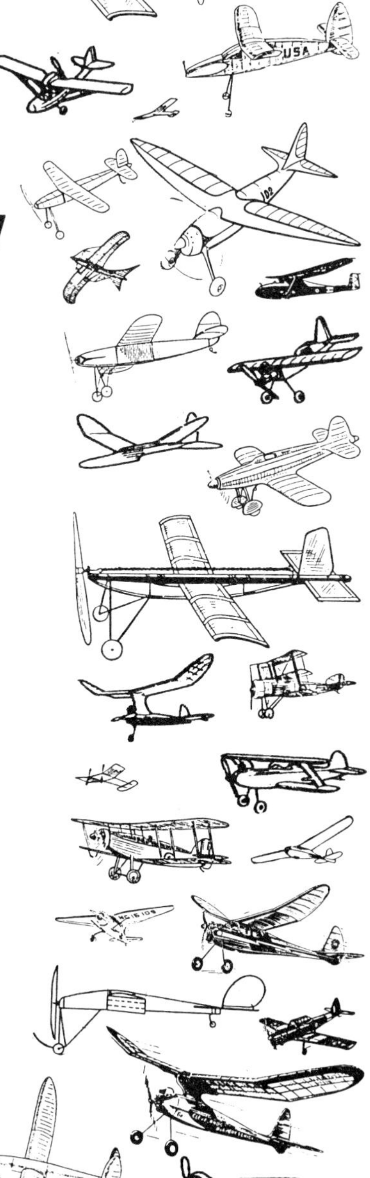

Compiled by
Vic Smeed

ARGUS BOOKS

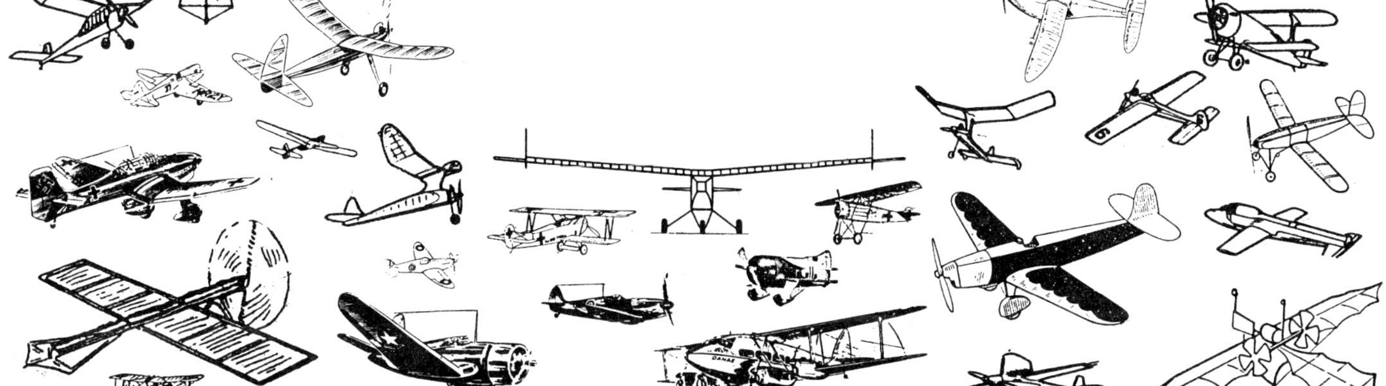

Model Flying — the First Fifty Years

Introduction

A comprehensive history of model flying is no more than a pipedream. Even with an international team of researchers working for several years it would still not be possible to guarantee complete accuracy or that no serious omissions had occurred, while what might be found would fill far more pages than could be produced at a price affordable by the average modeller. A book produced by an individual with constraints on his time and comparatively limited archival access, on the other hand, is bound to reflect his own prejudices and early experiences, to be less than complete, and, more seriously perhaps, to give a view biased towards his own country, if only because 'local' references are easier to come by. However, a book embodying such limitations which is *available* must be better than the ideal theoretical book which never actually appears?

There are people who know far more about developments in various aspects of model flying than this compiler, but fortunately modellers on the whole are perceptive people: show them a drawing of an early model and they can envisage the difficulties of designing, building and flying it. It is the models themselves which form the main focus of interest and which give the 'flavour' of a period. To a large extent they tell their own story and our task has thus been mainly to assemble a selection of illustrations likely to be of interest to the average (as opposed to specialist) modeller.

Most vintage model enthusiasts accept a cut-off date of around 1950 and with the beginnings of model activity for hobby purposes in the post-1900 period it seems logical to take the round half-century as the main theme of this book, with, however, a passing mention of the few earlier models built by pioneers during the previous century. The temptation to devote too much space to the first twenty years or so has to be resisted, fascinating though some of the models are, since in general vintage modellers are more interested in the years after balsa had become the established structural material, say from about 1932/3 onward; apart from early A-frame designs and an occasional *Kinglet* very few present day modellers are prepared to tackle spruce/birch/bamboo/wire and oiled silk construction. Perhaps another way of looking at this is that modern builders are interested in the better-performing models of the 1930s and later, or, if they are men of mature years, in models which they remember from their own early days.

Possibly the most consistent disseminator of information in the very early years was *Model Engineer*, published since 1898, which has always had an international input. Later, aeromodelling was to receive space in *Flight, Aero, Flugsport* and other full-size journals and in more general magazines such as *Practical Mechanics, The American Boy, Mechanix Illustrated* and many others. From the 1920s and 30s more specialised titles — *Model Airplane News, Flying Aces, Air Trails, Aeromodeller, Modellflug, Modele Reduit d'Avion* and *Hobby-boken*, to name only some — in which models were the main or sole content began to appear, and there were many smaller publications circulated in countries where the aeromodelling population was insufficient to support a full-blown magazine.

The Beginnings

Most histories of flight refer to the remarkable foresight of Leonardo da Vinci, but although there is a suggestion that spring-driven machines to his design were induced to rise off the ground, there is no general evidence to indicate that many of his ideas were — or could have been — put into practice at the time. Nowadays it is accepted that the first successful heavier-than-air machines were the work of Sir George Cayley, 'the father of aeronautics.' He was flying model gliders around 1800 and a man-carrying glider in 1809 (scaled up from models) and produced a remarkable three-volume work titled *On Aerial Navigation* in 1809.

Cayley established the need for cambered wings (or aerofoils), incidence angles, fixed wing and stabilising surfaces with movable controls, dihedral and many other basics now taken for granted; he even used a rotating test rig to measure aerodynamic forces. Among other of his inventions were wire-spoked wheels and caterpillar tracks, but his main life work was to develop hot air engines in a search for a lighter source of power than the steam engine to make flight practical, and in this success eluded him.

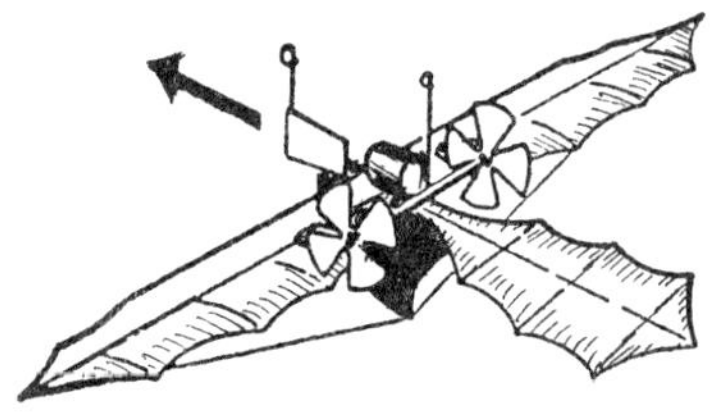

An impression of the Henson/Stringfellow model of 1848. The ¾in. bore × 2in. stroke single cylinder engine received steam at 100p.s.i. from a simple pot boiler.

The first engine-powered flight was achieved in 1848 by a 10ft., 8lb. model. built by William Henson and fitted with a lightweight steam engine by John Stringfellow, in the main shop of a disused lace factory at Chard. Stringfellow subsequently exhibited a 12ft. triplane at the Crystal Palace in 1868, but apparently the models lacked power and could not sustain flight for more than a few seconds; the longest flight of the 1848 model was 120ft.

Outdoor free flight with a power model capable of remaining airborne was *probably* first achieved by a Frenchman, Felix du Temple, in 1858, and it has also been indicated that a compatriot, Pierre Julien, flew a successful rubber model in that year. Rubber as a power source was, incidentally, another of Cayley's proposals. There is no doubt that Alphonse Penaud was achieving success from 1871, for he sold several hundred ready-to-fly models. Confusion surrounds his original model, the *Planophore*, which is usually shown with virtually flat wings and tail and no fin, following an early artist's impression, but another source suggests sharply turned-up wing and tail tips, which accord with a simplified drawing by Penaud in an article in *L' Aeronaute* in 1872. Penaud also refers to a small vertical rudder and some later designers employed 'the Penaud cruciform tail.' Possibly more than one model has become involved, since Penaud was prolific and ingenious. He produced quite a modern-looking rubber ornithopter in 1872 and in 1876 patented a design for a full-size twin-prop monoplane of very clean appearance, though it was never built.

Another Frenchman, Victor Tatin, was successful in 1879 with a compressed air model which would make short flights R.O.G. Many enthusiasts of many nationalities were exploring possibilities at this time, but documentation was often scant or superficial. Frank Wenham, a founder

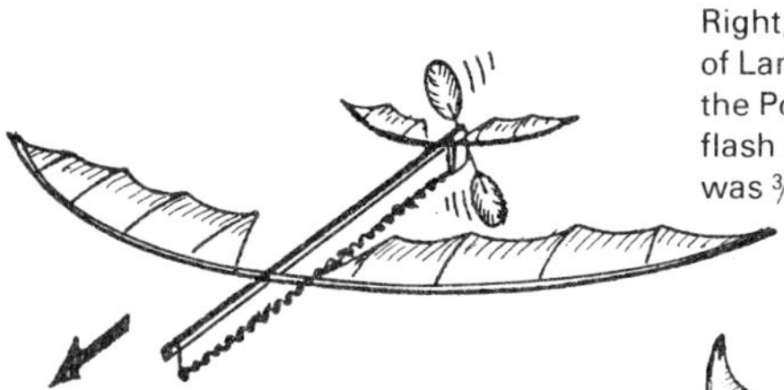

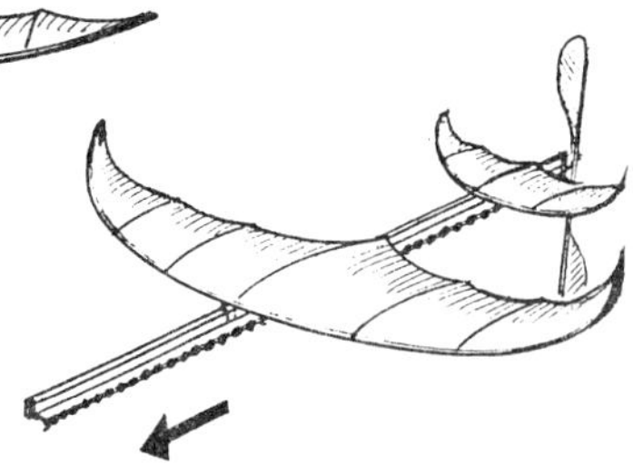

Two interpretations of Penaud's *Planophore*, the upper being the 'accepted' version, but without a fin it would be unlikely to fly. Span was 450mm (17.7in.) and weight only 16gms (0.56oz.)

Right, launched from a houseboat, one of Langley's two 1896 models flies over the Potomac river. These models used flash steam generators. Longest flight was ¾ mile in 1m. 45secs.

MODEL FLYING — THE FIRST FIFTY YEARS is published by

ARGUS BOOKS LTD.
1 Golden Square
London W1R 3AB

ISBN 0 85242 931 2

Phototypesetting by Multiform, Cardiff
Printed and bound by R. J. Acford, West Sussex

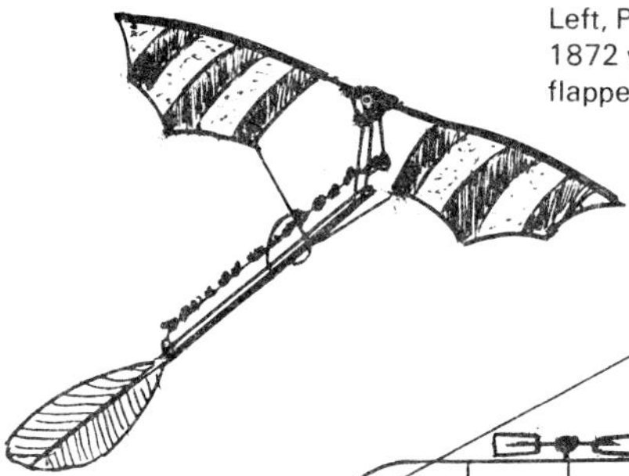

Left, Penaud's ornithopter of 1872 was very like present-day flappers.

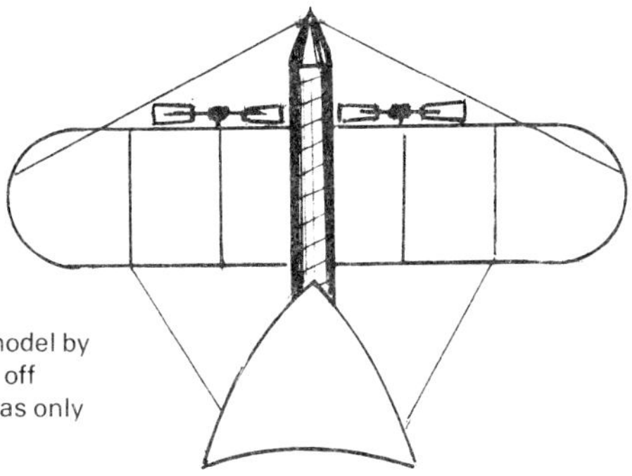

Right, compressed air model by Tatin, 1879, would take off ground but motor run was only about 5 seconds.

in 1866 of the Aeronautical Society of Great Britain (later the Royal Aeronautical Society) pioneered wind tunnels and influential work came from, among others, D. S. Brown (tandem wings), H. F. Phillips (wing sections — you've heard of a 'Phillips entry') and Australian L. Hargrave, whose box-kite developments formed a basis for many subsequently successful biplanes. He also built the first rotary petrol engine. Sir Hiram Maxim was flying tethered steam models in the 1880s.

The first documented free flights under power were those made in America by Professor S. P. Langley in 1896, who is believed to have been influenced by Brown's detailed accounts of his tandem-wing experiments. Langley, an astronomer and secretary of the Smithsonian Institute, spent ten years developing his *Aerodrome* concept and eventually achieved flights of up to 1400 yards with two 6ft., 25lb. steam-powered models. Later the U.S. Government asked him to build a man-carrying machine and he collaborated with a petrol-engine pioneer, Charles Manley. A quarter-scale petrol model was flown in 1901, the first-ever 'gas model'!

In October and December 1903 the full-size machine twice failed to launch from a houseboat top and there has been controversy since. Fouling of the launch gear was the likely cause, but a flight would not have been without problems since there were no movable controls on the machine. Failure to provide any form of control (other than pilot weight-shifting) or to realise that it would be necessary to learn to fly were two recurrent blind spots in the early days, though Cayley had stressed these two essentials.

The 1903 Langley attempts were overtaken in a matter of days by the Wright brothers making history at Kitty Hawk. Again, extensive work with models and gliders preceded their success and their story is well known. However, radio modellers may not appreciate that the Wrights were the first to use coupled aileron and rudder, developed on gliders in 1902 to overcome the drag-produced yaw in the opposite direction to the required turn.

Most early pioneers worked with models and in later years nearly all famous makers readily acknowledged that they had learned the basics from building and flying models, especially in the 1900-1914 period. One famous name, for example, won the top award at a competition organised by the *Daily Mail* in 1907 with prizes totalling £250 (something like £20,000 today), the contestant being A. V. Roe, later Sir Alliott Verdon Roe.

This competition provided the first major model aircraft article in *Model Engineer*, although there had previously appeared two model airships, plus reference to Zeppelin and other full-size airship notables, as well as Hiram Maxim's and Santos Dumont's aeroplanes and, in 1904, a quote from a Philadelphia newspaper describing a mains-powered electric aircraft which apparently made one successful flight indoors, for which the builder was unprepared; the model broke up when it hit the end wall.

The following drawings and snippets illustrate development but any indication that something was the first of its kind should not be taken as incontrovertible; in many cases the origin of a practice would be almost impossible to pin down. There are bound to be omissions too, often because of the difficulties of getting usable reproductions without redrawing. All that can be hoped for is a general look at what sort of models were popular when, and 'popular' is a word chosen with care.

As with *Fifty Years of Aeromodeller*, reproduction cannot be perfect or even, at times, reasonably good without involving processes which would make the price of the book prohibitive. This is compounded by having to work from rare bound volumes without damaging them and by the quality of paper and, indeed, reproduction in the original publications. The aim has been to provide as large a variety of models as possible at a size which can be scaled up and, generally, with annotations legible, even if a glass has to be used, or at least with a scale or a major dimension given. Access to an enlarging photocopier will be found helpful.

With similar limitations the previous book enjoyed a remarkably enthusiastic reception and it is hoped that this present work will prove of equal interest and value. If so, it is thanks to the publishers of all the source material and to the designers and enthusiasts of the past whose work has been included in an attempt to record some of the trials and triumphs and the evergreen spirit of model flying in its developing years.

A. V. Roe entered several models, that on the left winning the second prize with several steady flights, the longest some 30 yards. The model was rubber-powered, 9ft. 6in. span, of cane and birch construction covered with tracing paper. Roe's models appeared to be the only ones with springy (bamboo) landing skids.

Above, H. Burge Webb's machine, about 5ft. long, had a built-in ratchet winder at the tail. It impressed by being the only entry capable of climbing. Left, third prize went to the simplest entry,which flew into the crowd after a 108ft. flight. Clockwork powered, by W. F. Howard, who flew two identical models — first use of a reserve?!

Exhibition and Practical Trial of Model Flying Machines.

The *Daily Mail* put up prizes of £150, £75 and £25 for models at the same time as its famous £10,000 full-size prize offer. There were 128 entries exhibited for a week in April 1907 at London's Agricultural Hall, with trials at Alexandra Palace on the following Monday, where over 3,000 spectators paid to watch. The judges did not award the first prize.

Model Engineer May 2, 1907.

Left, another of A. V. Roe's models, a 10ft. 6in. span tandem wing, referred to as 'Langley type', which he did not fly. Note triangular motor housing used on both models illustrated. Right, ascensional screw machine or 'hélicoptere' by A. Chubb, on its starting platform, was one of the rare petrol-engined entries but failed to make an official flight.

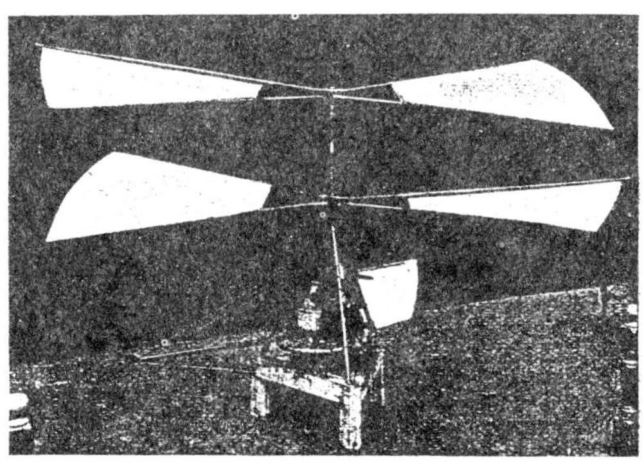

1903-1909

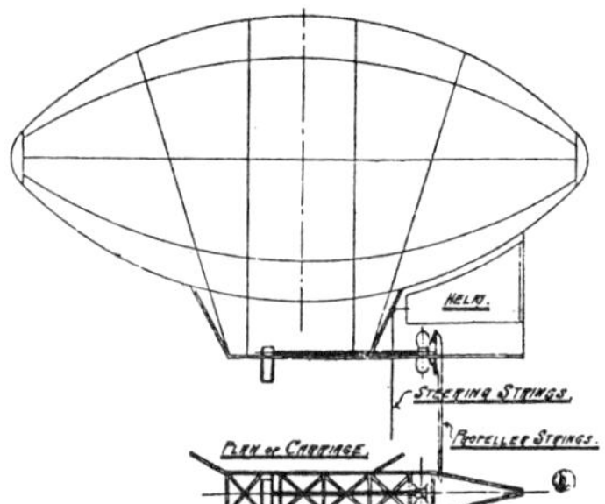

A Design for a Model Dirigible Air Ship.

This 'dirigible air ship' by D. Shannon (*M.E.* March 1903) is 3ft. 6in. long by 1ft. 9in. diameter with an 8-panel tissue envelope varnished with boiled oil, intended for hydrogen. Flying weight is 5oz. and propulsion clockwork, made from parts of an alarm clock and aluminium sheet. The propeller is 3in. dia., geared 200:1 giving 3mins. run at 300r.p.m. and it is driven via a long flexible shaft. A double string held by the operator allows the prop to be angled upward or downward thus giving control over climb and dive.

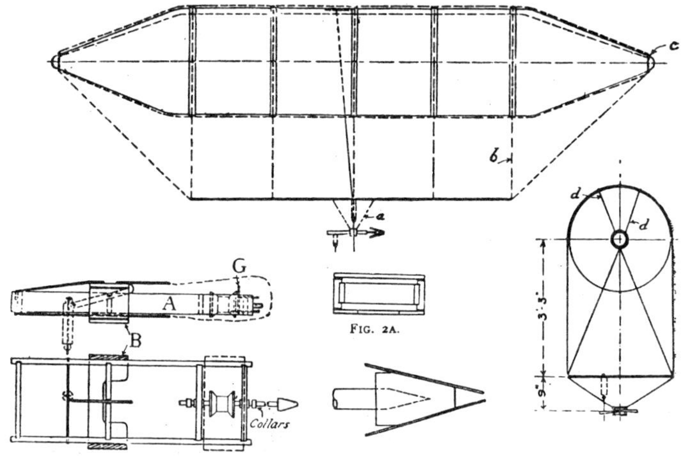

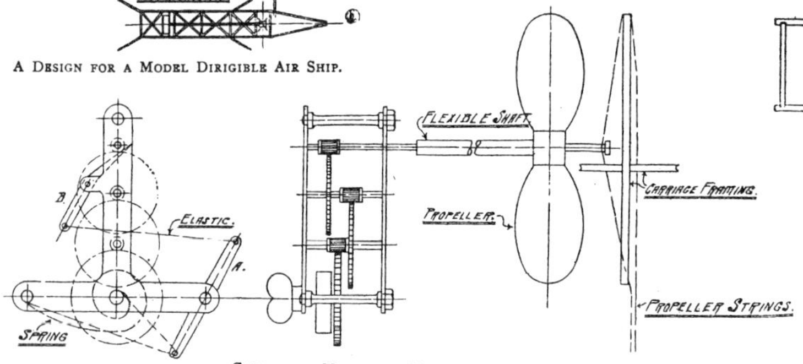

Clockwork Motor for Model Air Ship.

Also in 1903 (*M.E.* April) was this 'model flying machine' with a 15ft. × 2ft. 6in. envelope of waxed tissue, using coal gas and having a flying weight of nearly 4lbs. A rearward C.G. allows the machine to fly upward and forward and on pulling a string carriage A springs forward, moving the C.G. forward and releasing the propeller. The 12in. prop is spun by a 2oz. weight falling, attached to a line wound round the cotton reel shown; thrust is only needed to start the machine on its descending course. A separate line controls a gas release valve for use if needed. Designed by F. B. Scoffham.

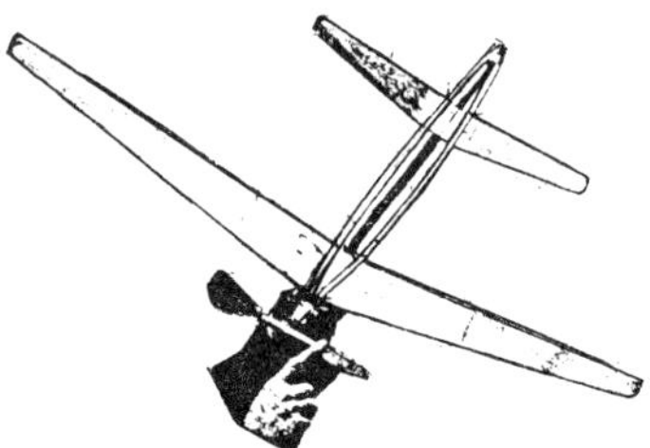

April 30, 1908.

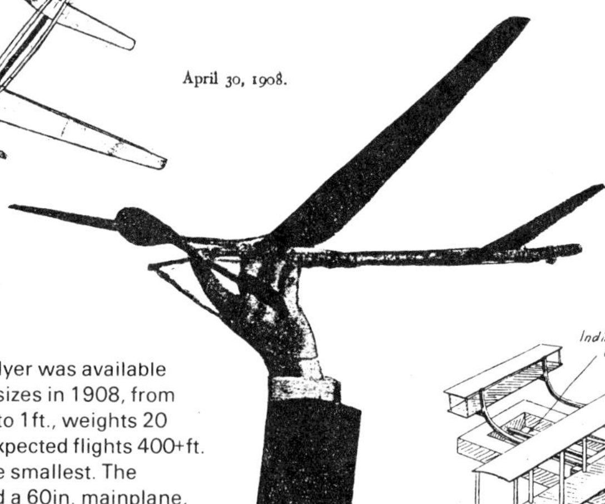

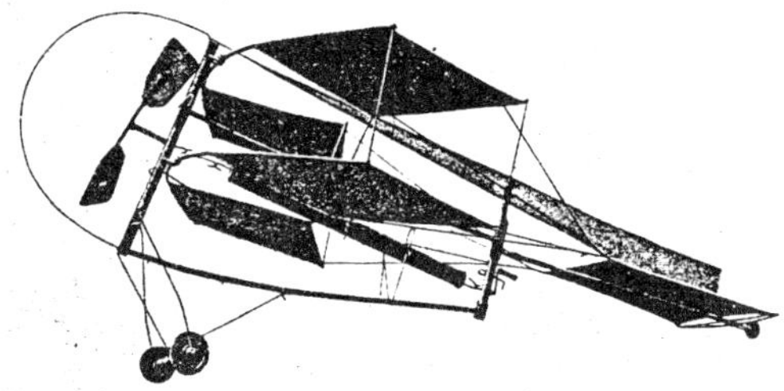

Commercially available were the up-market guaranteed-to-fly biplane above and a similar monoplane, by The New Things Ltd.
Sizes ranged from 28 to 48in. length. This firm offered a scale Wright biplane with a catapult-type launch pad, shown left.

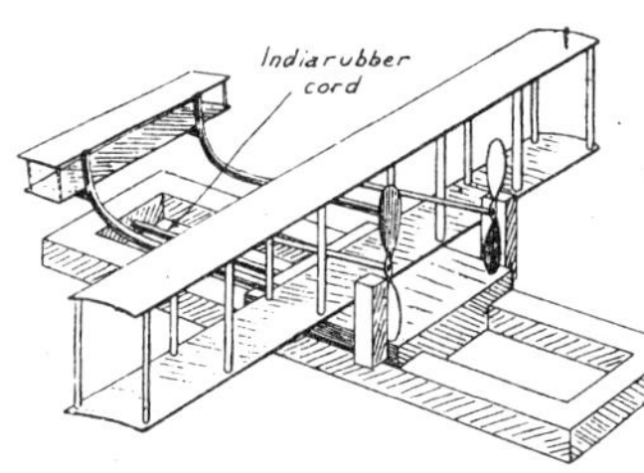

The Clarke Patent Flyer was available ready-to-fly in four sizes in 1908, from 3ft. in length down to 1ft., weights 20 down to 1oz., and expected flights 400+ft. to about 90ft. for the smallest. The prototype model had a 60in. mainplane, 4½in. chord down to 1½in., and all planes were cambered sheet spruce; power was 12 strands of ⅛in. *square* rubber.

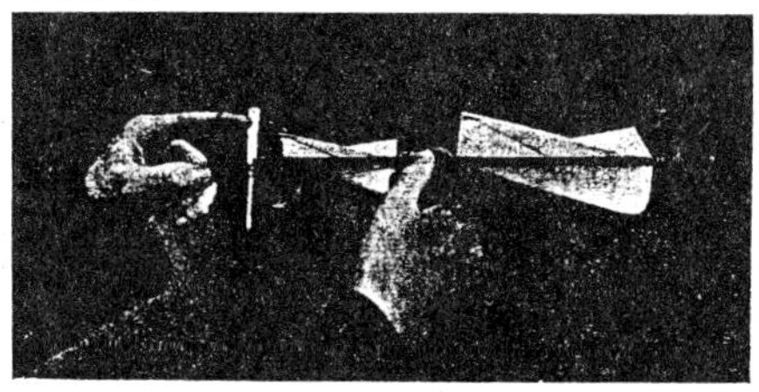

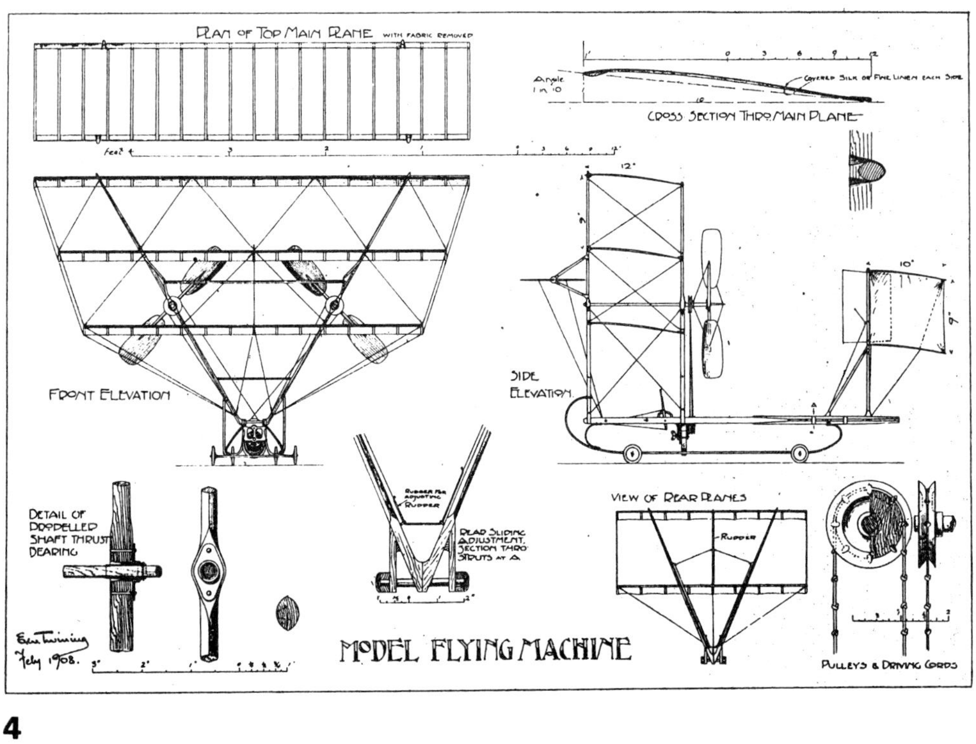

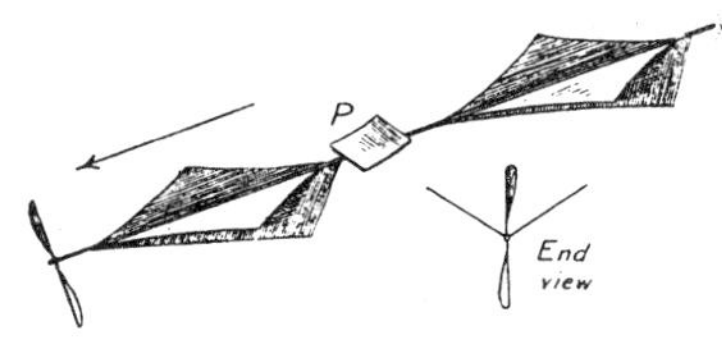

'Model flyers' offered by W. Howard in 1909 included the dart-style rubber model above (the square plate P is a movable trimming device) and the simple paper chuck glider, designed by Dr. Hele Shaw, F.R.S., below, which was marketed in boxes of 6 or 12 'for a few pence.'

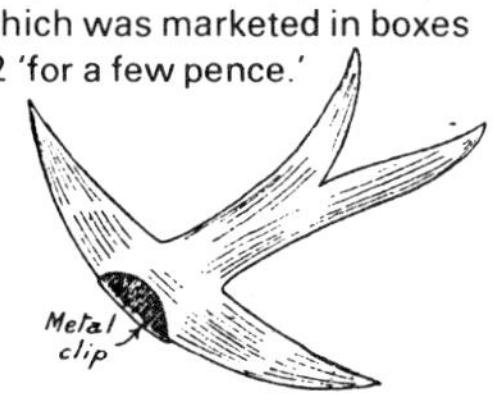

After experiments with gliders and propellers in the 1890s, E. W. Twining returned to model aviation in 1904 and the design at left, drawn in 1908, was published in June 1909. Intended for clockwork power; the motor could be mounted further forward as necessary to achieve balance 54in. span

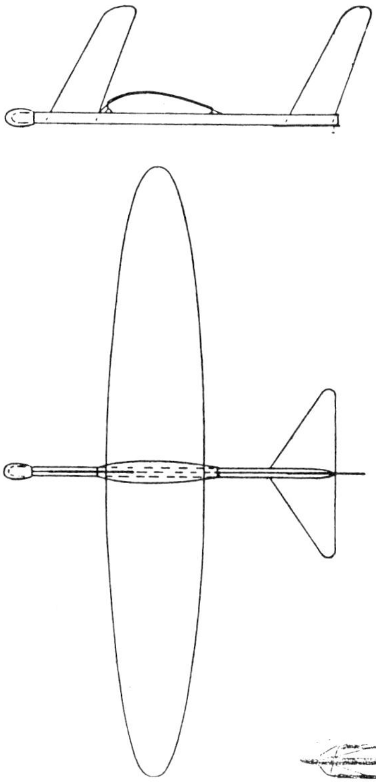

AERODONETICS
CONSTITUTING THE SECOND VOLUME OF A COMPLETE WORK ON AERIAL FLIGHT

BY

F. W. LANCHESTER

Another brilliant mind applied to flight theory was that of Dr. F. W. Lanchester, whose experiments in the 1890s led to the publication of *Aerodynamics* in 1907 and *Aerodonetics* in 1908. He established many of the principles on which subsequent work on airflows was based and coined words such as 'aerofoil' which are still in everyday use. His model experiments ranged from paper gliders dropped from a fixed height indoors (that on the left was 4½in. span and 0.25gm weight, up to the 26in. 12gm model on the right) to 1½lb. gliders catapulted at 60m.p.h. out of an upstairs back bedroom window, achieving flights of as much as 280 yards against winds of up to 30m.p.h., all carefully timed, documented and analysed; he termed the models 'aerodones'. Subsequently he built and successfully flew the rubber model below. Lanchester was more generally known as a car designer and manufacturer and inventor of such items as epicyclic gearing, overhead valves, live axles, cantilever springing and many other engineering developments.

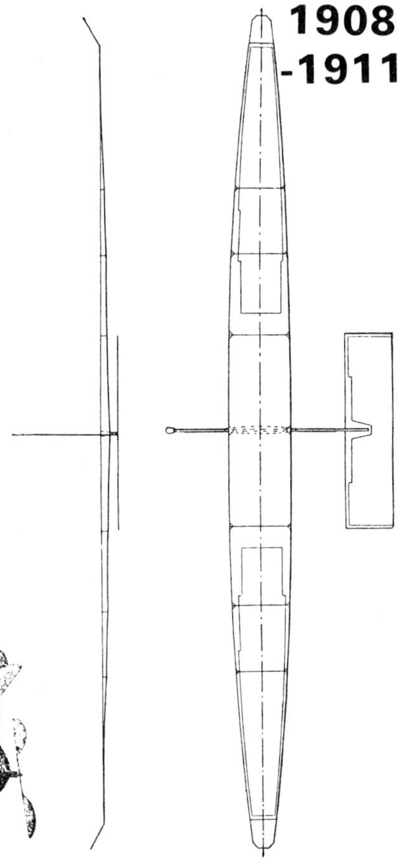

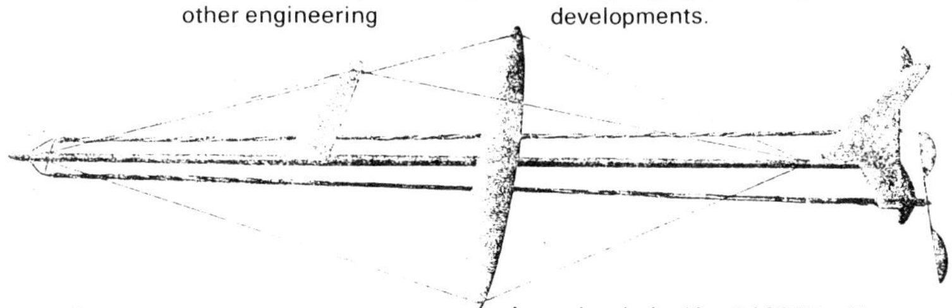

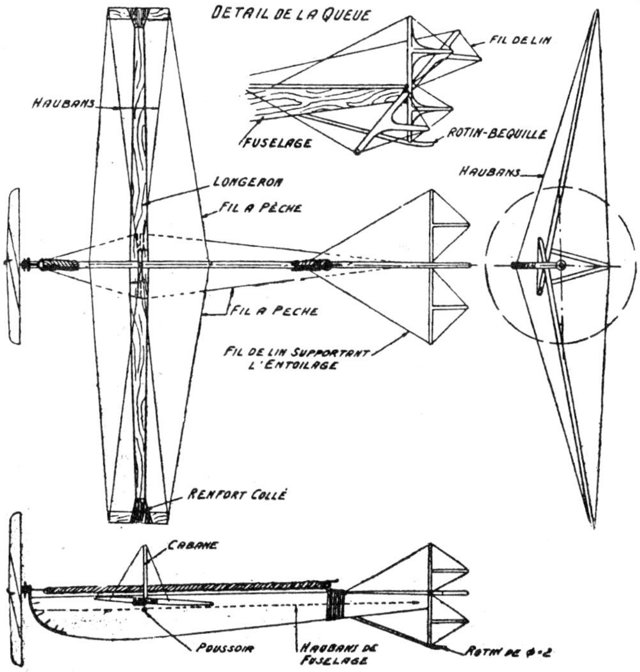

An early winder (April 1909) by H. Burge Webb used belt drive and a pair of pins between which the airscrew was engaged. The model shown was one of a range of single or twin pushers he manufactured; the 20in. span one would fly about 100ft. A sliding lead weight on the 'bowsprit' was adjustable and essential 'to compensate for alterations in the density of the atmosphere'.

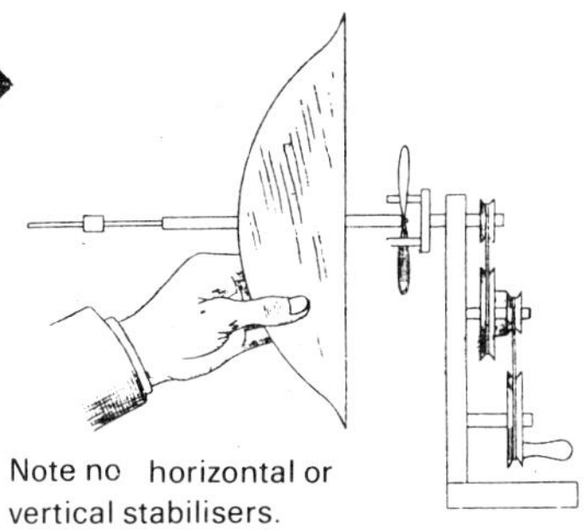

Antoinette-influenced model 40-45cm span by early French modeller Jean Delhaye had lots of dihedral and used a commercial Valentin propeller. Note fishing twine wing l.e. and t.e. and similar use of thread for tail surface outlines. The model made 'many good flights'. (Published M.R.A. 1950).

Note no horizontal or vertical stabilisers.

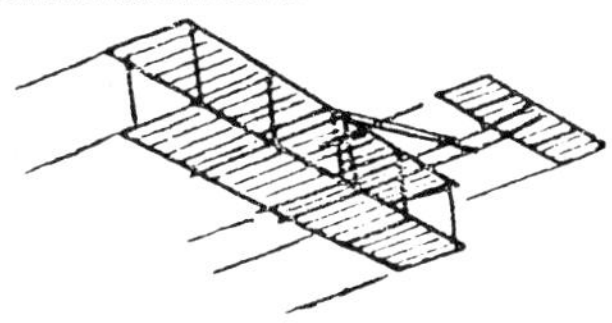

Pioneer Belgian modeller A. Watteyne began in 1908 with gliders of various configurations and built his first rubber-powered model (below) in 1911. Absence of fin and no dihedral must have made trimming tricky! (Published M.R.A. 1950).

From *Flight*, July 1912:

The first model aeroplane organisation in America was started in February, 1907, and was known as the Junior Aero Club of the United States. The first public exhibition of models that attracted any attention was held in New York, December, 1908. There had been prior to that a model flight contest. The first outdoor contest was held November 6th, 1909, the winning model travelling a distance of 92ft. The year 1910 ended with Percy Winslow Pierce winning the 'A. Les Stevens' Year Cup' with a flight (rising from the ground) of 22ft. This contest was held indoors; soon afterwards permission was obtained to fly models in Van Cortlandt's Park, New York, and on the Polo Ground, where every Saturday since (weather permitting) contests have been held. Percy Pierce is today the all-round champion of America (including Canada). The majority of the clubs are organised by students of high schools, and by a few of the Universities throughout the United States.

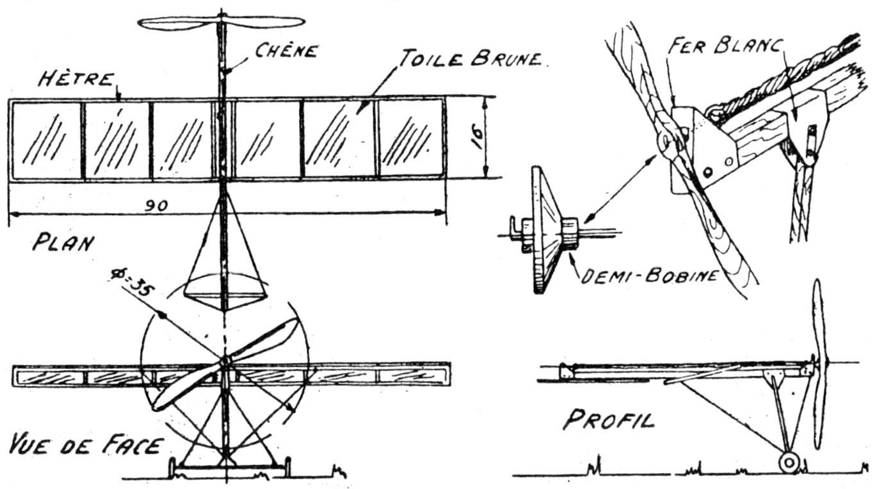

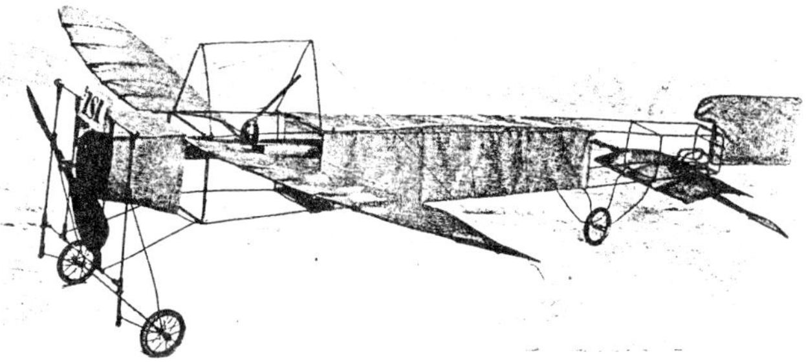

Two prize-winners at the 1909 Model Engineer Exhibition. First prize went to the modified Bleriot above left, by H. H. Hammond, and third to Capt A. H. W. Norgates' Bleriot, above right.

Suggestions re Model Aeroplanes.

To the Editor of *The Model Engineer.*

Dear Sir,—You often get readers asking for a small power plant for model aeroplanes. I have been experimenting with a slide valve engine, driven by carbonic acid gas derived from sparklet bulbs, and think that, with a bit of contriving, one could get a little power that way, as the bulbs contain liquid CO_2. I have made several model aeroplanes, not very successful I fear, and find that a splendid way to strengthen the woodwork is to varnish it well with celluloid varnish, made by dissolving celluloid in a mixture of 2 ozs. acetone, 1 oz. amyl acetate, 2 drms. glacial acetic acid. This makes a fine varnish. Being a chemist by trade, the long hours do not give me much time to do anything.—Yours truly,

G. H. Wiseman.

January 13, 1910.

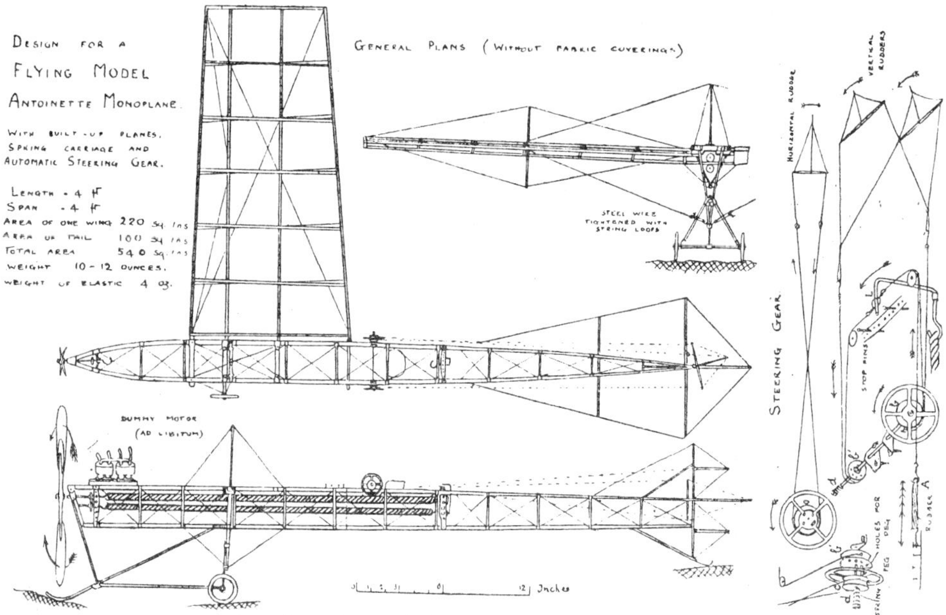

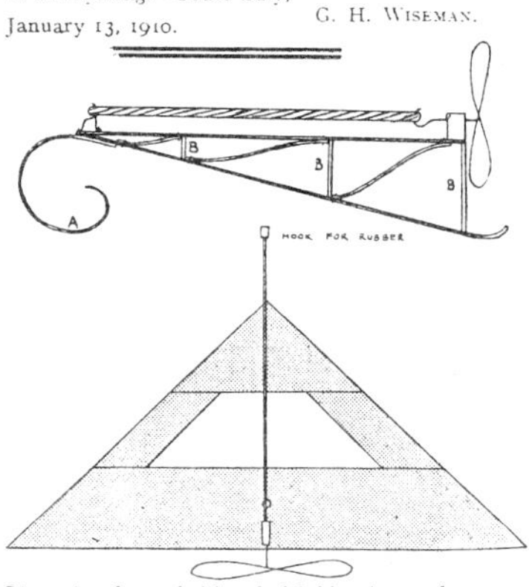

Sketch of model by A. V. Hughes of Canterbury, May 1910, which flew well (510ft.). No dimensions given: 2ft. span or less at a guess — the wing edges were ⅛ × $\frac{1}{16}$in. birch, silk covered, and the curly skid ¼ × $\frac{1}{16}$in. birch. Flying surfaces fit under upper frame member with no incidence.

Ingenious winner of a design competition run by *Model Engineer* in 1910 was this scale Antoinette by R. Koechlin, a Swiss reader who was confined to bed and drew the plans lying down. The most notable feature is the auto-rudder/elevator arrangement, operated by the pull of one of the geared rubber motors, to change trim for power on-power off. A light rubber tensioner (A) ensures that the operating lever (L) returns to the glide setting; rudder offset is controlled by adjustable stops on a cross-bar, elevator by a clever use of a sprung pulley with a locking pin engaging in one of several holes in its drive pulley.

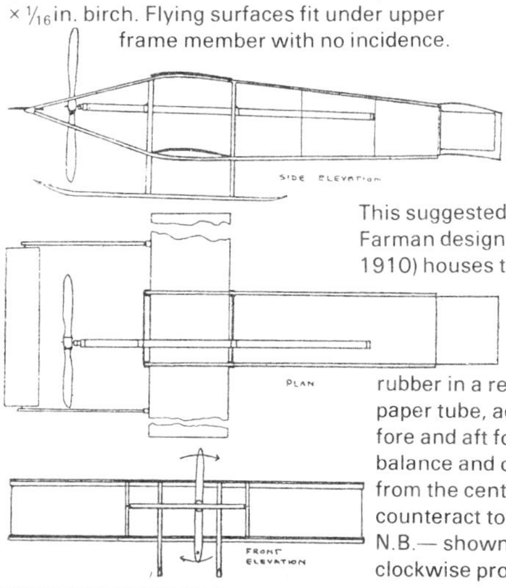

This suggested Farman design (May 1910) houses the rubber in a removable paper tube, adjustable fore and aft for balance and offset from the centre to counteract torque. N.B.— shown for clockwise prop!

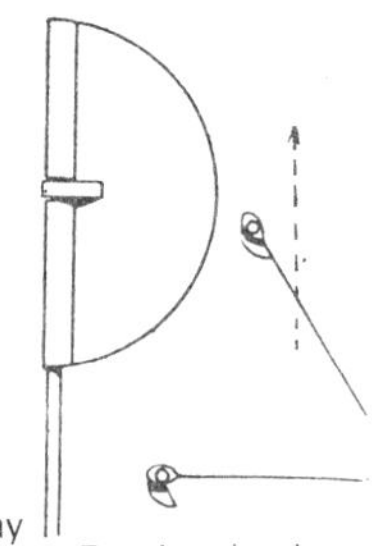

Freely- pivoting feathering prop and ratchet free-wheel prop, also 1910.

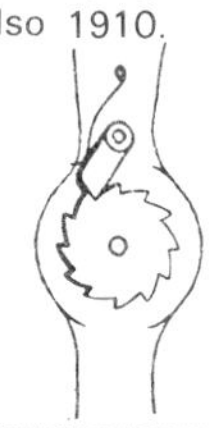

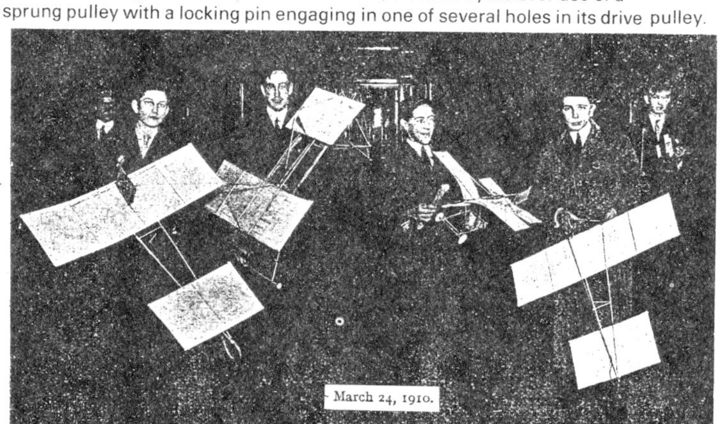

Model Aeroplanes in New York.

Model aeroplane building in America seems to be making very rapid progress, judging by the report of a recent competition held by the Aeronautical Class of the Y.M.C.A. at the Twenty-second Regiment Armoury. According to the particulars which have reached us, Dr. Carleton Dederer, with his odd darning-needle tandem monoplane, set a new record at 200 ft. for a straight-away flight, and his aeroplane raised itself from the floor, and did not have to be tossed into the air to get its start.

W. Morrill Sage ran a consistent second with a remarkable miniature model of the Wright biplane. With all the graceful lines of the Wright model, and less than 2 ft. across the planes, the tiny thing got off the floor under its own power and flew 98 ft.

Dr. Dederer and Mr. Sage were the chief competitors in the class for men. None of the flying machines brought out by the other grown members of the Aeronautic Class were able to make any sort of a showing.

The boys' class did better. Master Wilson Marshall, who has won two of these meets with his French flying machines, found himself strictly up against it on this occasion, finishing far outside the money.

Master Fred M. Watkins, with a machine of his own model and his own make, sailed away with the first prize. Watkins had one of the largest models in the contest, and it has been observed that the large models do not often fly well; but, driven by two strong ropes of rubber bands, the big Watkins monoplane, which is like and yet unlike the Blériot model, flew 126 ft. before it came to the floor.

Percy W. Pierce took second place with a monoplane which went 87 ft.

There were more than thirty flying machines on the floor when the tests began. One boy had a monoplane less than 10 ins. long, which would have won had the boy been able to fly it in a straight line. It persisted in circling the middle of the armoury.

By far the best toy model yet seen is the one handled by Dr. Dederer. It is a modification on the Langley idea.

LONG-DISTANCE FLYING MONOPLANE.

WINNER OF THE GAMAGE CHALLENGE CUP & GOLD MEDAL 1910.

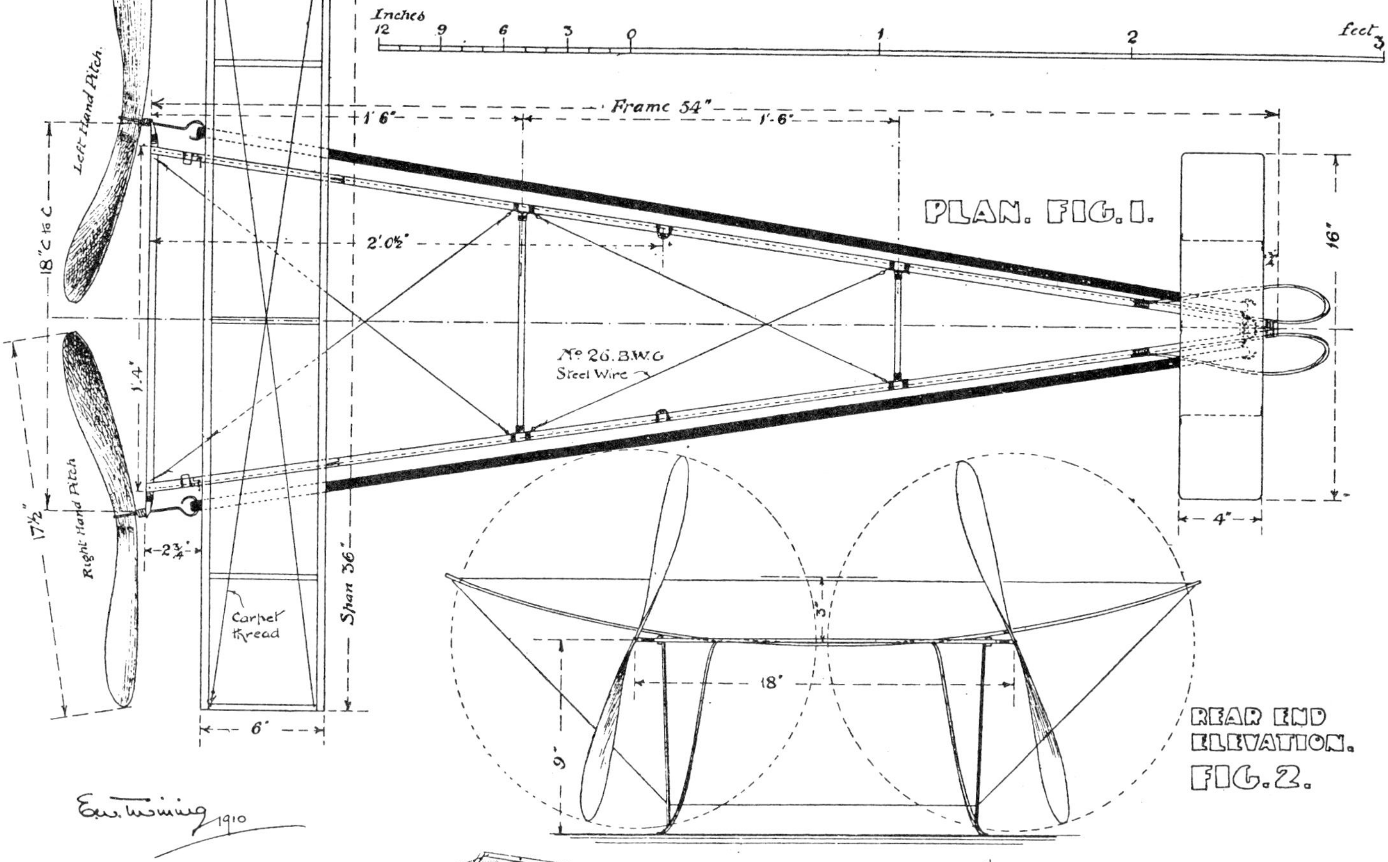

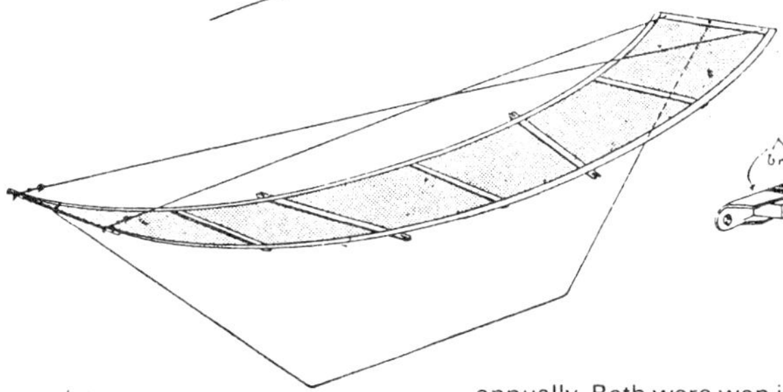

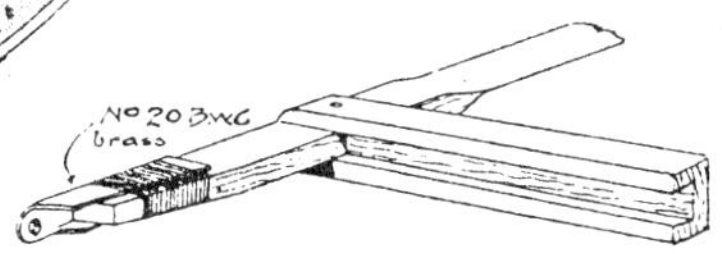

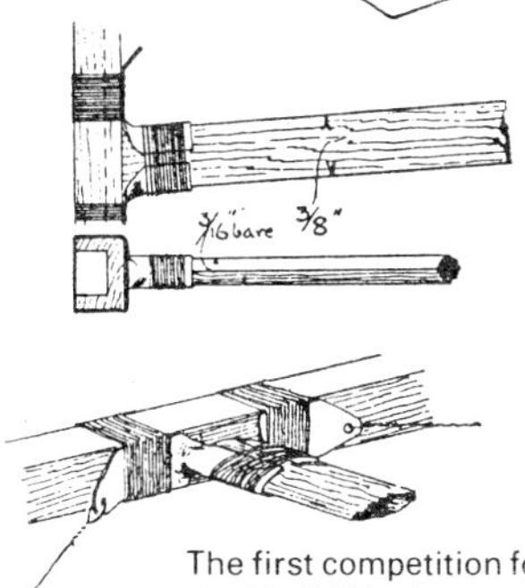

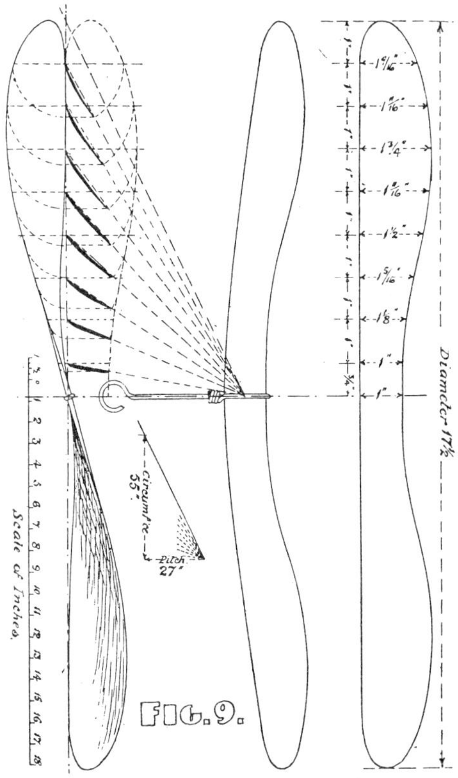

The first competition for the Model Engineer Challenge Cup was held in June 1910 and the first for the Gamage Cup in July 1910. Both these trophies are still competed for annually. Both were won in 1910 by E. W. Twining, with different models: that in the photo won the *M.E.* and that drawn the Gamage, despite having only been completed on the morning of the contest and extensively damaged on the first competition flight. The *M.E.* event, for duration, was won at 46secs. (see also page 8) and the Gamage, for distance, at 236 yards, though the model later proved capable of ca. 450yds. in still air and between 600 and 700 yards downwind.

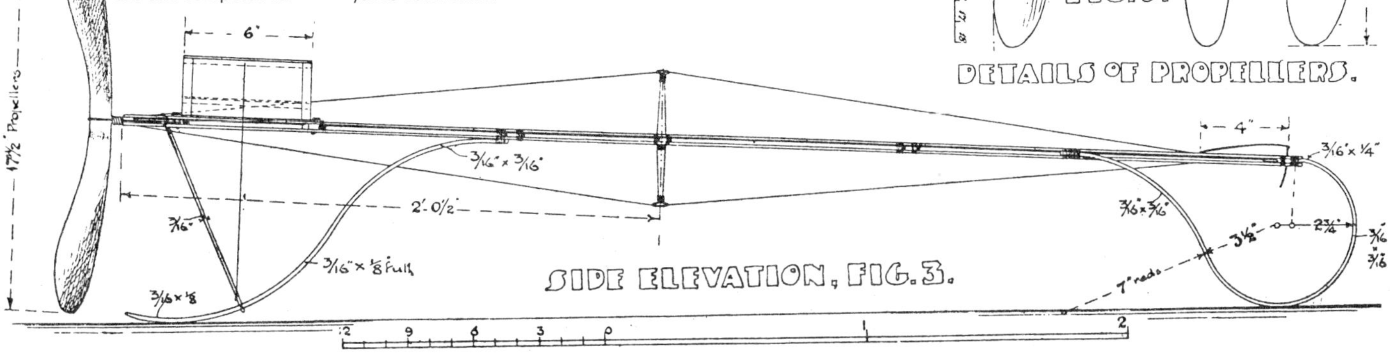

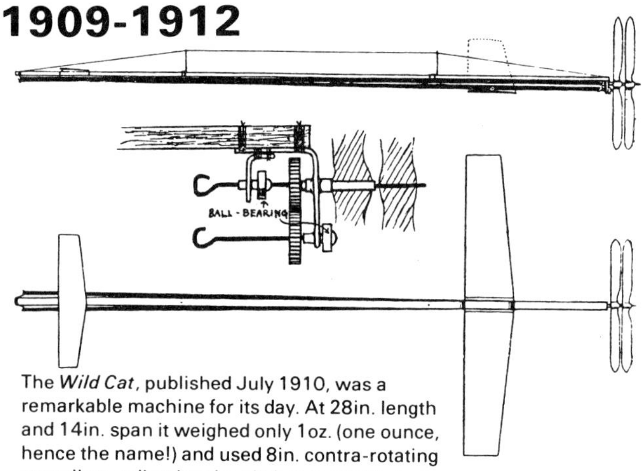

The *Wild Cat*, published July 1910, was a remarkable machine for its day. At 28in. length and 14in. span it weighed only 1oz. (one ounce, hence the name!) and used 8in. contra-rotating propellers, adjusting the pitch to cancel torque. Planes were piano wire outlines, spar wire-tensioned $\frac{3}{16} \times \frac{5}{16}$in. walnut, props paper-thin American whitewood. Builder was W. G. Aston.

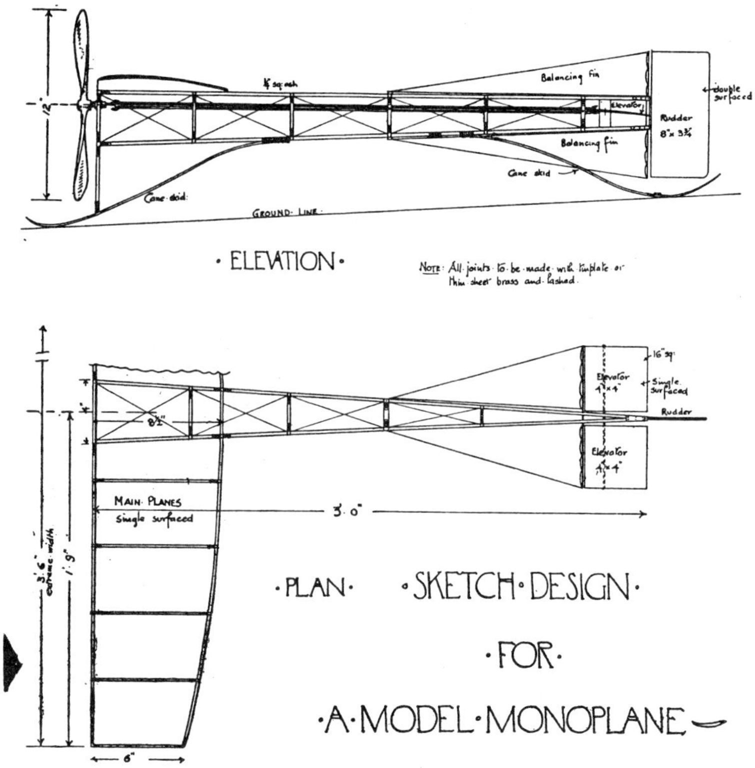

Design project by 'H.A.B.', an architect, also published in July 1910, shows influence of full-size shapes but has a nice simplicity.

Full-size pilot of some note, Louis Paulhan with one of his early models (1909).

H. H. Groves' successful 1912 steam model, 60in. span double-surfaced (Jap silk proofed with celluloid dissolved in amyl acetate) with 525sq. in. wing area and $1\frac{1}{2}$lbs. airframe weight. Steam plant was an additional $1\frac{1}{2}$lbs. and was a twin opposed single-acting rotary valve engine, $\frac{5}{8}$in. bore, 1in. stroke ($3\frac{1}{4}$oz.) with a benzoline blowlamp boiler using compressed air water feed.

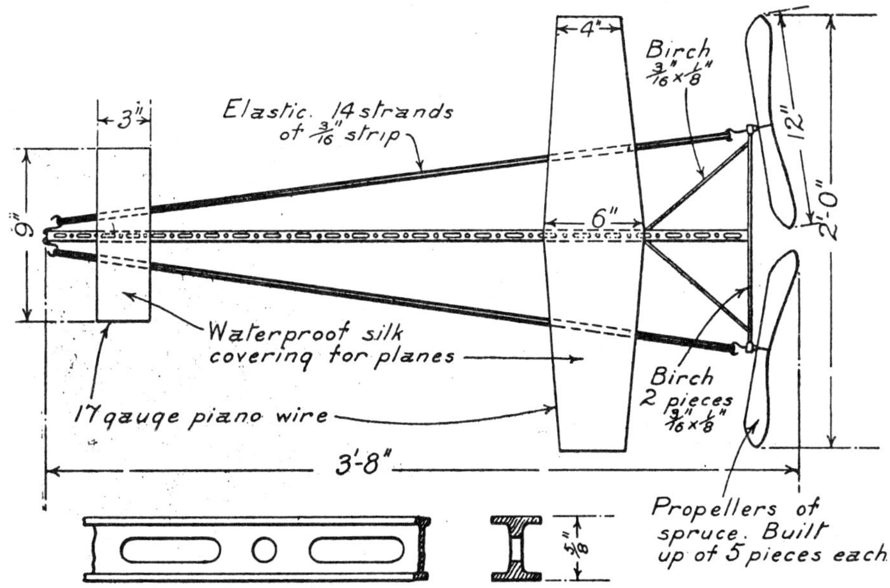

Sketch of part of main frame showing method of slotting to reduce weight, all the slots being covered in to reduce head resistance

Longest flight in the 1910 *M.E.* Cup was by a junior, Cyril Ridley, timed at 61secs. O.O.S. behind a 'mountain railway' (roller coaster) at a height of about 100ft. However, the model was under the 4oz. minimum and had to be disqualified. Twining won (previous page) and next of the 19 entrants was A. E. Cartlidge with 42secs. with the model above. Entrants included well-known names — C. R. Fairey, G. P. Bragg-Smith, R. F. Mann, R. P. Grimmer, H. B. Sayers and V. E. Johnson among others.

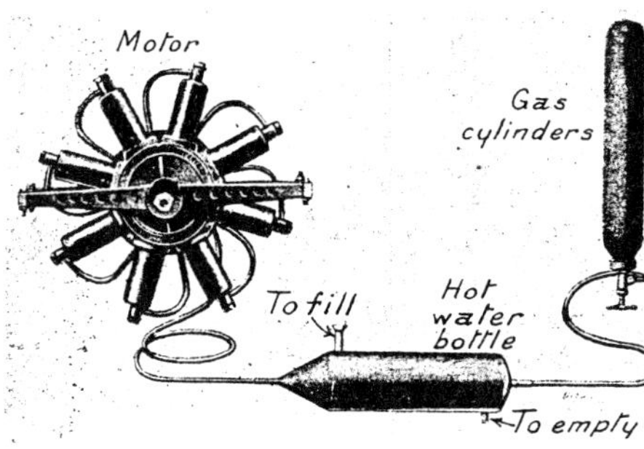

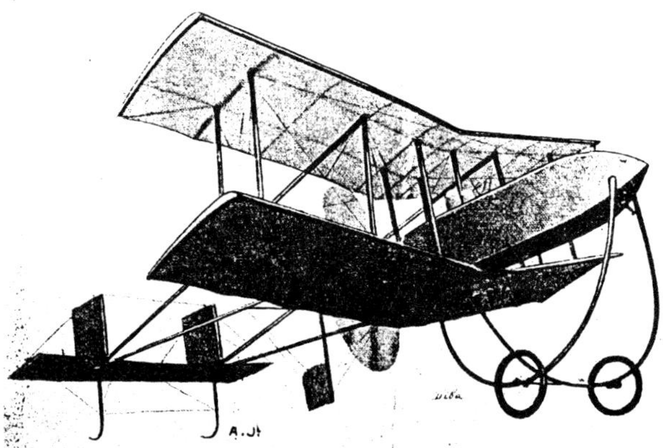

CO_2 motors were much in evidence in 1911/12. This French Cetonia plant weighed $8\frac{1}{2}$lbs. (motor weight 27oz., bore and stroke 19 and 28mm, overall diameter $7\frac{1}{4}$in.) and on official tests produced 1.03h.p. at 1710r.p.m. for one minute. It flew the large biplane (left) 1,178m in 92secs. to win the Gordon Bennett Cup and a gold medal. Model was 2m span, 3m long with no less than a $33\frac{1}{2}$in. prop and weight of $38\frac{1}{2}$lbs.

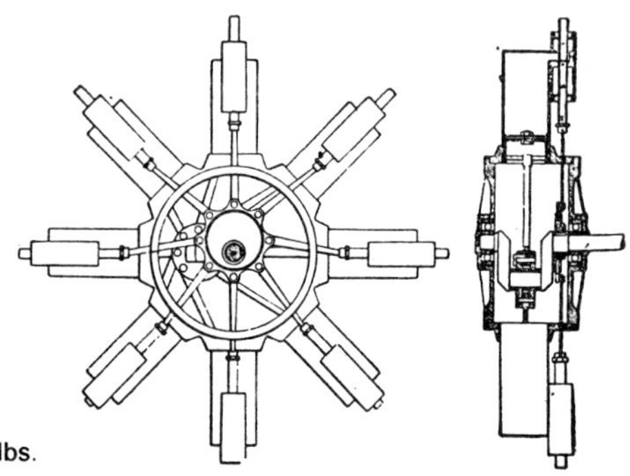

Alex Houlberg, later for many years Chairman of the S.M.A.E. and awarded the M.B.E. for his services to modelling, established a rubber duration record of 89secs. in 1912 with this machine and raised it to 128secs. with a similar but larger model in 1913.

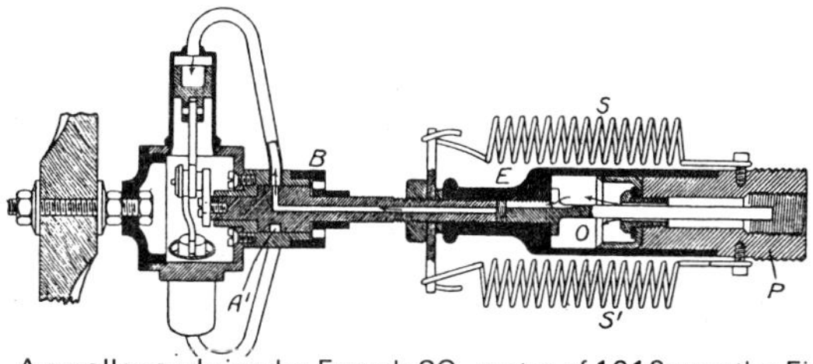

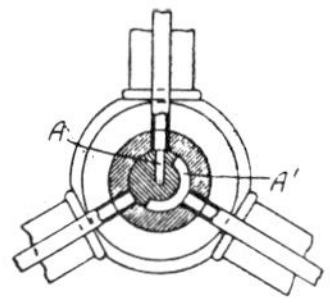

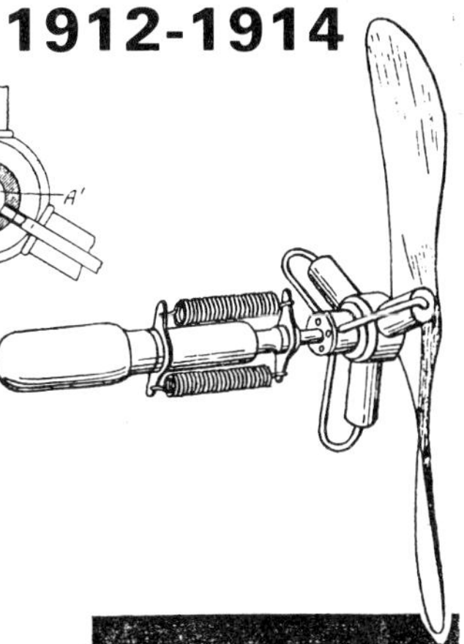

A smaller and simpler French CO_2 motor of 1912 was the Fieux. There were three types — 2, 3 or 5 cylinders, weighing 1½, 2 and 2½oz. and all with 10mm bore and 25mm stroke. The supply was a Prana Sparklet bulb (yes, they existed even then!) with an ingenious pressure balanced on/off gas feed. Total plant weight was 7oz.

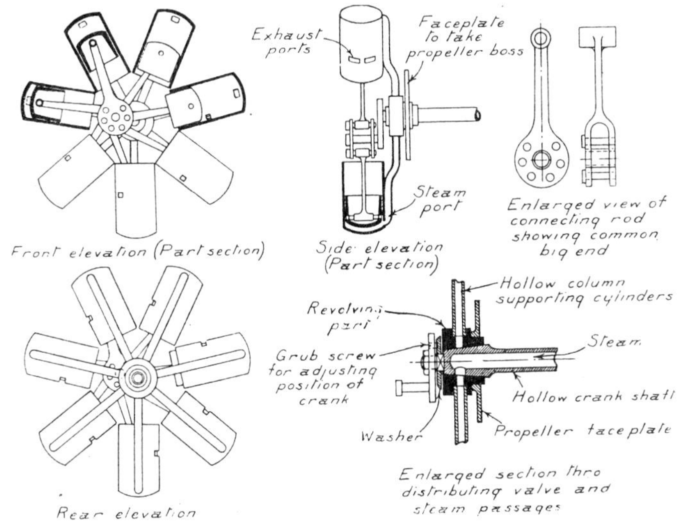

Published in June 1912, this 7-cylinder rotary steam engine by Arthur Thornton was a simple, valveless concept, ⅝in. bore and ½in. stroke, overall dia. a little over 4in., with no crankcase, the cylinders being supported only by the steam pipes.

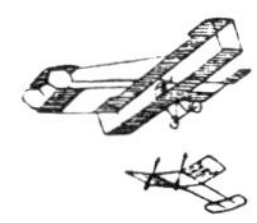

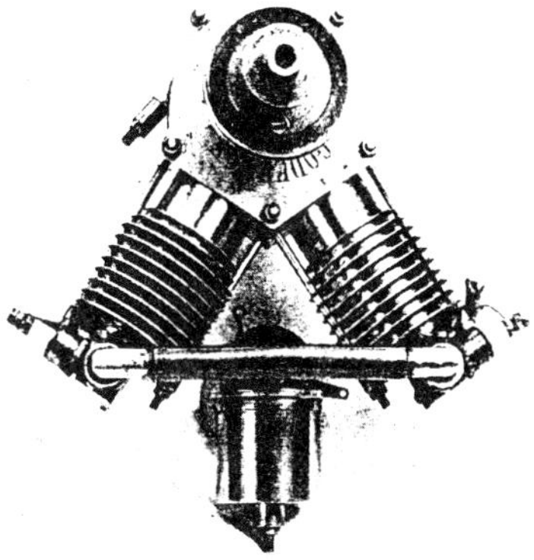

One of the earliest model petrol engines was this V-twin built by the Godfroy brothers (France), which flew an Antoinette-type model for five laps tethered to a pole before the line broke and the model crashed into railings. This occurred in 1911,

The first officially-observed flight (by Royal Aero Club officials) by a petrol-engined model in the U.K. was by D. Stanger in 1914, setting a time of 51secs. which was to remain unchallenged until 1932 and winning for its builder a prize of £10. Span of the 10¾lb. model was 7ft. and the engine a V-twin.

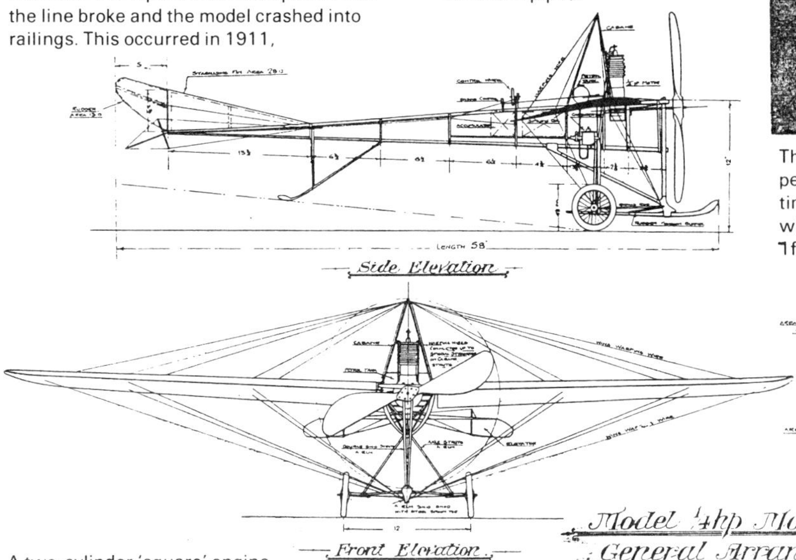

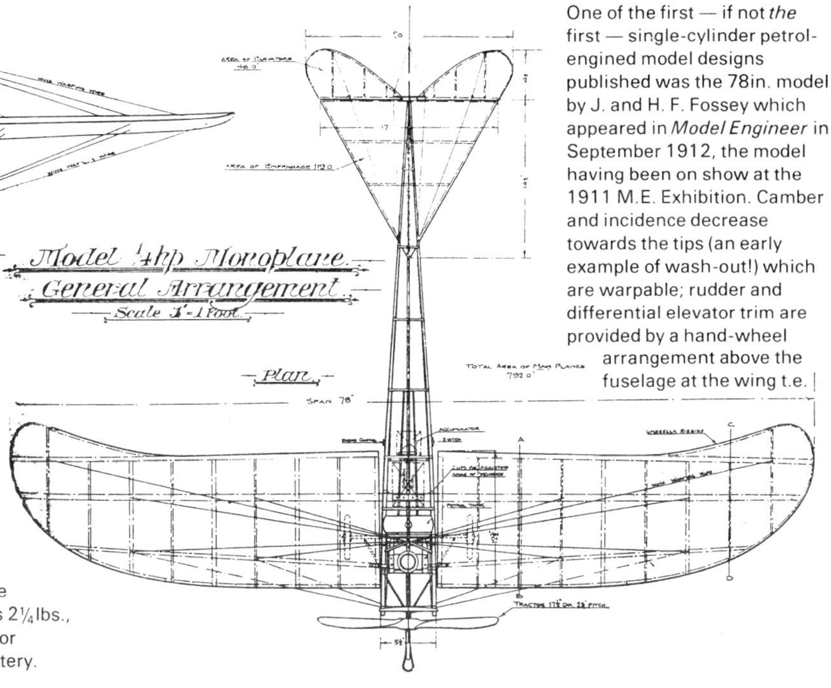

One of the first — if not *the* first — single-cylinder petrol-engined model designs published was the 78in. model by J. and H. F. Fossey which appeared in *Model Engineer* in September 1912, the model having been on show at the 1911 M.E. Exhibition. Camber and incidence decrease towards the tips (an early example of wash-out!) which are warpable; rudder and differential elevator trim are provided by a hand-wheel arrangement above the fuselage at the wing t.e.

A two-cylinder 'square' engine, $1\frac{3}{16}$in. bore and stroke, with a variable-pitch propeller, by W. J. Smith, July 1913.

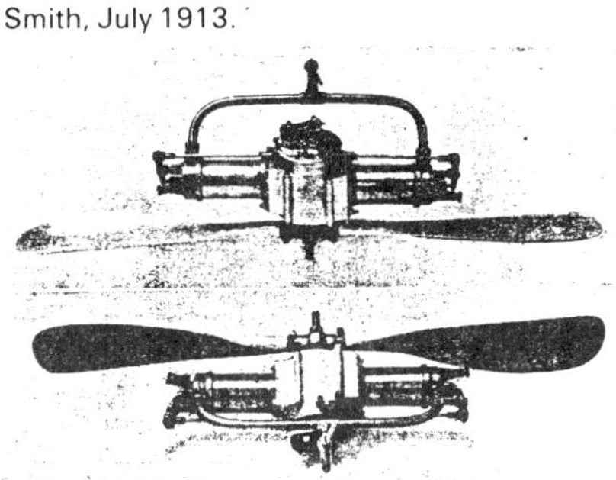

Bare engine weight was 2¼lbs., plus 9ozs. for coil and battery.

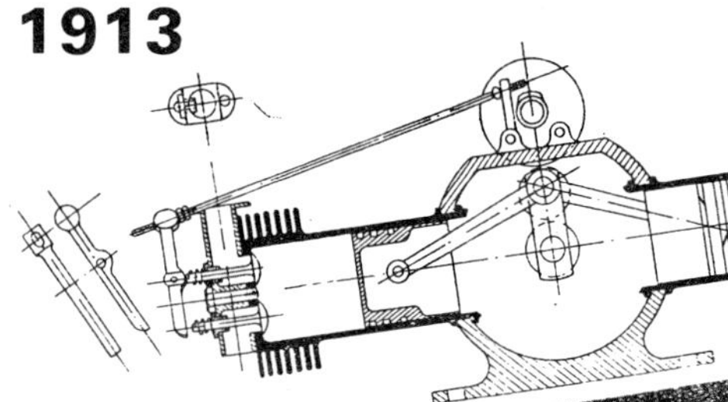

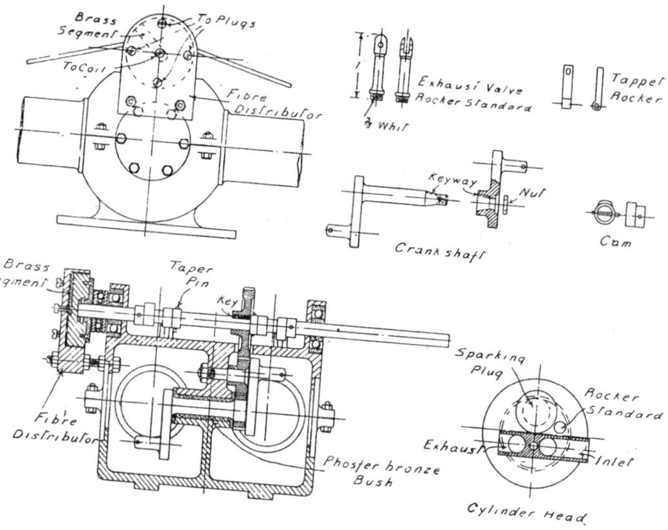

This four-cylinder four-stroke (shown with a contemporary full-size sparking plug) appeared in *M.E.* in February 1913 and was described as '1 h.p.' but few dimensions were given other than overall width, top fin to top fin, was 6½in.: it is probably 1in. bore × ¾in. stroke. Made on a 4in. Drummond lathe by its designer, A. T. Hawkins, who wrote that each cylinder, turned from the solid, required six hours of steady treadling!

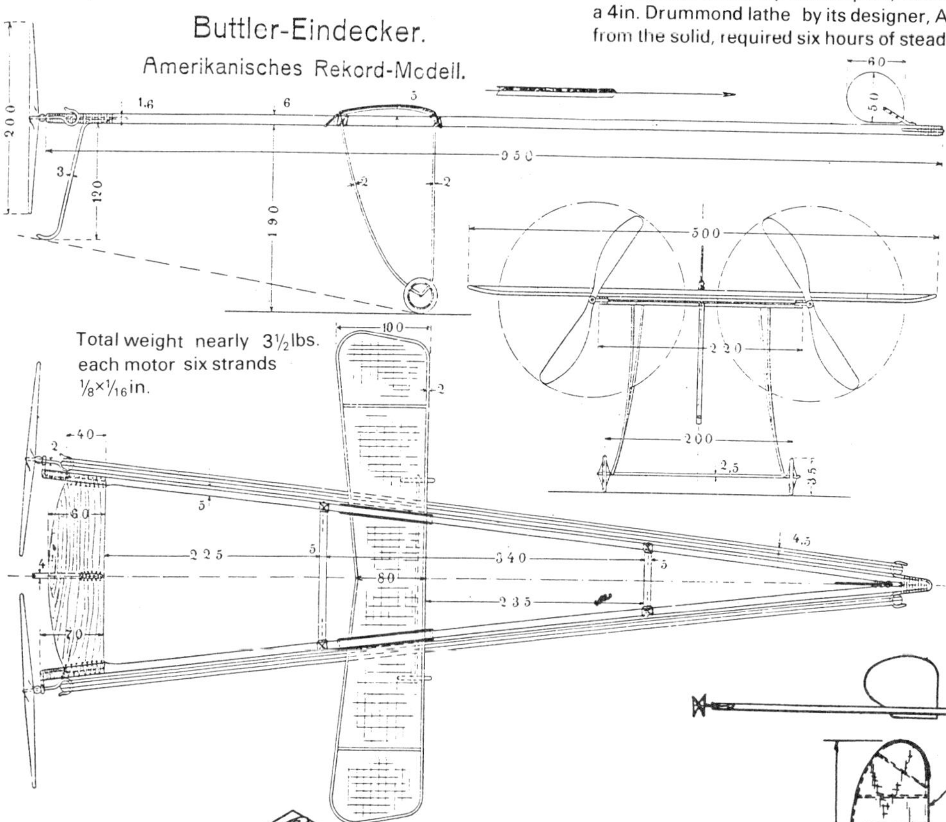

By 1913 the A-frame or 'twin pusher' had established itself as the best and most consistent performer among rubber models, a situation which was to continue for another ten years. Competitions were frequently for distance flown, which was to remain a criterion of performance until well into the 1930s, though flight duration (usually the best of three timed flights) was also used from about 1910; we have seen examples of each in the Gamage and M.E. Cups, page 7.

The model below, shown courtesy of Frank Zaic (it appeared in his 1938 Yearbook) was clearly intended for hand-launching and was familiar to the compiler, who enjoyed correspondence with Bill Butler for several years in the 1940s/50s. That on the left, published in *Flugsport* in 1913, was a surprise: it seems a reasonable surmise that it was the same modeller, though perhaps an earlier model. The record was for distance, R.O.G., 345 yards, the model reaching a height of approximately 150ft.

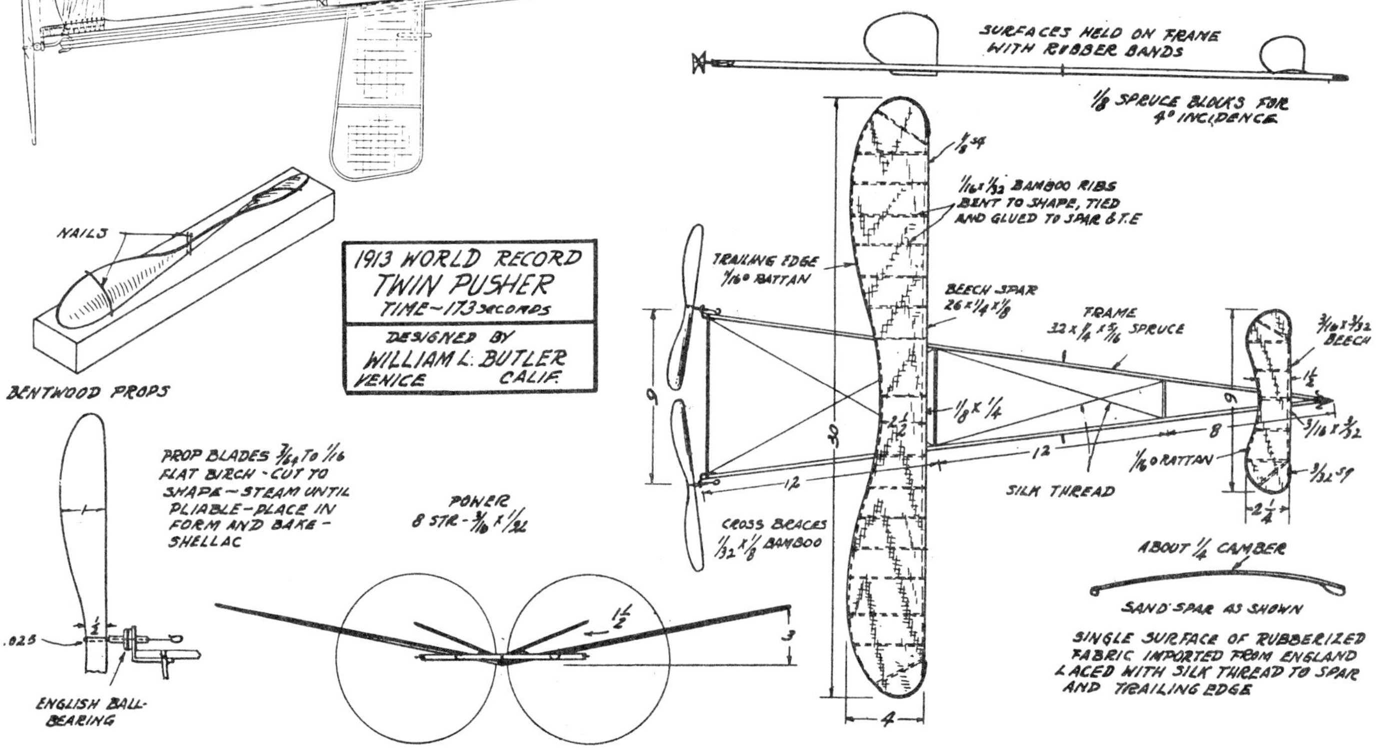

Excerpts from *Flugsport*, 1913, courtesy John Blake and Ken Macdonough.

Four of the five models on this page are by English-type names and may have been reproduced from other journals.

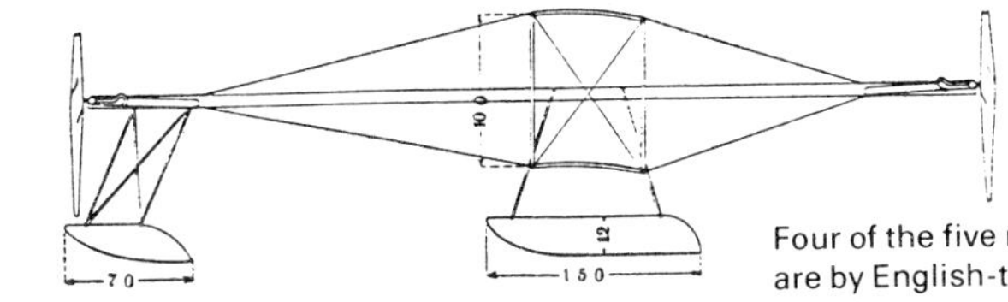

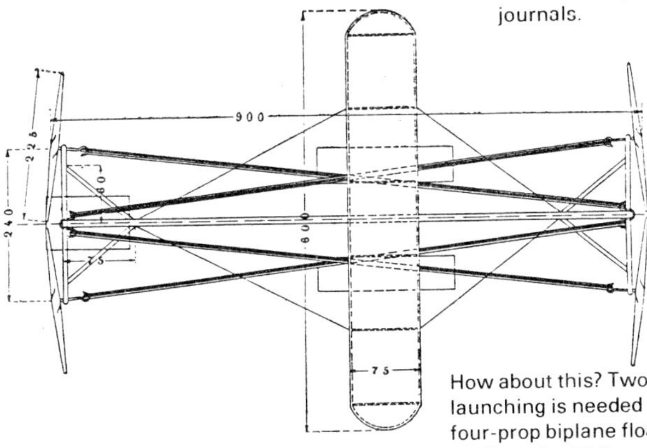

How about this? Two-man launching is needed with this four-prop biplane floatplane, attributed to 'Johnson'. Span 600mm, length 900mm, weight about 8½oz.

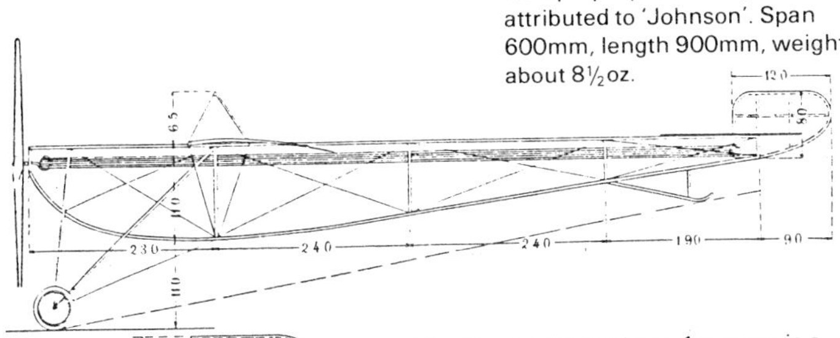

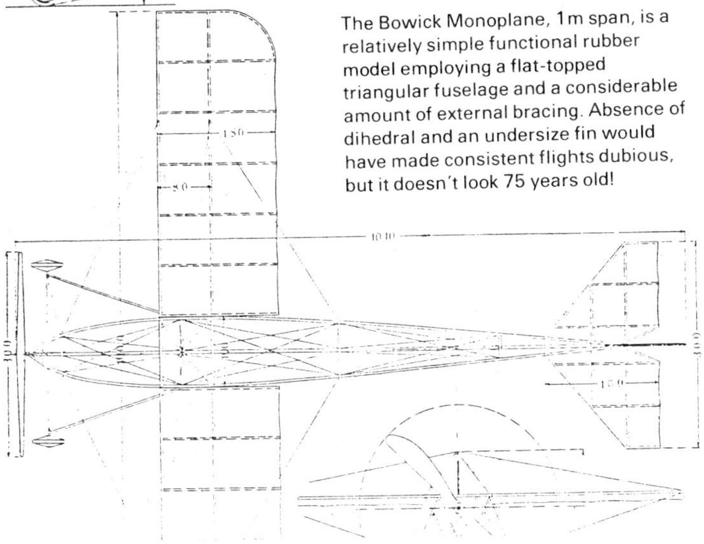

The Bowick Monoplane, 1m span, is a relatively simple functional rubber model employing a flat-topped triangular fuselage and a considerable amount of external bracing. Absence of dihedral and an undersize fin would have made consistent flights dubious, but it doesn't look 75 years old!

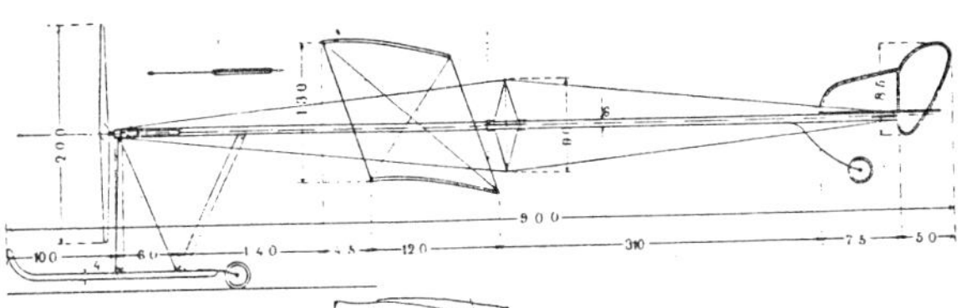

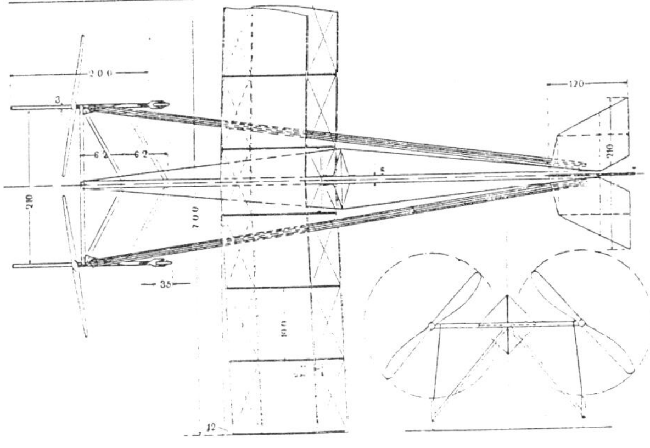

Unusual tractor A-frame biplane 27½in. span and 35½in. long, using a central fore-and-aft spar. Attributed to 'Smith'.

1913

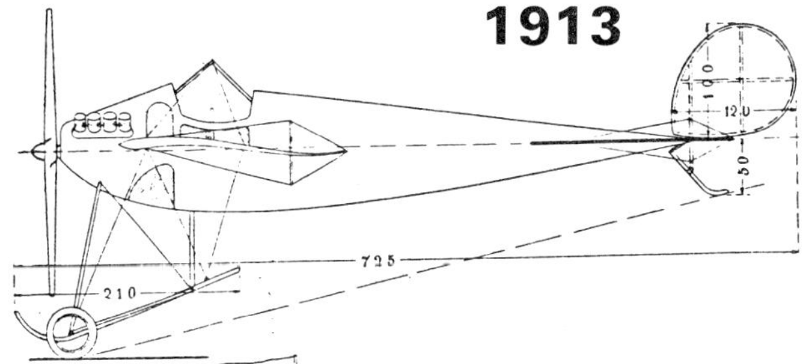

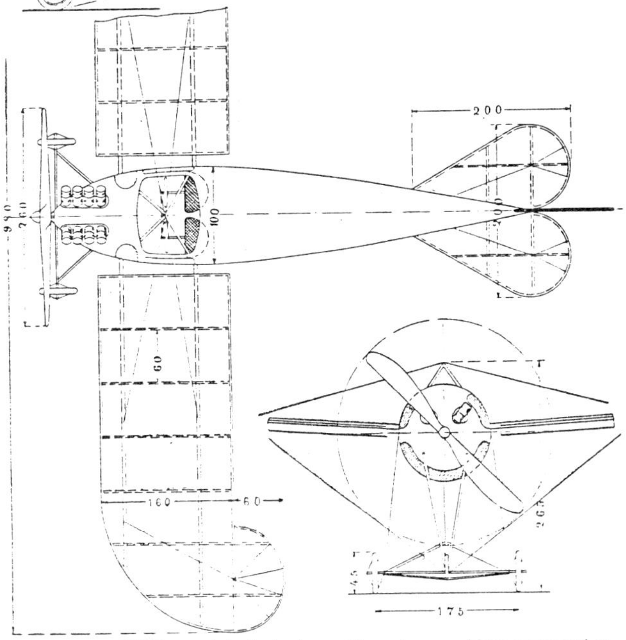

Attractive Wilkinson 'eindecker' must have been rubber-powered at 39in. span and appears to have been an early attempt at a monocoque fuselage. Note the addition of a V8 engine and side-by-side cockpit seats. The wing root gaps must have lowered efficiency but at least some dihedral was incorporated.

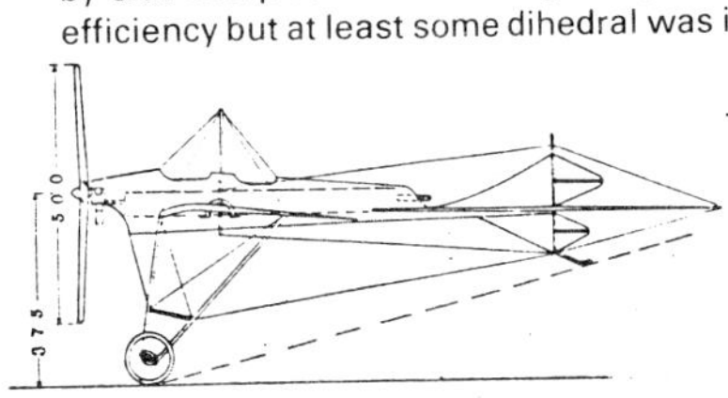

There were at least a dozen clubs in Germany before the 1914-18 war.

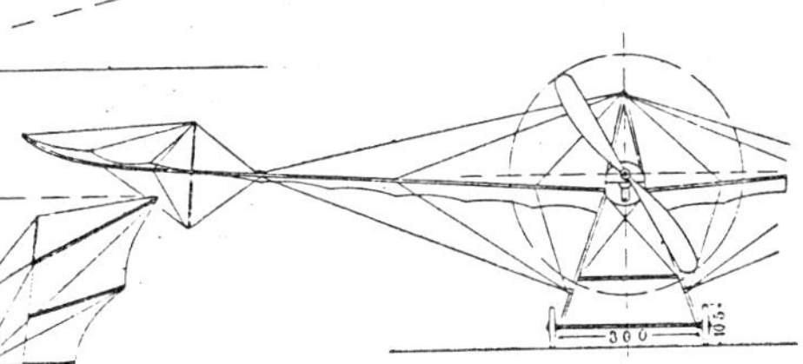

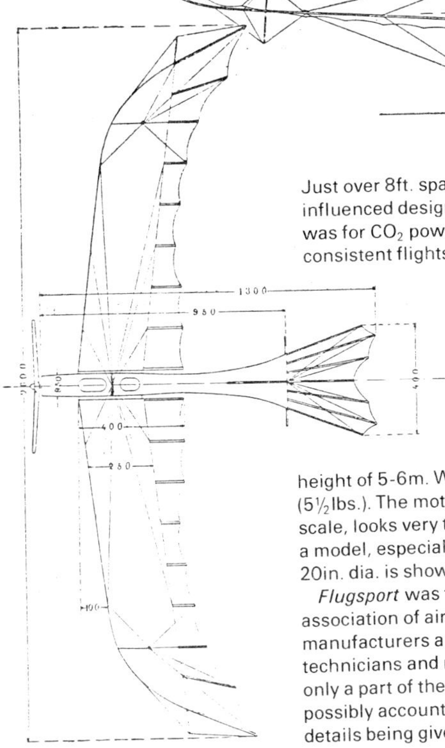

Just over 8ft. span, this Taube-influenced design by Wollenstein was for CO_2 power and would make consistent flights of 120-150m at a height of 5-6m. Weight was 2.5kg (5½lbs.). The motor, if drawn to scale, looks very tiny to power such a model, especially as a prop about 20in. dia. is shown.

Flugsport was the organ of the association of aircraft manufacturers and flight technicians and models formed only a part of the content, which possibly accounts for only bare details being given in some cases.

Span 600mm, length 500mm.

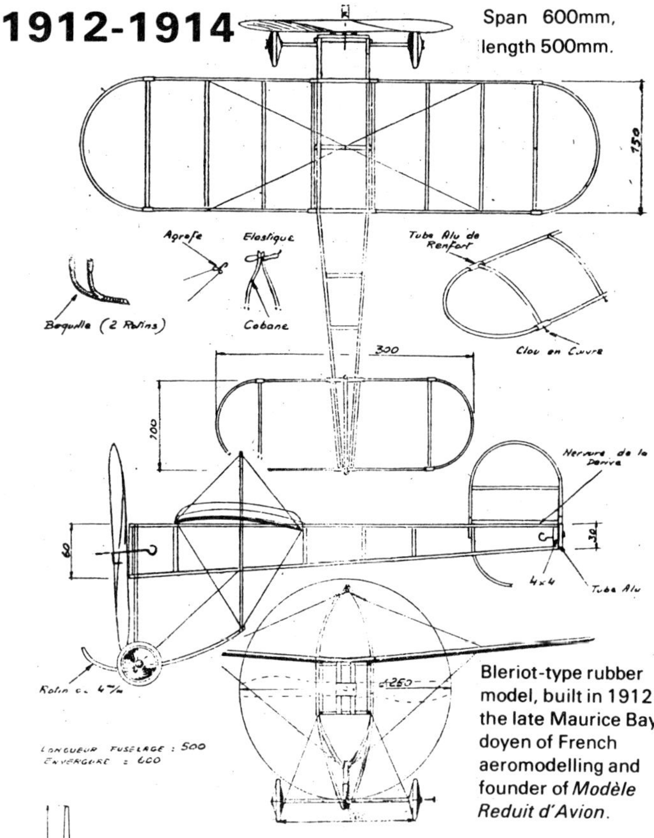

Bleriot-type rubber model, built in 1912 by the late Maurice Bayet, doyen of French aeromodelling and founder of *Modèle Reduit d'Avion*.

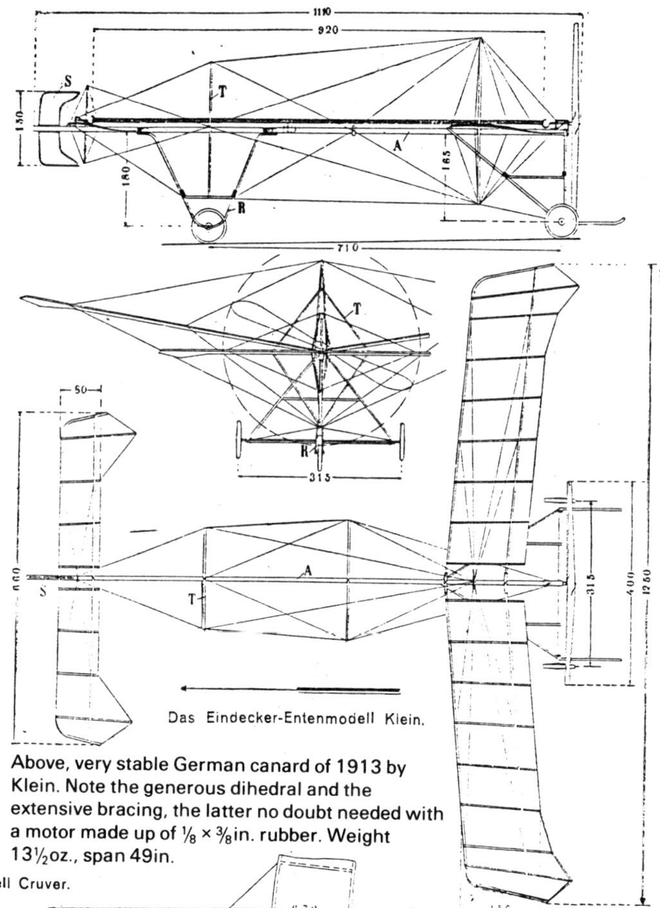

Above, very stable German canard of 1913 by Klein. Note the generous dihedral and the extensive bracing, the latter no doubt needed with a motor made up of $\frac{1}{8} \times \frac{3}{8}$in. rubber. Weight $13\frac{1}{2}$oz., span 49in.

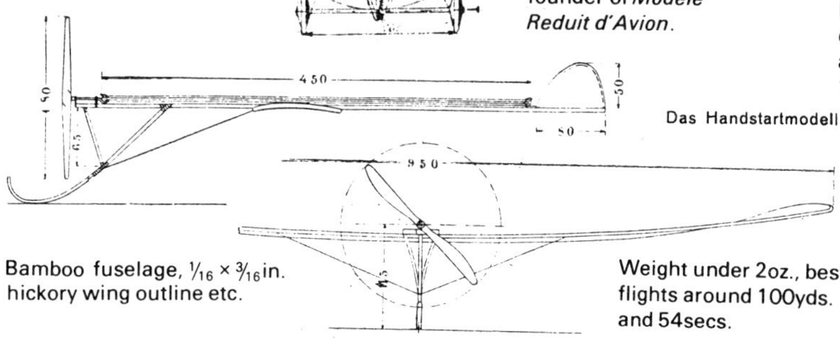

Bamboo fuselage, $\frac{1}{16} \times \frac{3}{16}$in. hickory wing outline etc.

Weight under 2oz., best flights around 100yds. and 54secs.

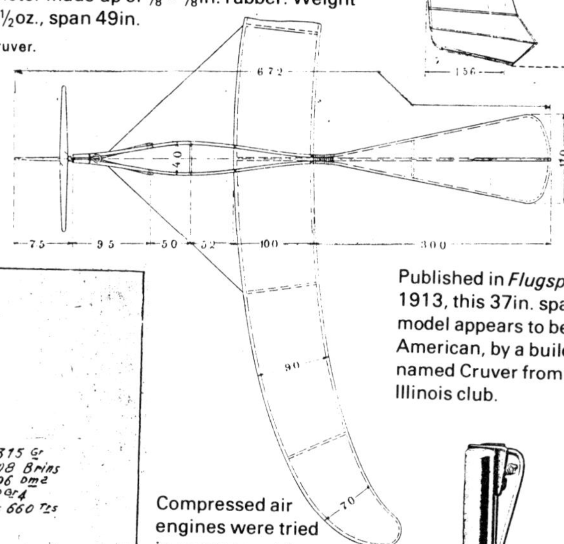

Published in *Flugsport*, 1913, this 37in. span model appears to be American, by a builder named Cruver from the Illinois club.

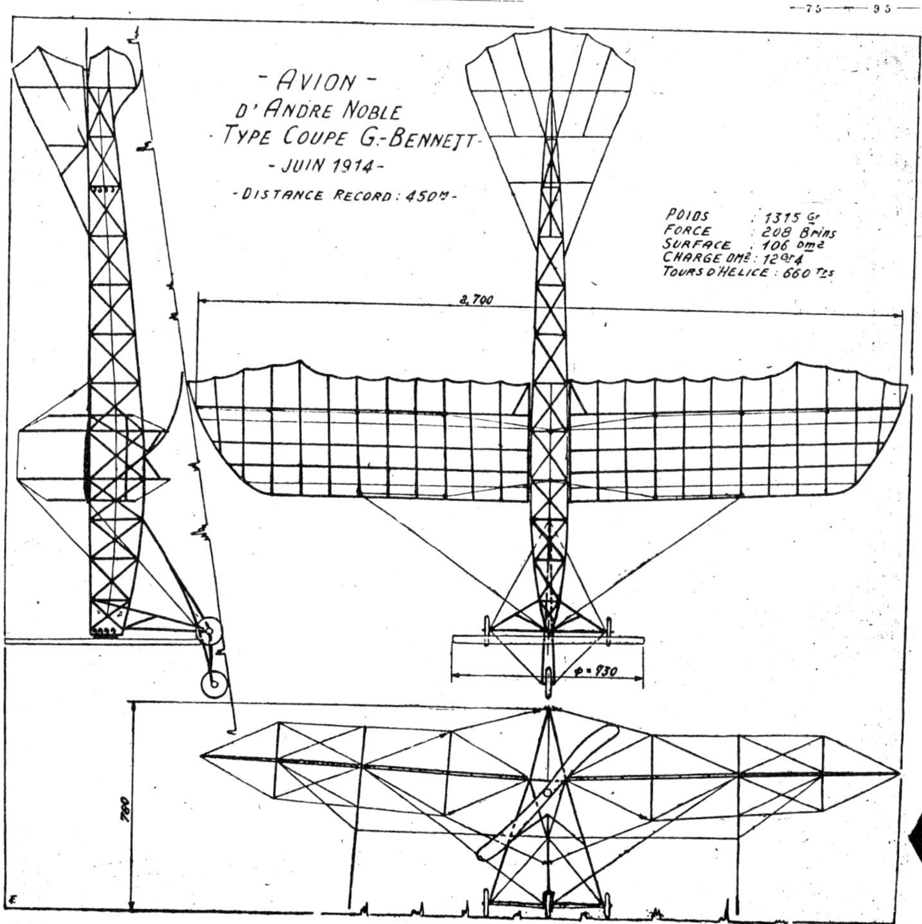

Compressed air engines were tried in many countries. This German example appears to use the simple single-acting oscillating principle found in basic steam engines from around 1800 and still used today.

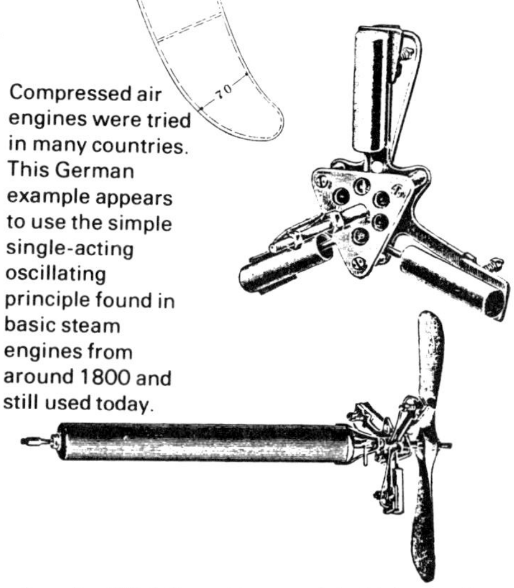

Amazing 8ft. 11in. rubber model which won the 1914 Gordon Bennett Cup for André Noble with a record distance of 492 yards. It could have been more, since the model ended up about 20ft. up a tree. Triangular fuselage, 208 strands of rubber (approx. $15\frac{1}{2}$ozs.) arranged in four motors geared to drive a 29in. prop, total model weight 46ozs. Noble's prizes included a flight in a Blériot, a 5-cyl. Comby compressed air engine and 20 francs.

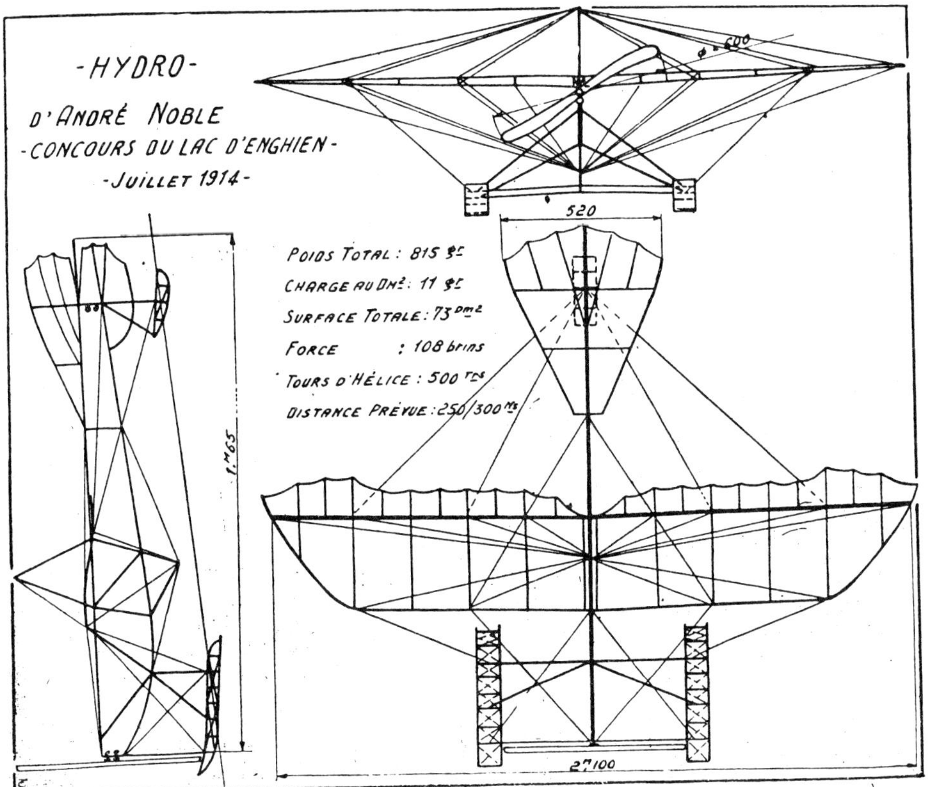

A smaller and lighter André Noble model was this profile-fuselage floatplane, just under 7ft. span and 28oz. weight, powered by 108 strands grouped into two motors. Built for a contest which was never held. Tests had shown promise and proved the practicability of the variable incidence tailplane, operated at the end of the motor run. (Published M.R.A. 1950).

Advertisements from *The American Boy*, 1916.

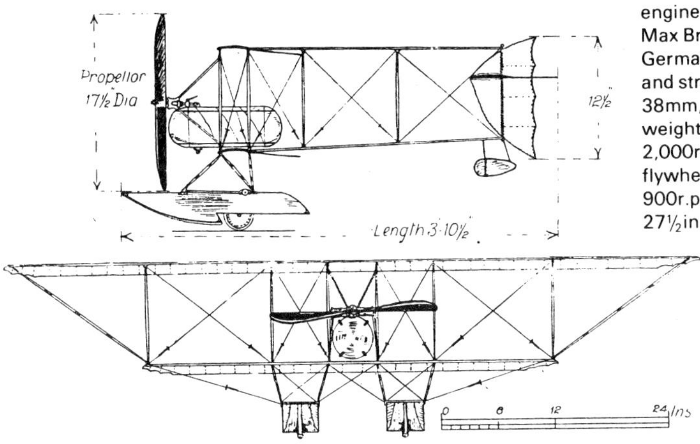

V-twin petrol engine, 1913, by Max Braune in Germany. Bore and stroke 26 × 38mm, 4½lb. weight, 2,000r.p.m. on a flywheel, 900r.p.m. on a 27½in. prop.

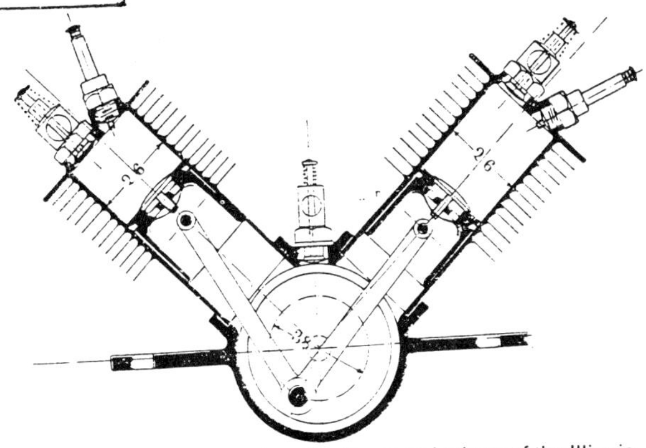

Below, 'world record tractor monoplane' by Donovan Lathrop of the Illinois club, drawn and described by William B. Stout in *The American Boy*, May 1916. This model was lost at 5:00min. at a distance of 4,000ft. The concept of a small prop and long motor on tractors was ascribed to a clubmate, Hittle. Stout, a prominent member of the aviation side of Ford (he designed the Trimotor) was a major influence on American model flying.

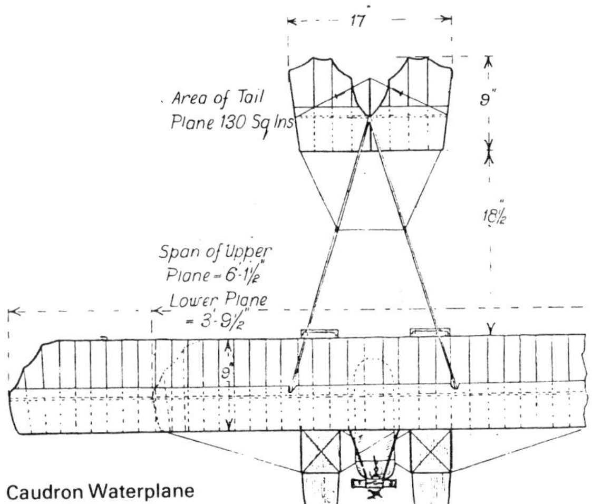

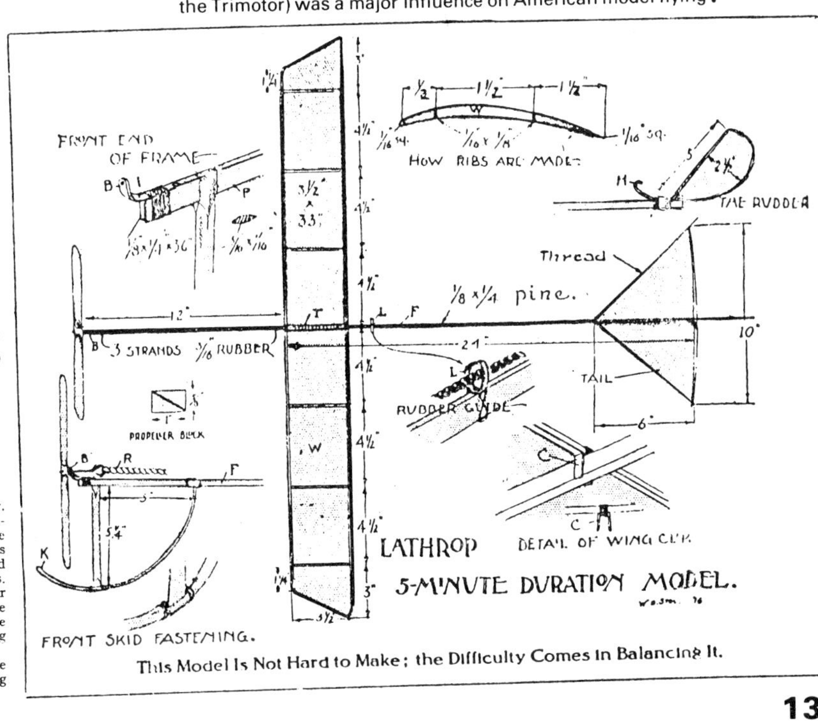

Caudron Waterplane model featured in *Model Engineer*, August, 1914. The brief extract, right, gives details of the motor, etc., the model itself being by D. Hiscox of Bexhill, who deplored in the article 'elastic driven flying freaks'.

The engine, which weighs 1½ ozs., is of Mr. Charlie Desoutter's excellent workmanship, comprising two cylinders horizontally opposed. The pistons are fitted with leather cups, and the valve is a rotary one. The engine is driven by compressed air stored in a steel cylinder 12 ins. long 4 ins. diameter, also made by Mr. Desoutter. A tractor propeller of 17½ diameter supplies the thrust. The motor runs for two minutes when 300 lbs. pressure is pumped into the reservoir, but a real flying thrust only lasts for a little over half a minute.

The complete machine weighs 2 lbs. 3 ozs.; the reservoir 11 ozs.; and the planes, before being braced together, 8 ozs.

1919-1924

"Flying-Stick" or Scale Model Aeroplanes?

By A. R. Walkley.

Brief excerpts from a 1919 article. Scale *vs.* 'freaks' has been a perennial argument!

To one who has had no previous experience, it is useless to attempt to build a scale model straight away. The first thing is to obtain a working knowledge of modern theory and practice—then buy one of the many model tractor monoplanes now on the market. We will name no names, but if the purchaser cannot see room for improvement "all over" the machine he has bought he must be singularly lacking in talent and ingenuity. In common fairness it must be said that there are a few machines obtainable which bear evidence of thought and originality in design; some of these, however, are not water-proof, and become distinctly soggy after a few landings on wet grass.

It will be found that any machine above 30 ins. or so in length becomes very inconvenient to transport from place to place, and everyone is not fortunate enough to have an "aerodrome" at the door of their workshop. It must be borne in mind that clear open spaces are not plentiful in the neighbourhood of large towns, and even a small machine that flies 200 yards needs a field of 400 yards square if collisions with trees or houses, etc., are to be avoided. This means that the long-distance flying-stick, with a flight of 600 yards and upwards, is in the nature of a white-elephant to most people. A good flight of comparatively short duration with a realistic aeroplane will surely be of more interest than any amount of long flights with "floating freaks," such as the twin pusher affairs that wander about aimlessly and eventually get lost!

When one decides on a given length for a machine, length of flight (time in the air × distance covered), depends chiefly on three factors: 1. number of strands in motor, *i.e.* number of turns possible; 2. weight of machine in flying trim; and 3. resistance. Therefore, by fitting a geared motor we add extra weight; so, unless the power (as well as number of turns) is much increased, the advantage is only apparent. The air-screw may take more turns, but, being heavier, the machine absorbs more power and the actual distance covered may not be much greater than a single skein motor would give. Also, a geared motor means complication, trouble in replacing rubber and other disadvantages, so that if a good flight can be obtained without its aid—so much the better.

The great thing to remember is that weight must be cut down to the minimum. Strength depends on lightness—the lighter a machine is the stronger it will be! At first this sounds contradictory, but a moment's thought will show it is so. The flying strains on a model are negligible compared with the landing-shocks it has to undergo, so reduced weight means less landing-shock—further, if the weight is cut down in suitable places resistance is reduced as well.

September 18, 1919

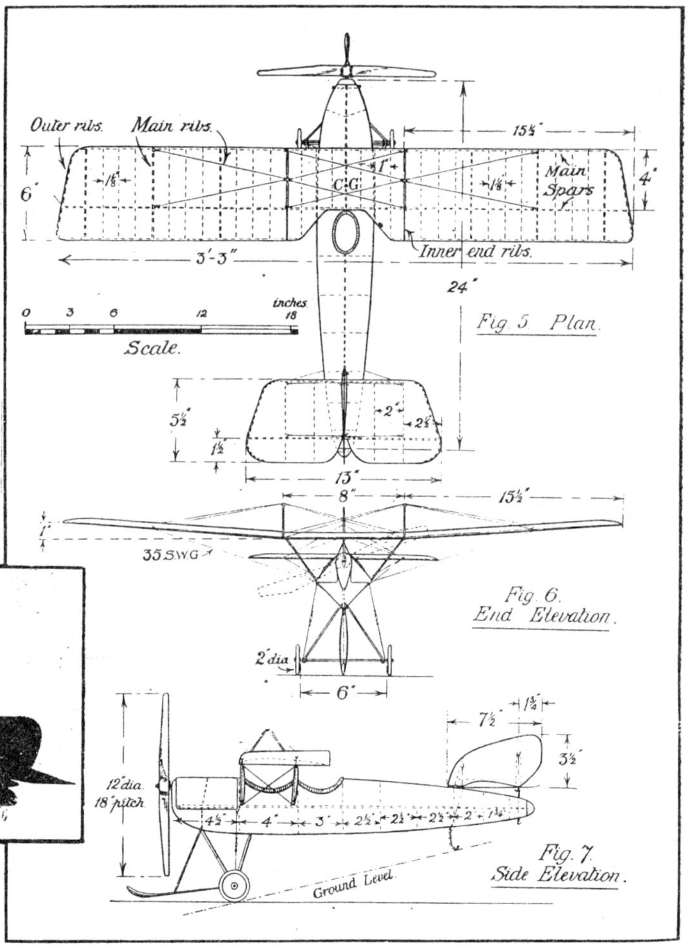

Realistic appearance and double-surface (RAF 6) wing were features of this parasol design by F. F. Jackson, published in November 1921, though the model first flew in 1917.

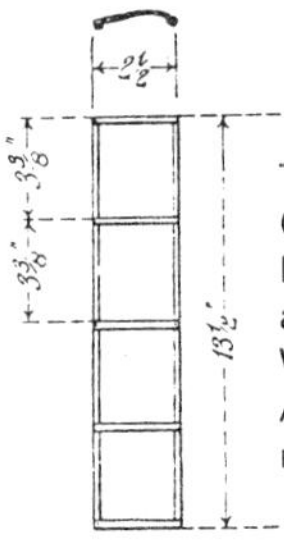

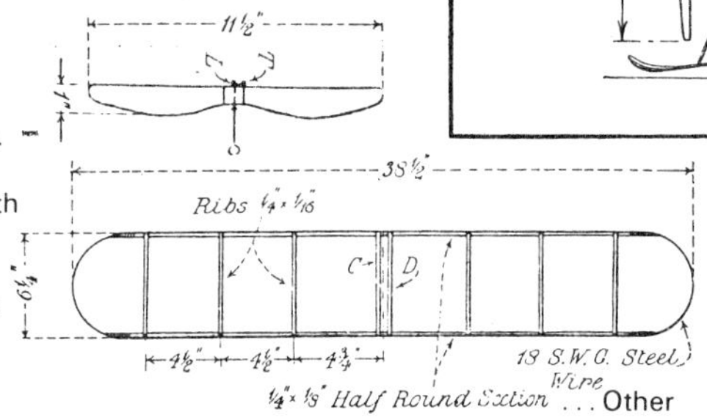

This twin-pusher model by S. C. Hersom set up a world R.O.G. record for the type with a flight of 4min. 7secs. on Wanstead Flats, London, in August 1922, one of the first recorded thermal flights . . . Other recorded twin pusher times were R. Jaros (U.S.A.) 3mins. 29secs. and Heuvelinck (Holland) 1min 17.6secs.

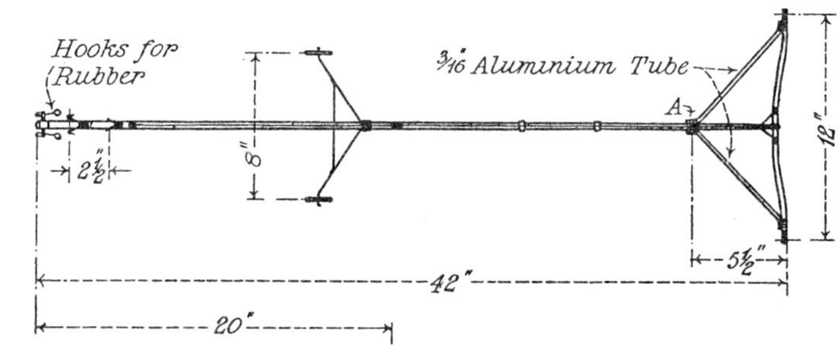

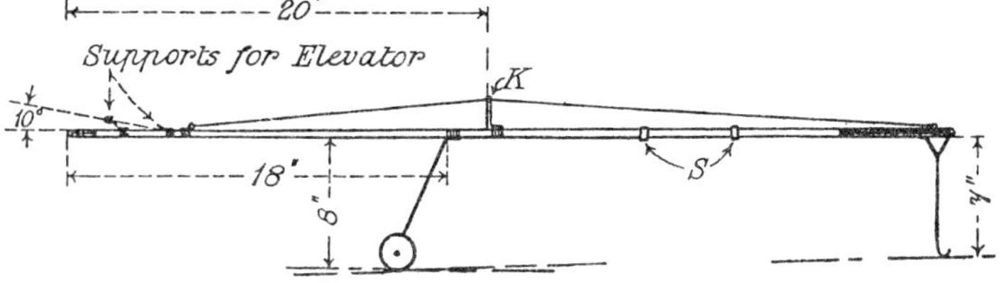

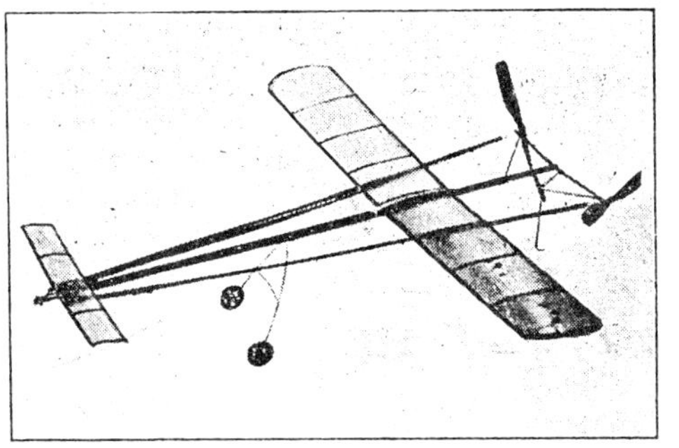

Below, an early attempt at a lifting fuselage,by W. E. Evans (Paddington D.A.C.); with double-surface wings.

Fuselage was sheeted with 0.5mm ply (weight 3¼ozs.) and ribs cut from ⅛in. sheet spruce.

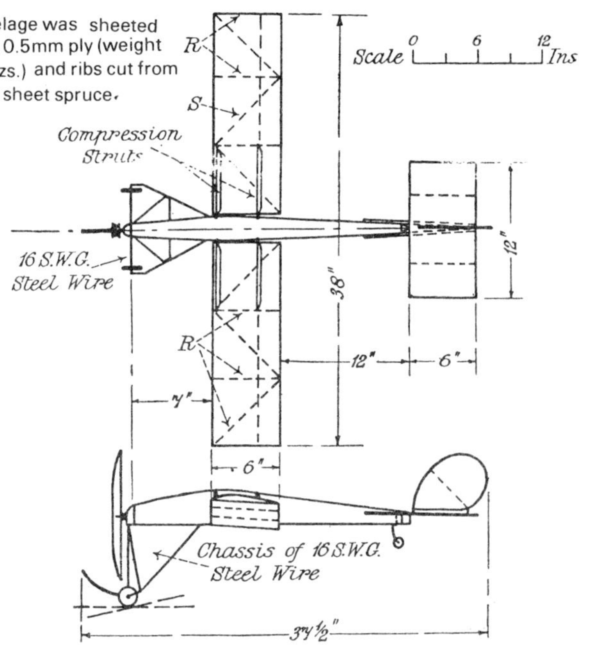

Another fuselage model was the 53in. 'light plane' illustrated at left, plans for which appear on the opposite page. This too has a ply-covered fuselage (1mm) and was intended to explore the effects of a 'deep keel' fuselage and to produce a rugged model; one result was the opportunity of dispensing with a normal wire undercarriage. Resemblance to the A.N.E.C. machine flown at the Lympne light aircraft trials in 1923 was coincidental. Exhibited at the 1923 Model Engineer Exhibition, published July 1924.

This model was by B. K. Johnson, Technical Secretary of the Society of Model Aeronautical Engineers, which had been founded in 1922, largely by the efforts of A. E. Jones, its first Secretary, to take the place of the old Kite and Model Aeroplane Association and the London Model Aeroplane Society. The title of the new body (which some later modellers have thought a shade pretentious) was suggested by F. J. Camm, a founder member, later editor of *Practical Mechanics*.

1924

Fuselage gliders were a different class and a leading exponent was C. W. Howes, who established a H.L. record of 40.4secs. in 1923

Interest in model flying flagged world-wide in the years following the First World War. Strangely, model boating prospered and many of the developments in model petrol ('gas') engines were for boats, though often with fairly large (25-30cc) engines. Rubber remained the chief power source for aeromodellers but gliders began to attract experimenters.

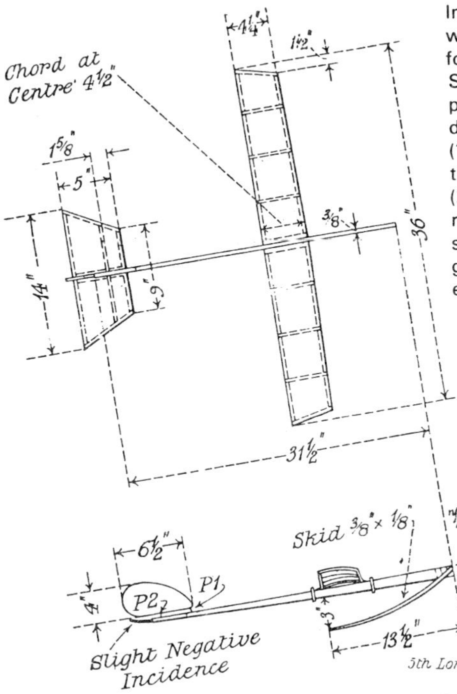

Left, 4.6oz. 'spar glider' by C. J. Burchell held the British H.L. record (44secs.) in early 1924. B. K. Johnson with a similar model discovered slope lift near Reculver, Kent, making numerous 4min.-plus flights and 12 consecutive flights averaging 2:04.

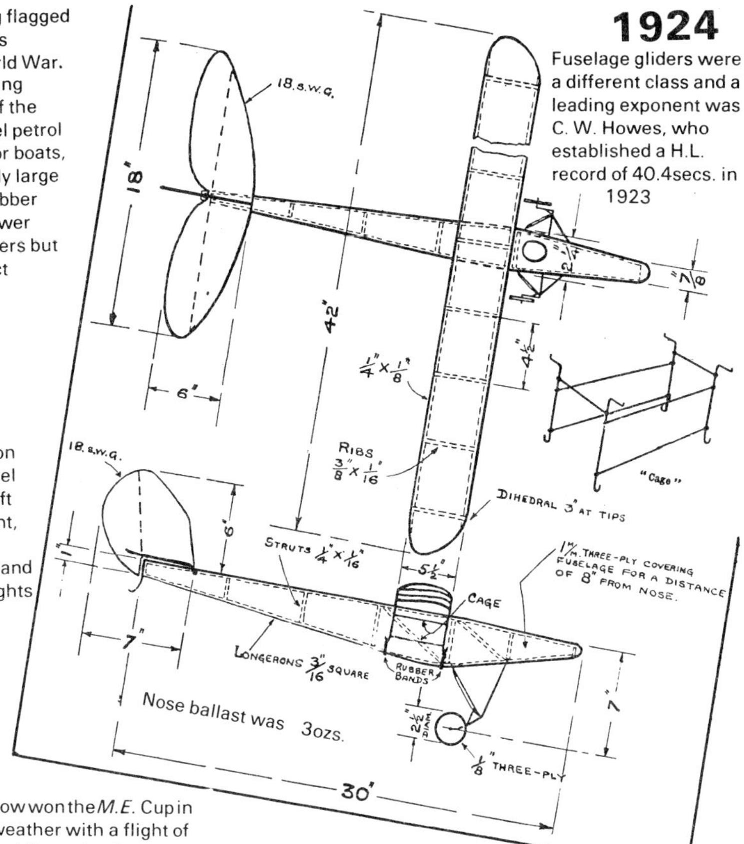

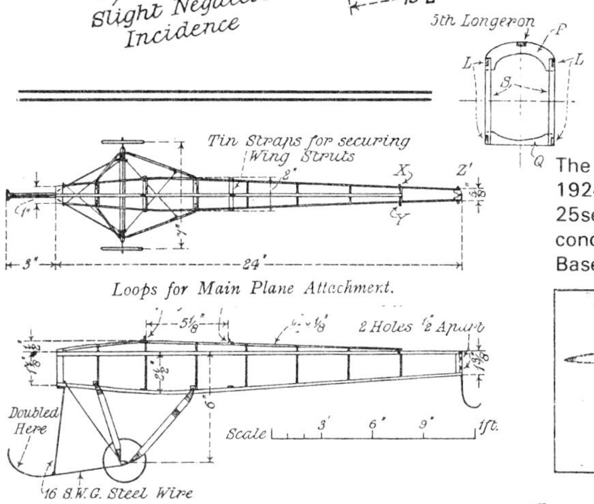

The model below won the *M.E.* Cup in 1924 in bad weather with a flight of 25secs. but had flown in all conditions for three years previously. Based on a design by L. A. Gray, it was built by F. de P. Green, who was the man whose approach to Lord Wakefield in 1927 resulted in the establishment of the Wakefield Cup.

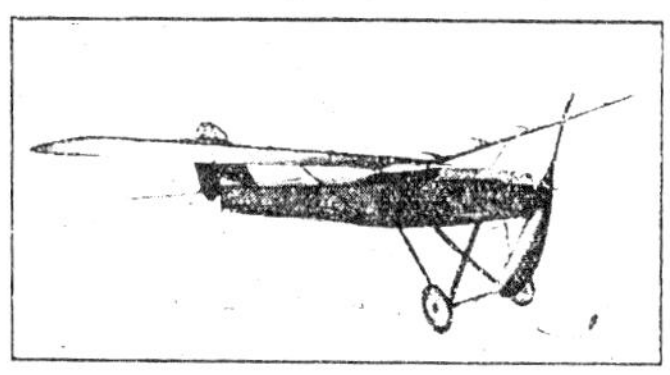

B. K. Johnson's model (illustrated opposite) used a single-surface wing covered with 'oiled silk'

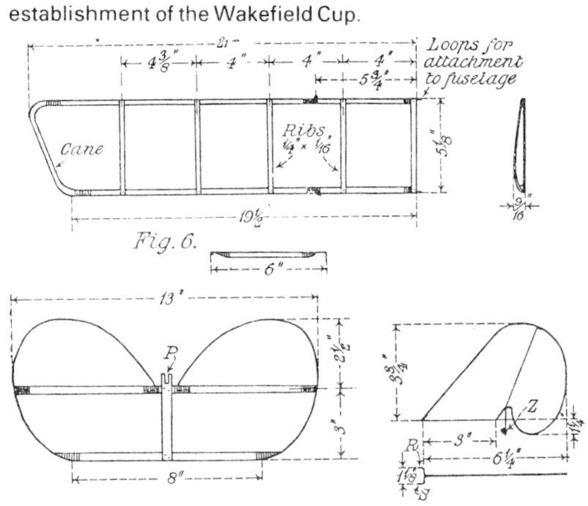

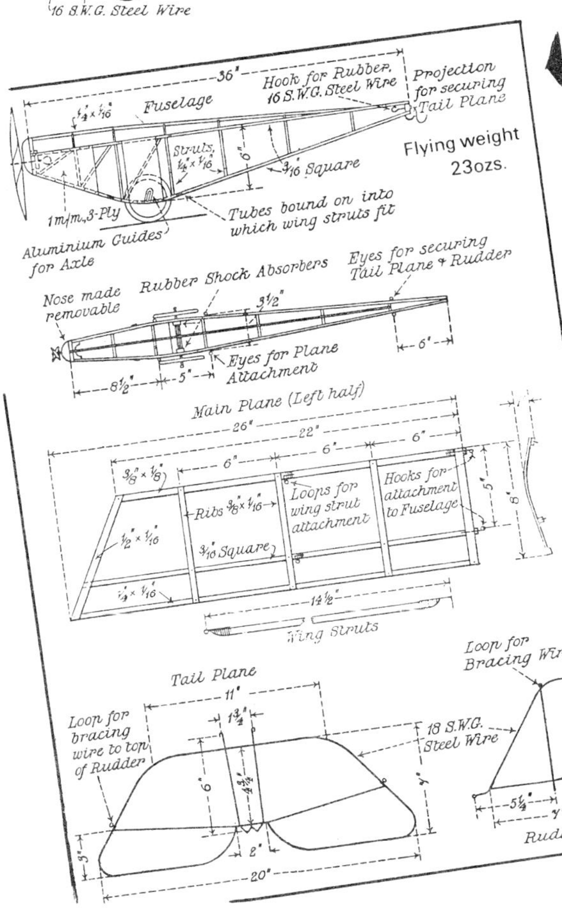

Consistent 50sec. flights were achieved by D. A. Pavely in the 1924 *M.E.* No. 2 and Gamage Cup competitions with the spar tractor below, right. The model weighed 5¼ozs., 1¼ of which was ¼in. rubber driving the 12in. dia., 18in. pitch prop.

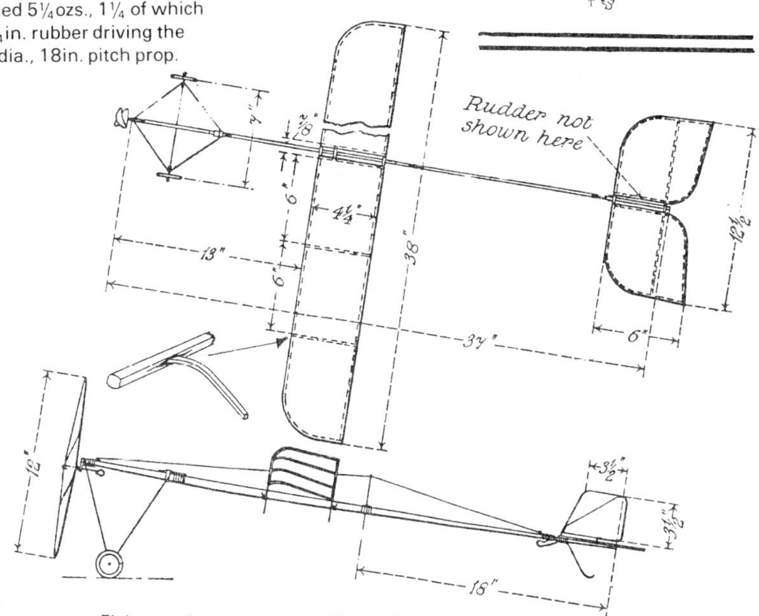

Flying surface outlines bamboo

Scale 0 6" 12"

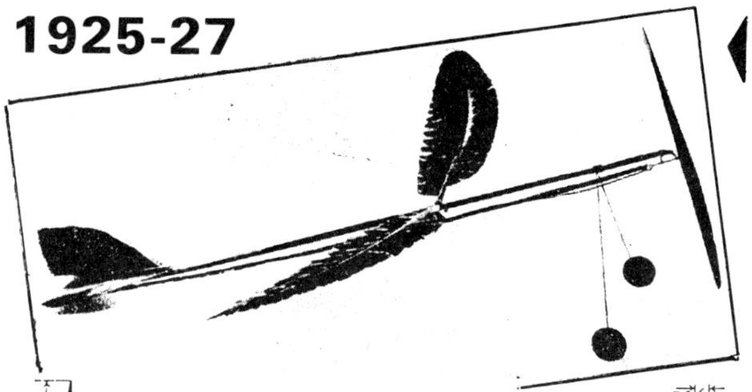

R. N. Bullock, one of the great British flyers of the 1920s and 30s, produced a series of lightweight models in 1925 using mainly turkey feathers; model shown flew for 77secs. R.O.G.

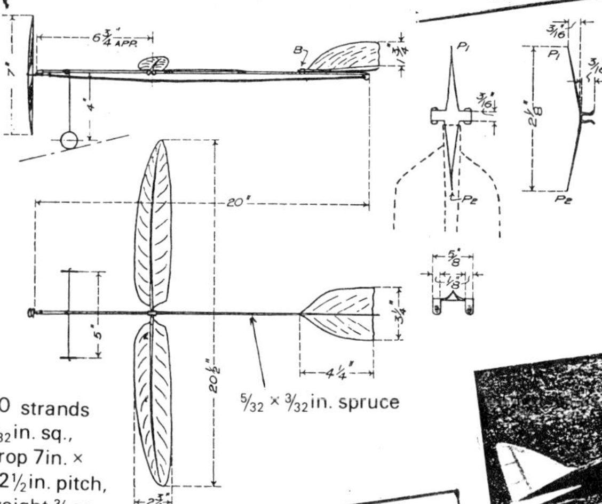

10 strands 1/32in. sq., prop 7in. × 12½in. pitch, weight 3/8oz.

The S.M.A.E. Cup was put up for the fastest model over a 50 yard course (into wind) in 1926. Ralph Bullock won with this advanced machine, which covered the course into a 12m.p.h. wind in 6secs.

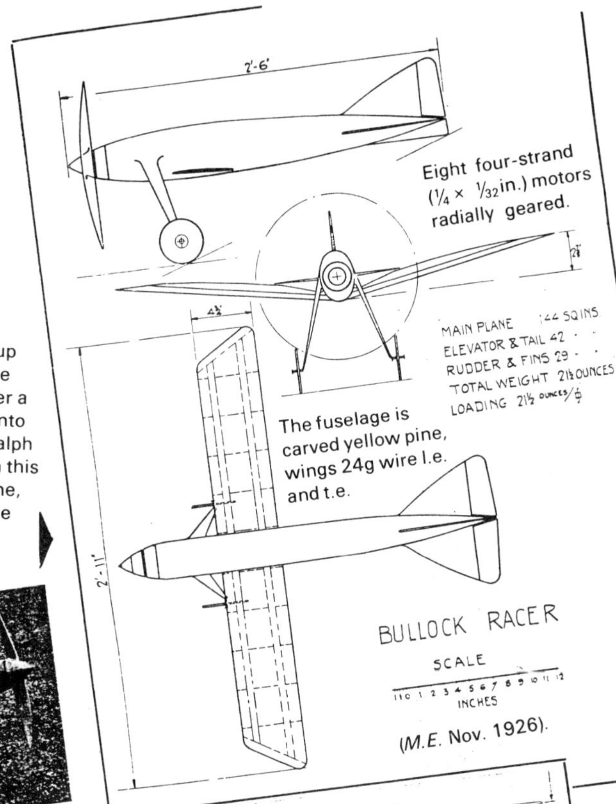

Eight four-strand (1/4 × 1/32in.) motors radially geared.

The fuselage is carved yellow pine, wings 24g wire l.e. and t.e.

(*M.E.* Nov. 1926).

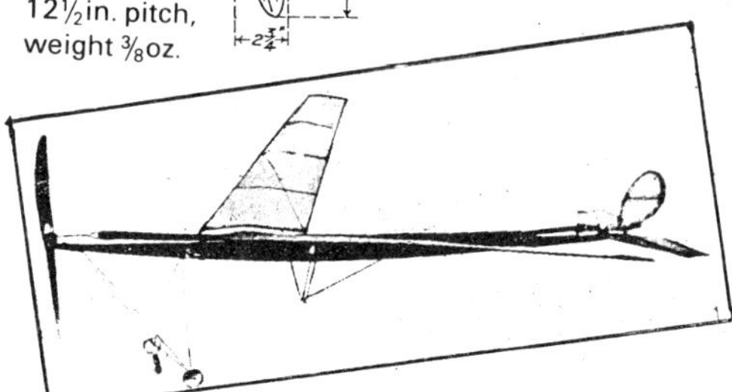

The 1925 Lady Shelley Cup was offered for 'self-righting' tractor models with an included sweep angle of 90-140 deg. and the winning machine (left) was by D. A. Paveley. Single-surface wing, 10 strands of 1/4in., weight 8ozs.

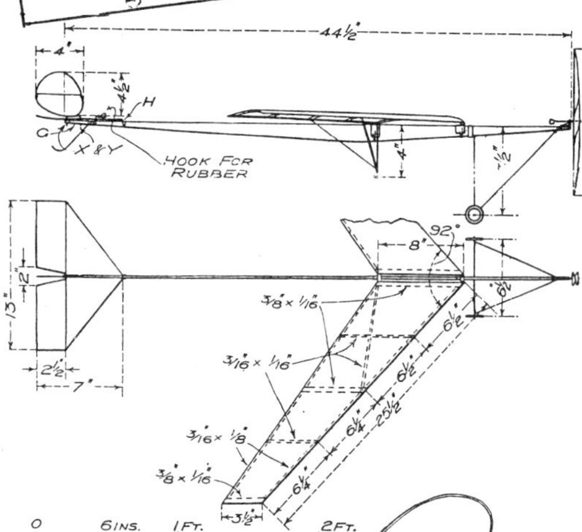

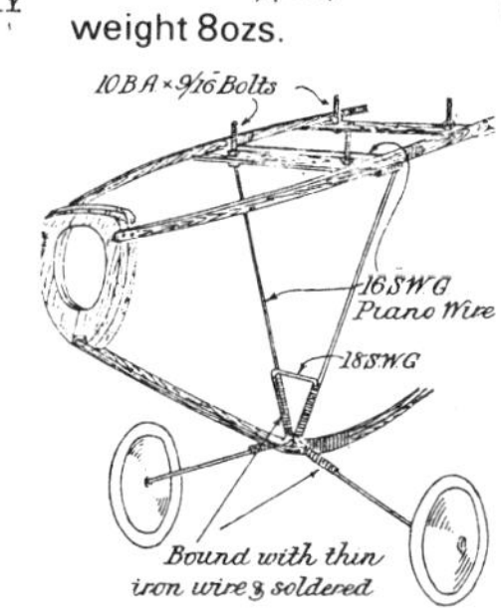

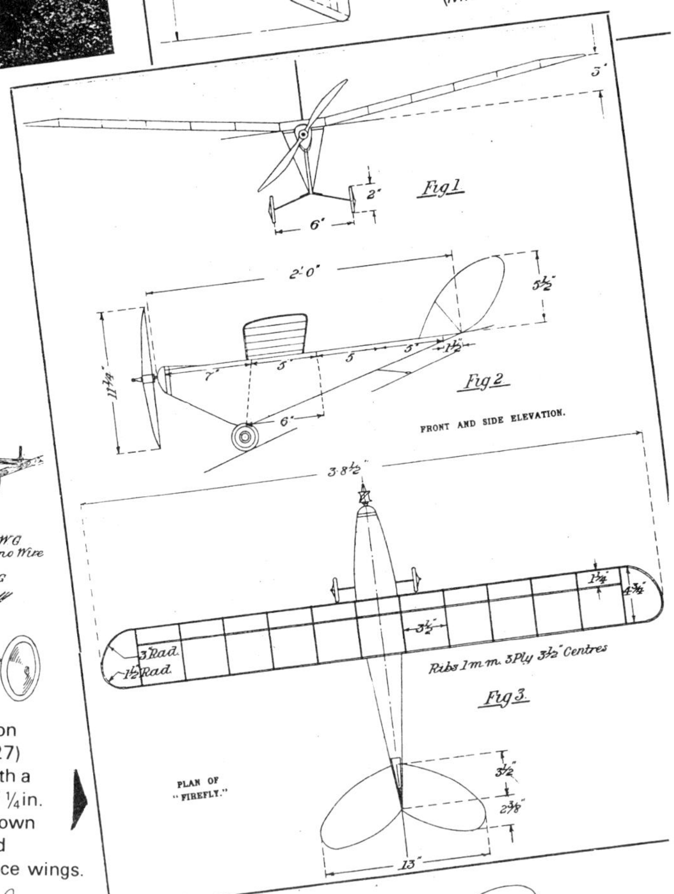

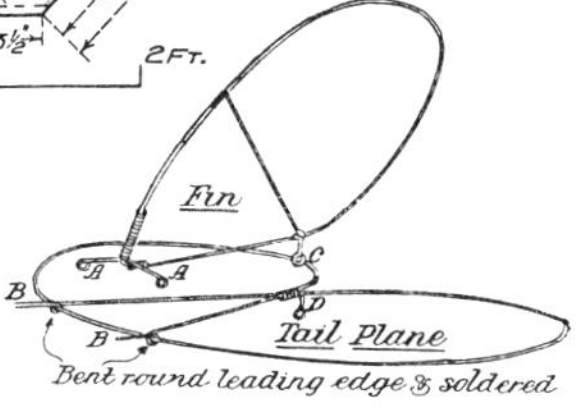

1925 *Firefly* by H. Jackson (published *M.E.* July 1927) originally flew R.O.G. with a 7in. prop and one loop of 1/4in. rubber at 6ozs. but as shown had geared twin 5-strand motors and double-surface wings.

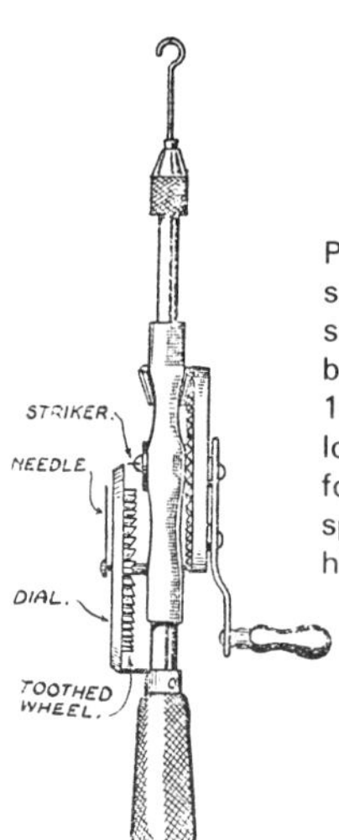

Plater monoplane, Type 8, Orville series No. 9 (!) was the ninth model in a series. Span 39in., very strongly built but capable of H.L. flights of around 1min. Fuselage is 1/8 × 3/16in. spruce longerons on six 1/8in. and one 3/16in. ply formers, wing 1/8 × 5/32in. l.e., 1/4in. sq. spar and 3/32 × 1/4in. t.e., with 1/16in. hickory ribs and half-ribs.

Power two 10 strand motors of 1/4in. rubber 54in. long.

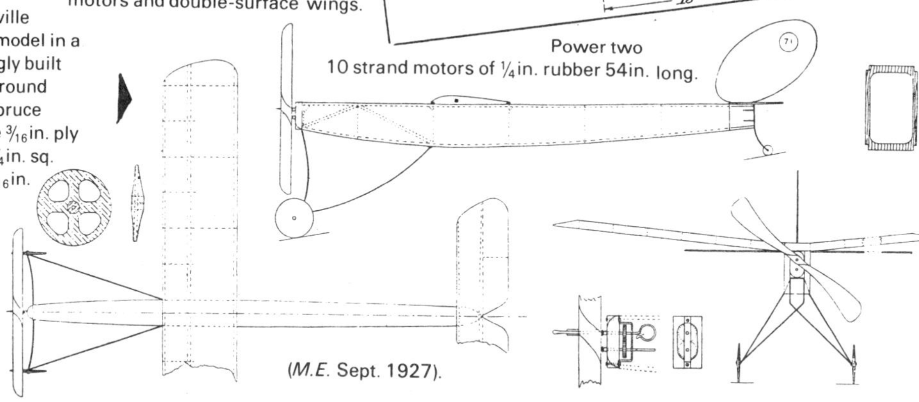

Winders with automatic counters seemed to occupy a lot of thought. This one uses a striker to drive a gearwheel one tooth per rev, (Newnes).

(*M.E.* Sept. 1927).

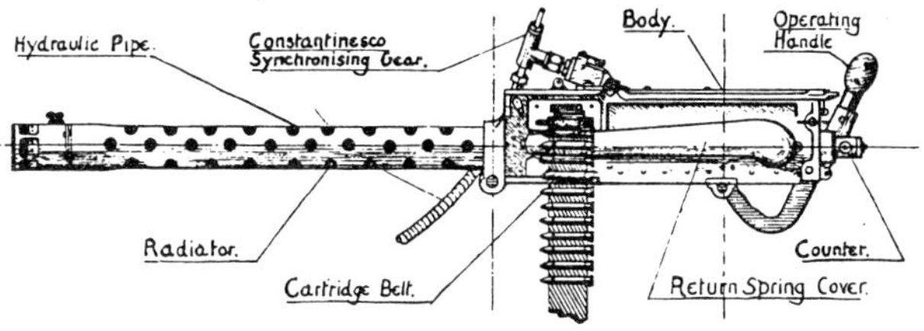

The Vickers Machine Gun (Aero Type) .303.

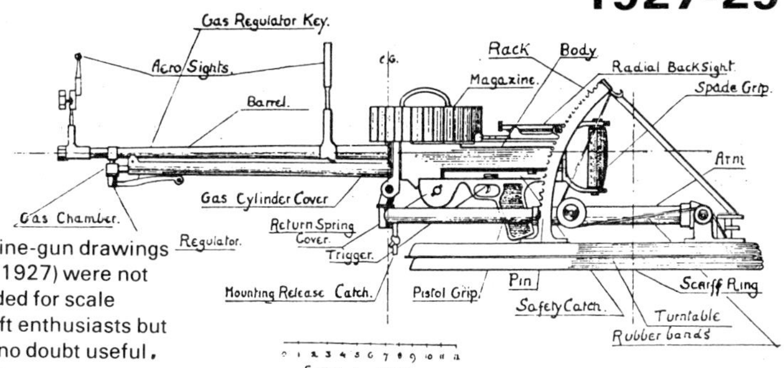

The Lewis Gun, Aero Pattern, .303.

Machine-gun drawings (July, 1927) were not intended for scale aircraft enthusiasts but were no doubt useful.

I WAS very much surprised to learn quite recently that there are 200,000 boys in the United States who are actively interested in building and flying model aeroplanes. A gigantic competition for these enthusiasts was organised some months ago by *The American Boy*, the very enterprising and entertaining organ of youthful American life and interests. Mr. Merrill Hamburg, a member of the staff of that journal, conceived the idea that interest in aviation and some practical knowledge of its methods and possibilities could best be inculcated in the rising generation by getting them busy on building model aeroplanes. He published articles and designs in his paper, set up a department for supplying materials at practically cost price, and organised the competition I have above referred to. The principal prizes were two free trips to Europe. Mr. Merrill has just been over here with the happy winners of these free trips and • • •

Percival Marshall, publisher of *Model Engineer*, wrote supportively of American efforts in August, 1928.

Below, another 'M.E.' editorial piece, in September 1929,

The Revival of Interest.

Readers may perhaps remember that in the early days of flying a large number of model aeroplane clubs were started all over the country, and thousands of people built and flew model aircraft. Then came the war, and during the period of hostilities most of the clubs died out. An effort was made after the war to revive interest in model aircraft, but much uphill work had to be done. New clubs were formed, the object being to design, build and fly models. The type of model most popular in the early days was the "spar," which was purely a flyer and bore little resemblance to the full-size machine. Model enthusiasts now wanted something more like the real thing, so they set about designing fuselage models, mostly monoplanes of the tractor type. This type of model has been developed, and is now capable of a really high performance, and its flying qualities are much better than the earlier "spars."

Model aircraft returned to *The American Boy* with this 'Baby R.O.G.' in October 1927, followed by a model virtually every month through to mid-1934. Encouraged by William B. Stout, the magazine formed the Airplane Model League of America. Merrill Hamburg wrote up all the published models.

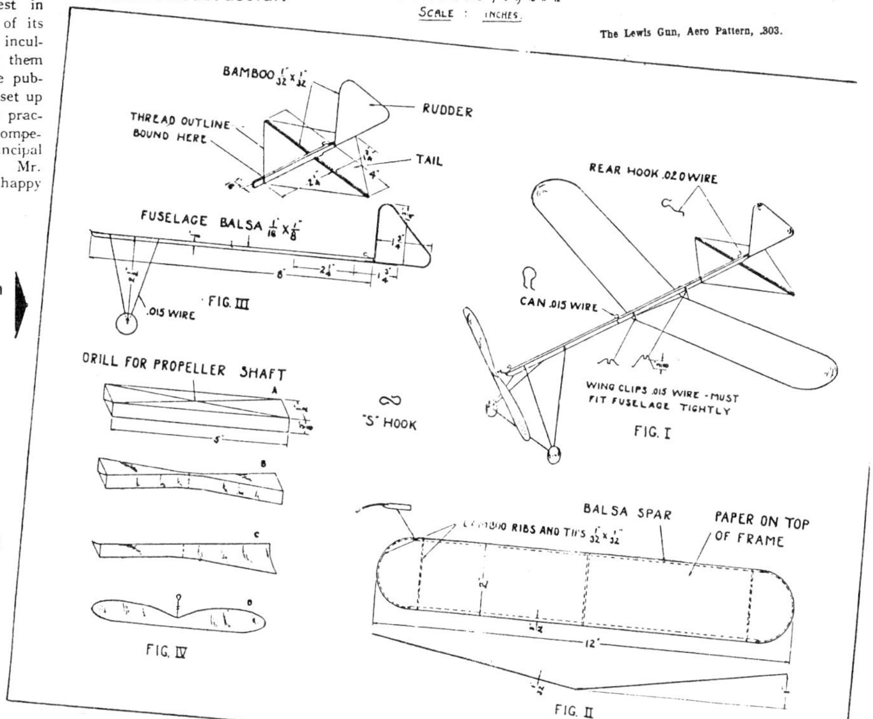

News of rocket-propelled models by B. K. Johnson (then Tech. Sec. of the S.M.A.E.) appeared in *M.E.*, December 1928.

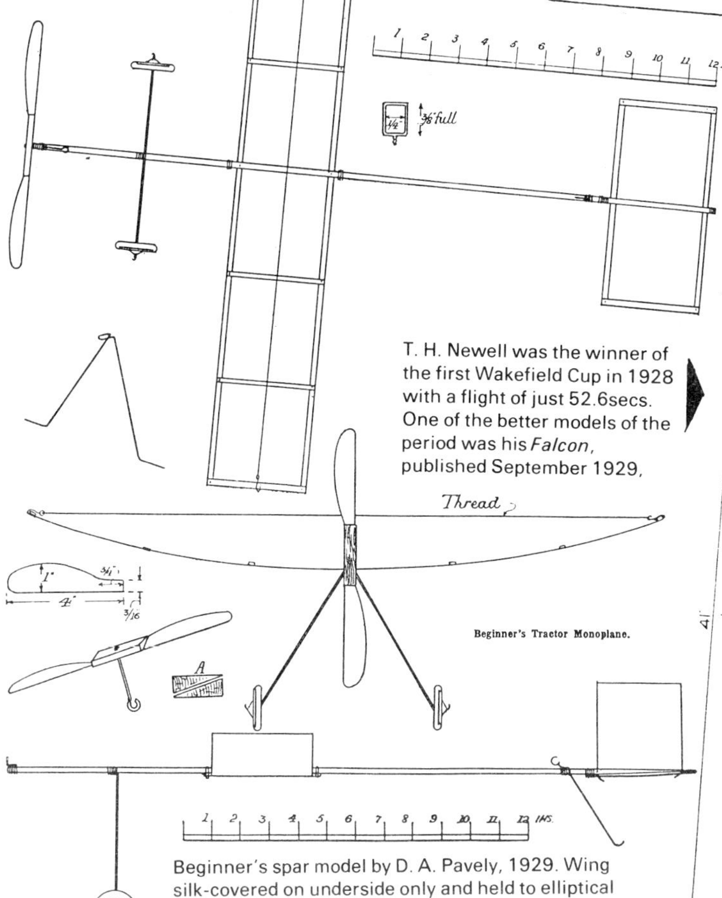

Beginner's spar model by D. A. Pavely, 1929. Wing silk-covered on underside only and held to elliptical dihedral by thread. Spar $\frac{1}{4} \times \frac{5}{16}$in., l.e. and t.e. $\frac{1}{4} \times \frac{1}{16}$, ribs $\frac{3}{16} \times \frac{1}{16}$, all birch.

T. H. Newell was the winner of the first Wakefield Cup in 1928 with a flight of just 52.6secs. One of the better models of the period was his *Falcon*, published September 1929.

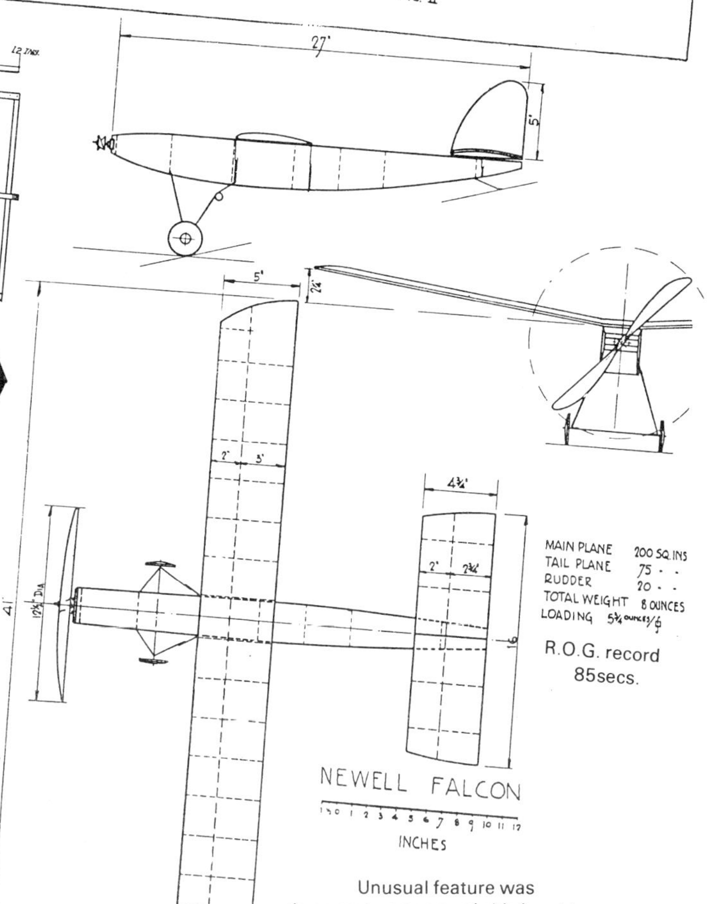

R.O.G. record 85secs.

Unusual feature was three motors arranged side by side.

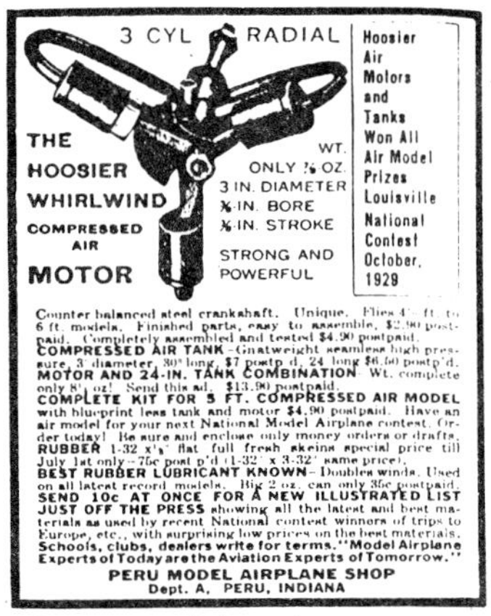

The late 1920s were perhaps the hey-day of compressed air engines although they spanned more than 30 years and were available commercially for much of that time. The 1928 *American Boy* adverts above show two, plus a book which was 'official' for the National Playground Miniature Aircraft Tournament, sponsored by Col. Lindbergh, Orville Wright and Cdr. Byrd, among others. The author was Paul E. Garber of the Smithsonian Institute. . . . . Lindbergh's epic transatlantic flight was responsible for a wave of airmindedness in, particularly, the U.S.A. and by happy coincidence balsa had just been 'discovered' together with Jap tissue, setting the scene for enormous expansion of model-building.

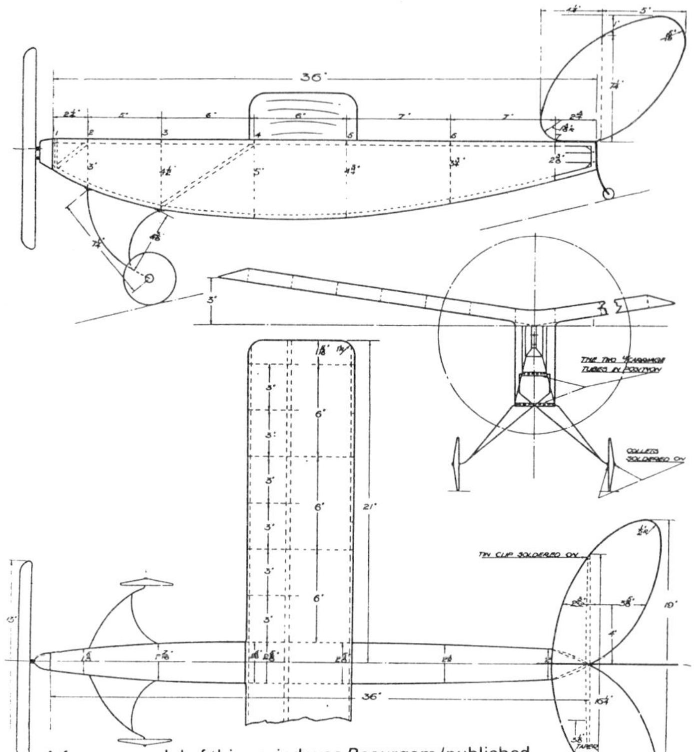

A famous model of this period was *Resurgam* (published *M.E.* November 1929) which was a consistent and reliable all-weather flyer if, to modern eyes, rather ugly. Twin geared and heavy at 17½ozs. it would nevertheless produce a regular 70secs. h.l. Designer was W. J. Plater.

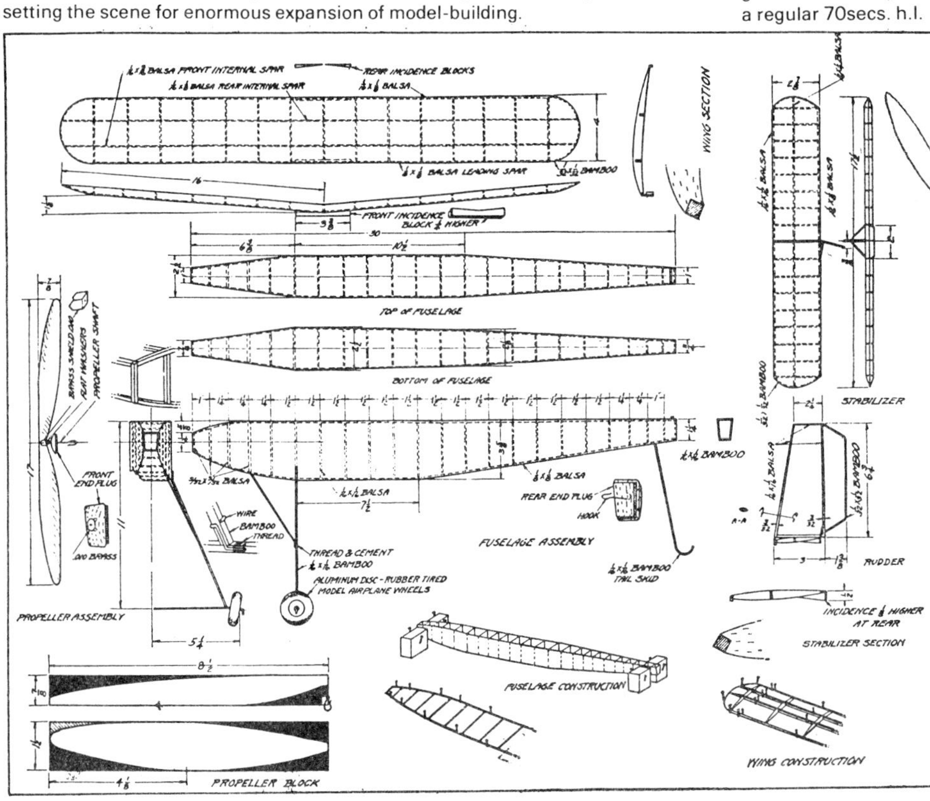

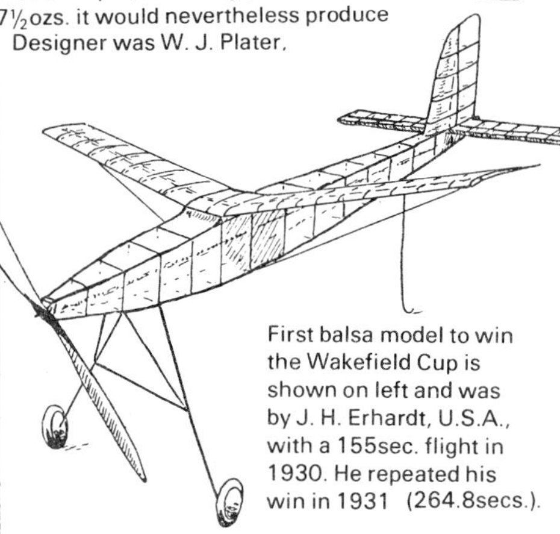

First balsa model to win the Wakefield Cup is shown on left and was by J. H. Erhardt, U.S.A., with a 155sec. flight in 1930. He repeated his win in 1931 (264.8secs.).

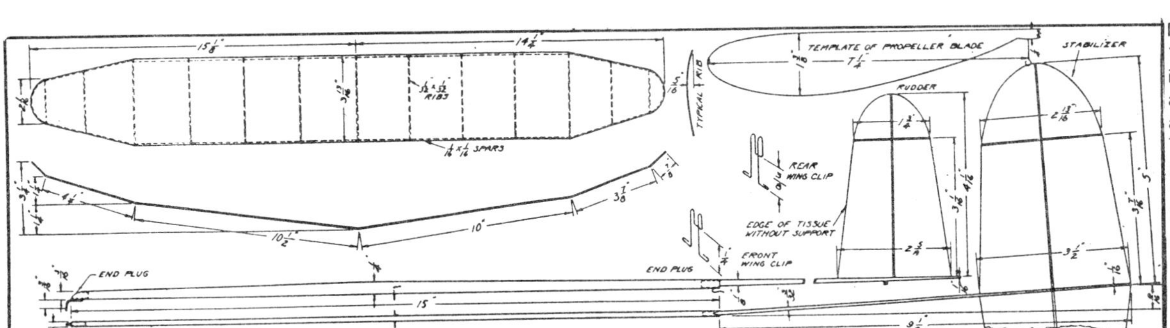

Most documented pre-1928 American flying was indoor, by clubs having access to the huge armouries in various cities. Three indoor models produced flights of over 11mins. in June 1929 in Detroit; that at left was second with 11:38, by Carl Goldberg, who was to become one of the greatest names in American model flying.

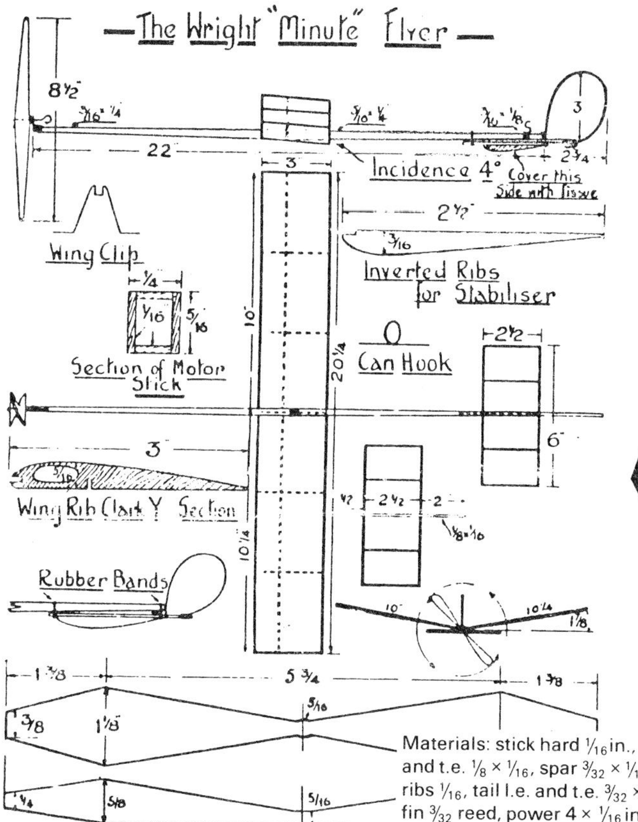

Materials: stick hard $\frac{1}{16}$in., l.e. and t.e. $\frac{1}{8} \times \frac{1}{16}$, spar $\frac{3}{32} \times \frac{1}{16}$, all ribs $\frac{1}{16}$, tail l.e. and t.e. $\frac{3}{32} \times \frac{1}{16}$, fin $\frac{3}{32}$ reed, power $4 \times \frac{1}{16}$in. sq. rubber.

This simple 'spar tractor' by H. Warrell of the Nottingham club was published in July 1930 and is fairly typical of the type of machine most people thought of as a model aeroplane.

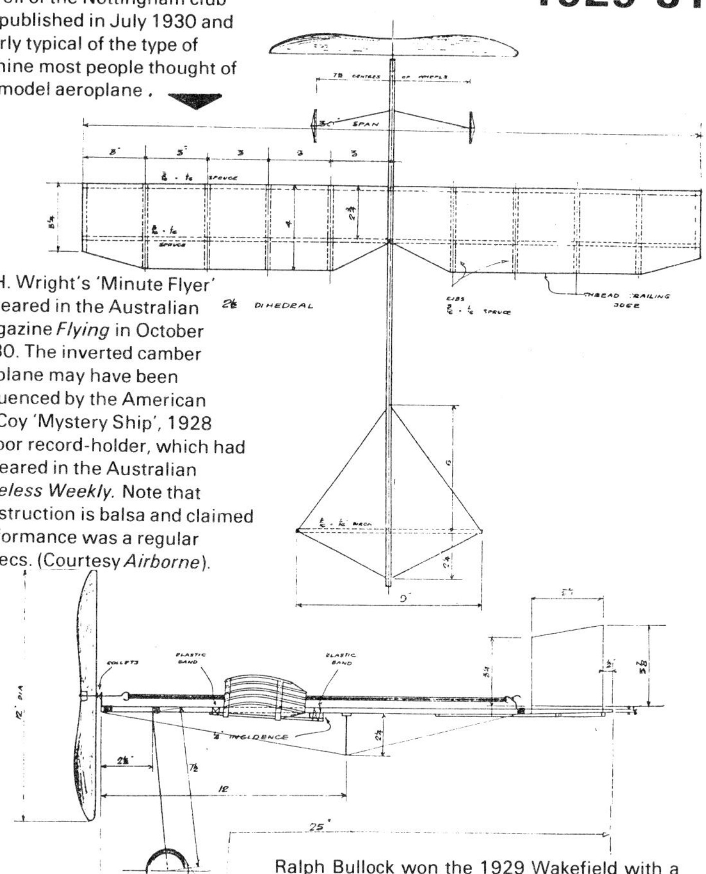

N. H. Wright's 'Minute Flyer' appeared in the Australian magazine *Flying* in October 1930. The inverted camber tailplane may have been influenced by the American McCoy 'Mystery Ship', 1928 indoor record-holder, which had appeared in the Australian *Wireless Weekly*. Note that construction is balsa and claimed performance was a regular 90secs. (Courtesy *Airborne*).

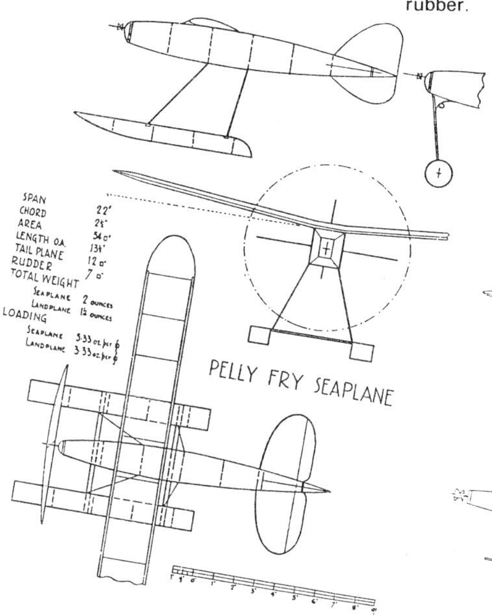

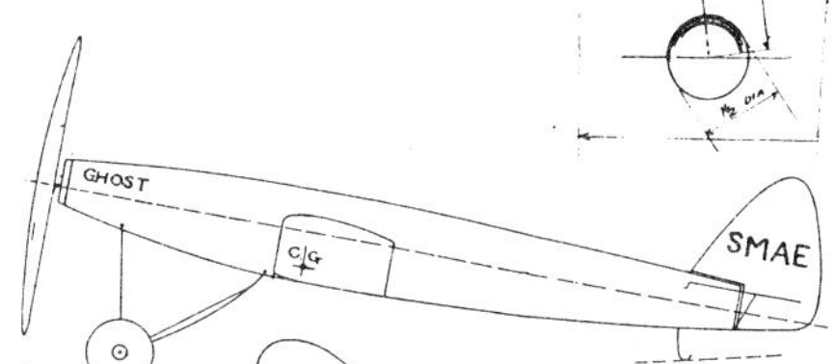

Ralph Bullock won the 1929 Wakefield with a low-winger (70.4 secs.) but this was the only win with such a layout. Juste van Hattum's V.H. 36 'Ghost' was his 1931 16oz. Wakefield; American models were around 2oz. The Cup stayed in America. For 1932 the best-of-three score was changed to average of three, a system at least reducing the impact of one long thermal flight, but the event was declared void because its date had been changed without adequate notice to overseas entrants.

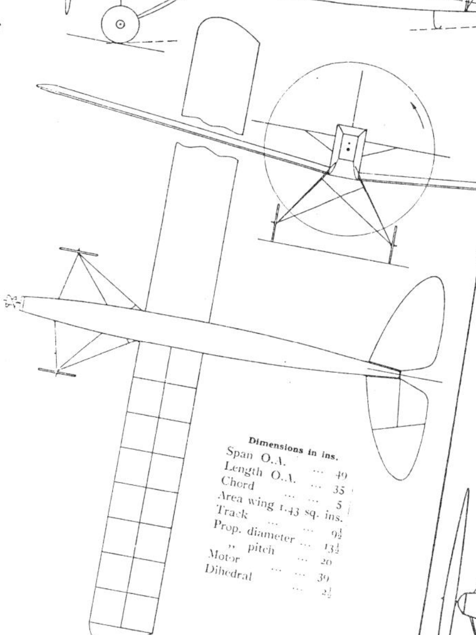

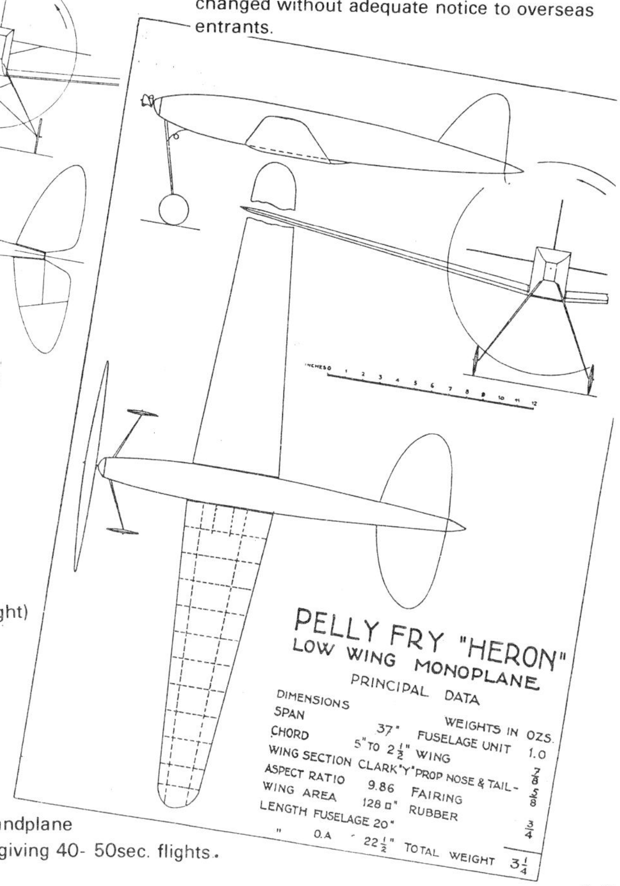

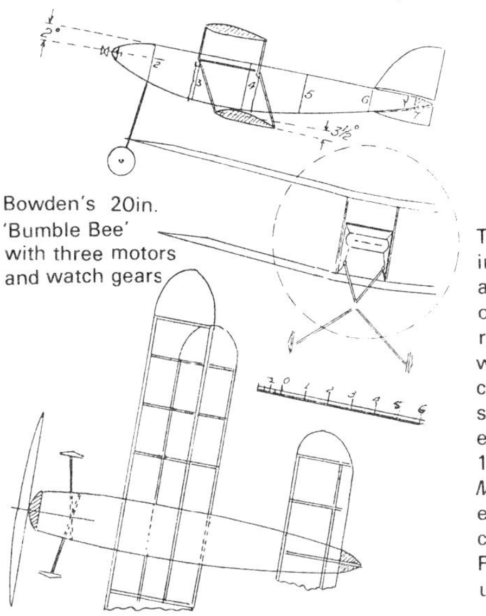

Bowden's 20in. 'Bumble Bee' with three motors and watch gears

Two major British names to emerge in the late '20s were J. E. Pelly Fry and C. E. Bowden, both career officers in the RAF and Army respectively. The elegant 'Heron' (right) was one of the most successful competition models of the 1929/30 seasons, winning among other events the 1929 Pilcher Cup, the 1930 Weston Cup and the 1930 *M.E.* No. 1 Cup. Both these enthusiasts built 'vest pocket' silk-covered models in late 1929, Pelly Fry's a 22in. convertible seaplane/landplane using two motors and watch gears, giving 40- 50sec. flights.

1931-32

Ever-growing aviation enthusiasm plus the revolution in construction brought about by balsa, tissue and cellulose cements led to the beginnings of a boom in kit production in the U.S.A. Most of these kits were for smallish scale models, at prices from 20 cents up (equivalent to 5p U.K. at the then rate of exchange) and if a new builder completed one successfully he was a modeller for life!

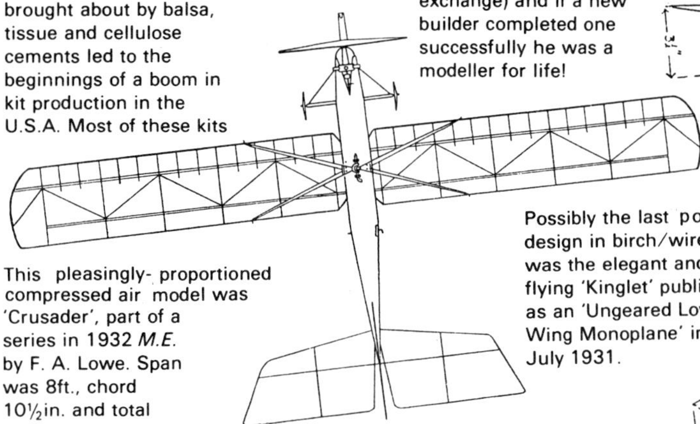

This pleasingly-proportioned compressed air model was 'Crusader', part of a series in 1932 *M.E.* by F. A. Lowe. Span was 8ft., chord 10½in. and total weight 2½lb., giving less than 6oz./sq. ft. wing loading, a remarkably low figure for the time.

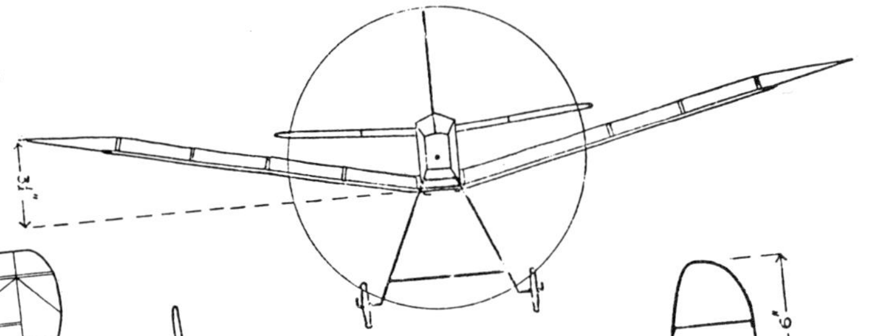

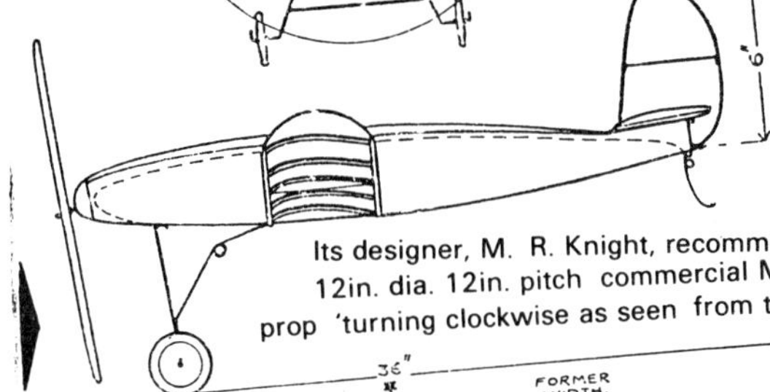

Possibly the last popular design in birch/wire/silk was the elegant and fine-flying 'Kinglet' published as an 'Ungeared Low-Wing Monoplane' in *M.E.* July 1931.

Its designer, M. R. Knight, recommended a 12in. dia. 12in. pitch commercial Mosstum prop 'turning clockwise as seen from the front.'

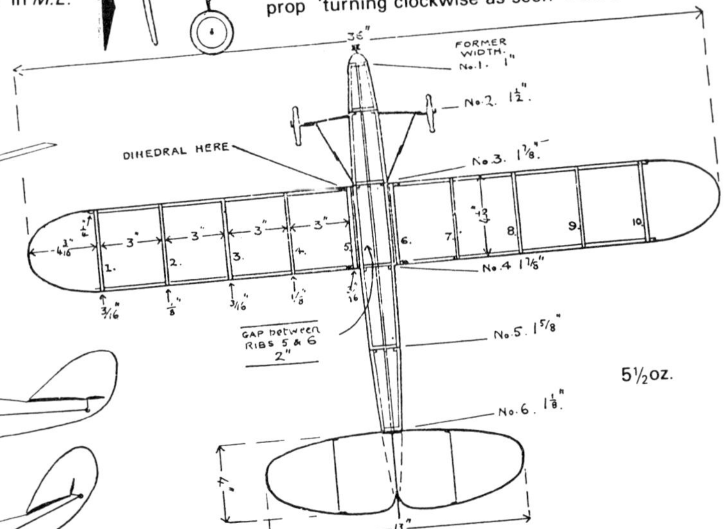

5½oz.

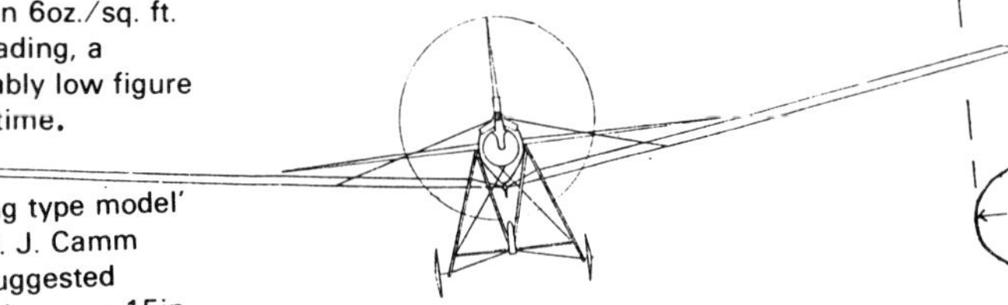

'Twining type model' in an F. J. Camm book: suggested dimensions are 15in. length, 12 and 6in. planes (of thin balsa)

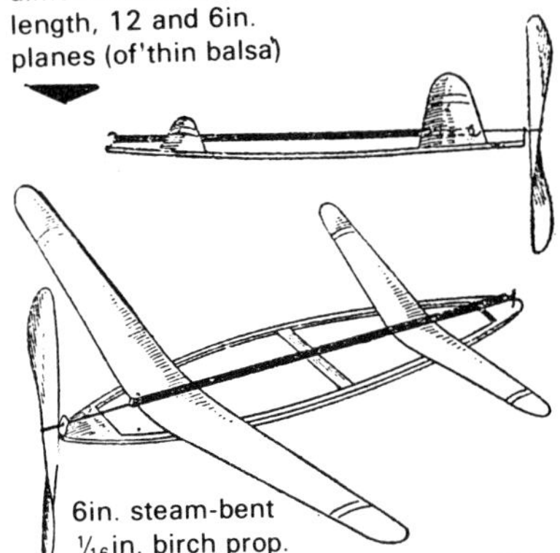

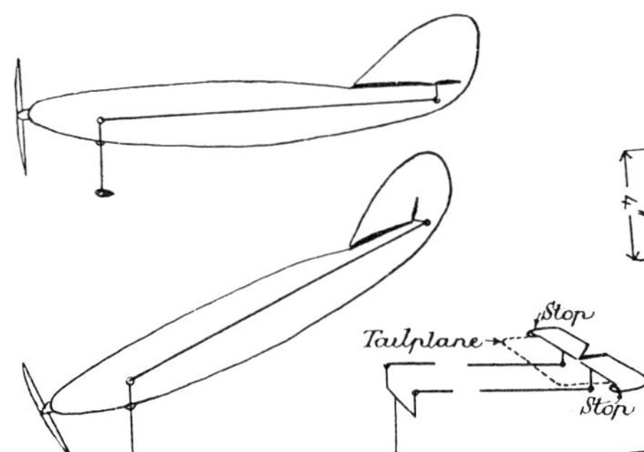

Pendulum control employed by L. Mann (Aug. 1931) though not a new idea, he said.

6in. steam-bent 1/16in. birch prop.

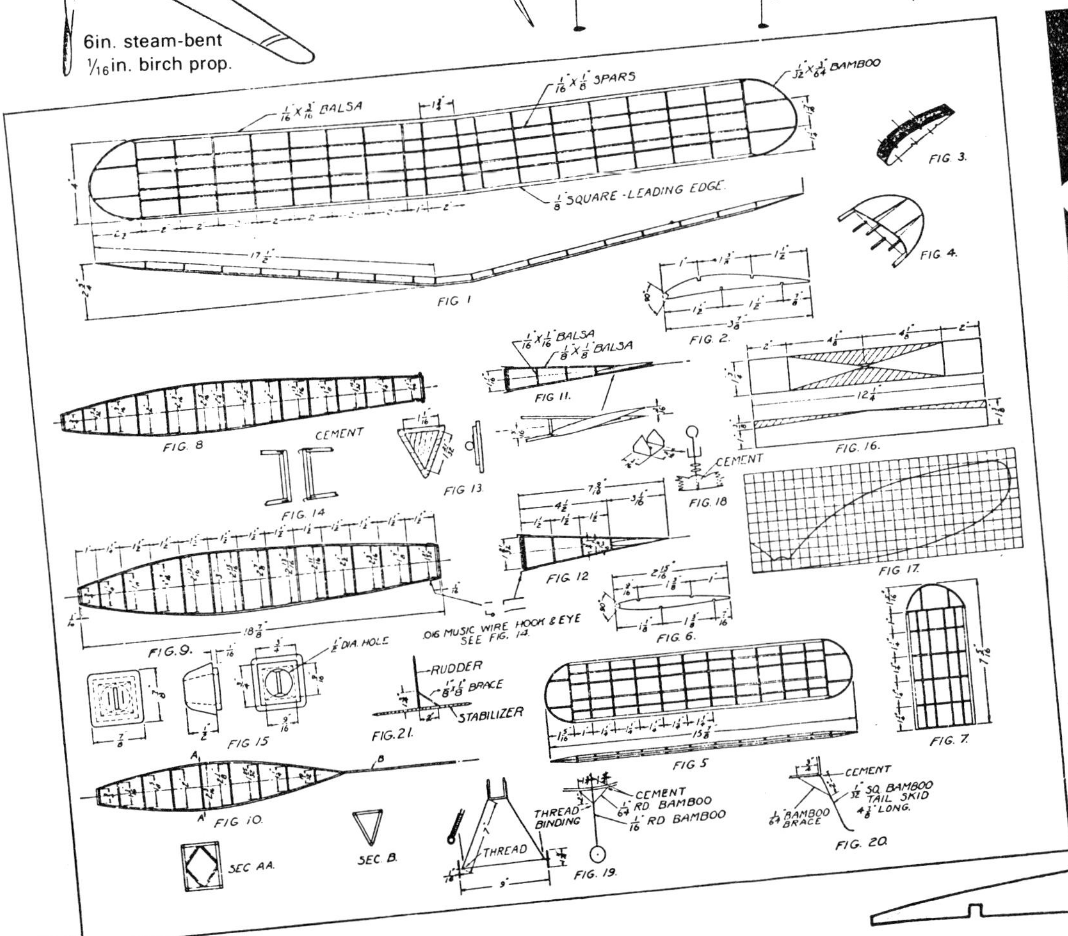

Outstanding Australian model by N. H. Wright of Sydney (another of his models on page 19) made numerous remarkable flights, the longest of 24min. 14.6secs.

The best ship of 1931, said *American Boy* of this 37in. Emanuel Feinberg design which flew 17 and 29½min. at the 1931 Nationals at Dayton. The same style of square-to-triangular fuselage reappeared years later.

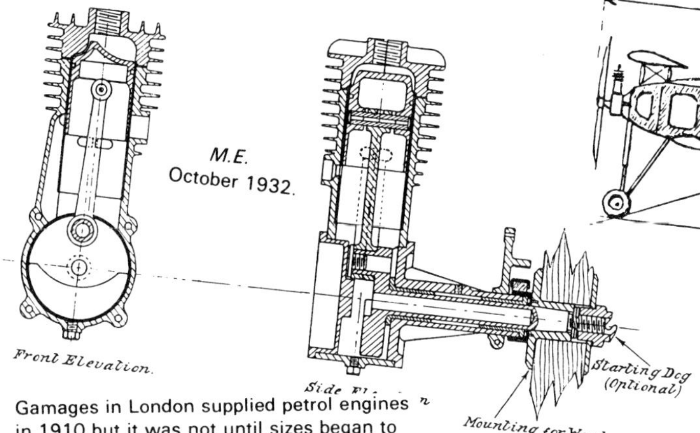

Gamages in London supplied petrol engines in 1910 but it was not until sizes began to come down that much interest was shown. One pioneer was Edgar R. Westbury, whose original 'Atom' was 52cc, reduced to 30cc and then to the engine above, the 'Atom Minor' at 14.2cc,

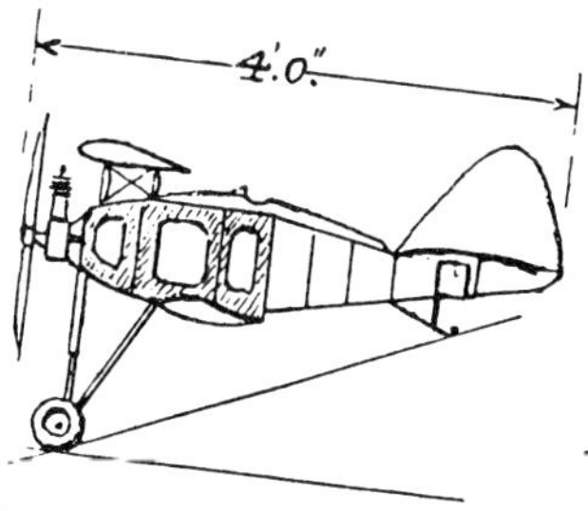

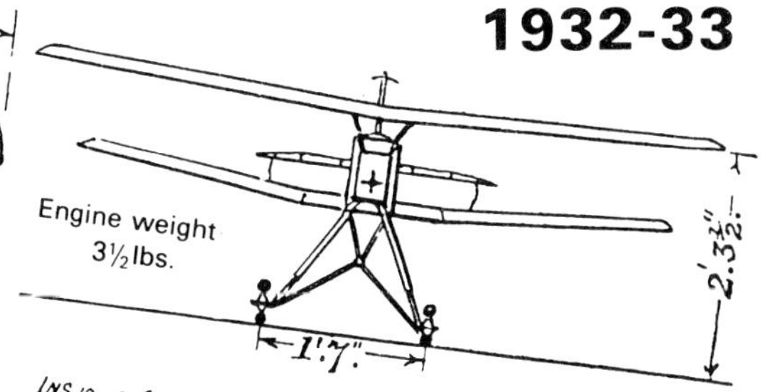

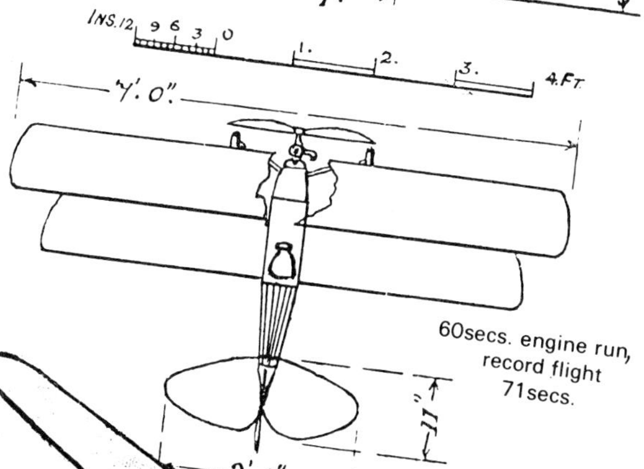

Col. (then Capt.) Bowden set the first petrol-engine duration record since Stanger's in 1914 with the 7ft., 8lb. 14oz. 'Kanga' using a Wall Junior 28cc two-stroke hydroplane engine.

60secs. engine run, record flight 71secs.

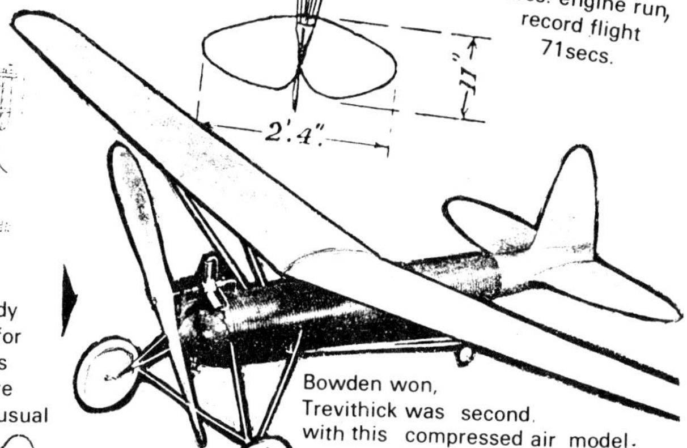

The 1933 Lady Shelley Cup for power models attracted more entries than usual — five.

Bowden won, Trevithick was second with this compressed air model.

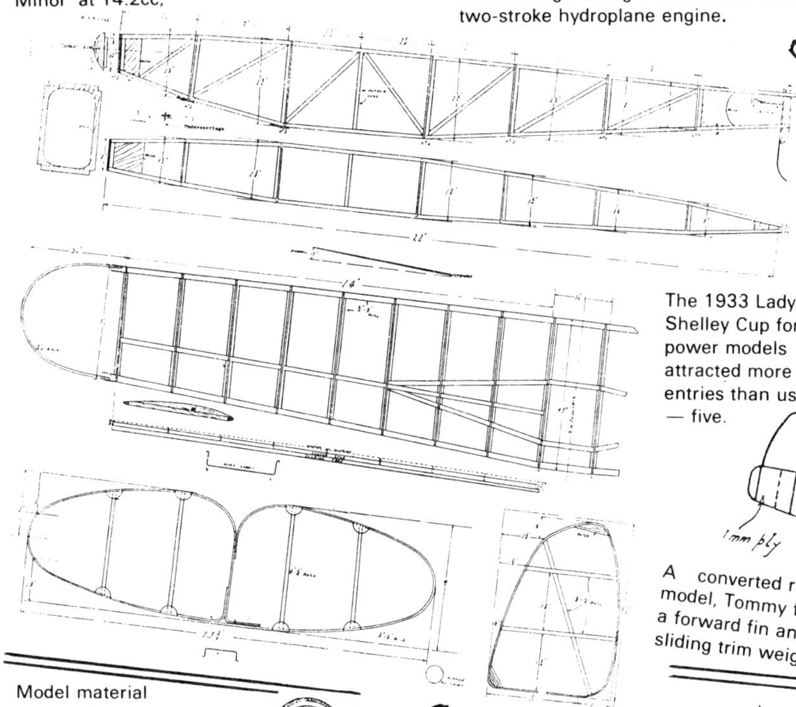

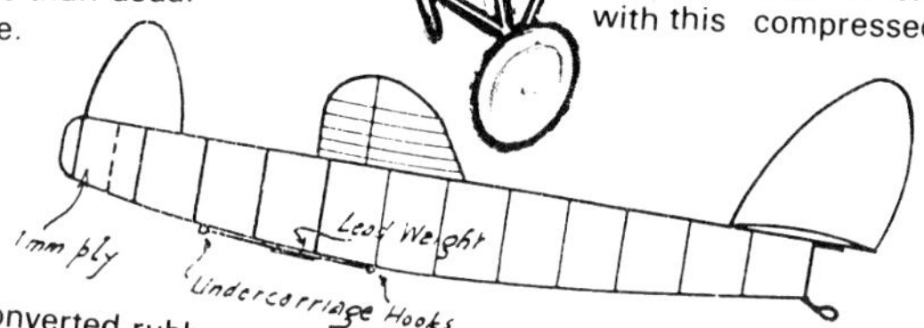

A converted rubber model, Tommy fitted a forward fin and a sliding trim weight.

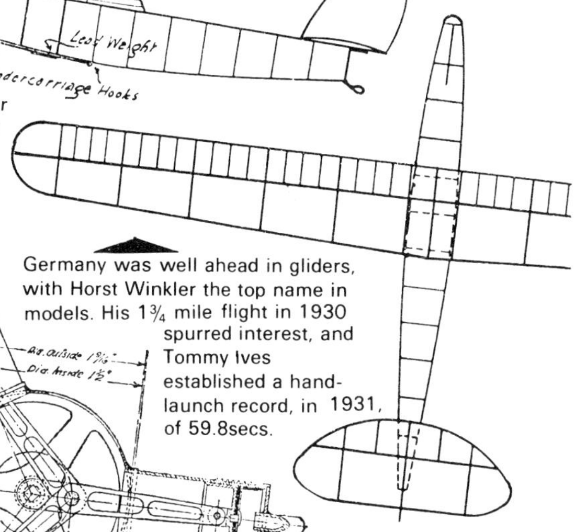

Model material appeared in George Newnes Ltd. publications, usually under the editorship of F. J. Camm. This rotary cone-valve compressed air engine was among the last of this form of power, rendered effectively a curio by 1934.

One of the first all-balsa English designs to achieve wide popularity was the 'Lincol' of 1932, designed by Linfoot and Cole; it was said that a Lincol was in 1935 the first all-balsa model seen in Germany. Flights of up to 12½mins. were reported.

Germany was well ahead in gliders, with Horst Winkler the top name in models. His 1¾ mile flight in 1930 spurred interest, and Tommy Ives established a hand-launch record, in 1931, of 59.8secs.

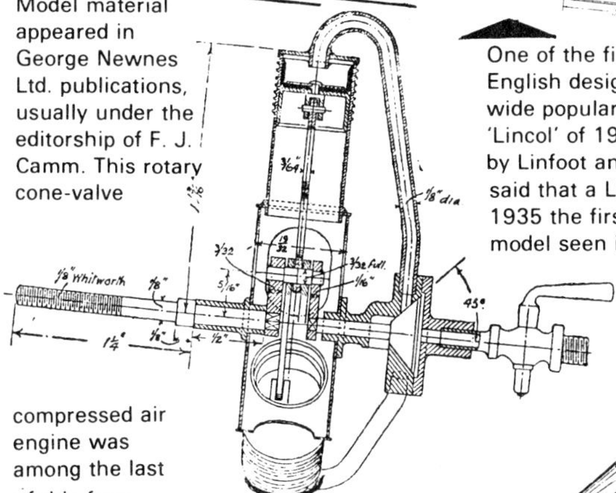

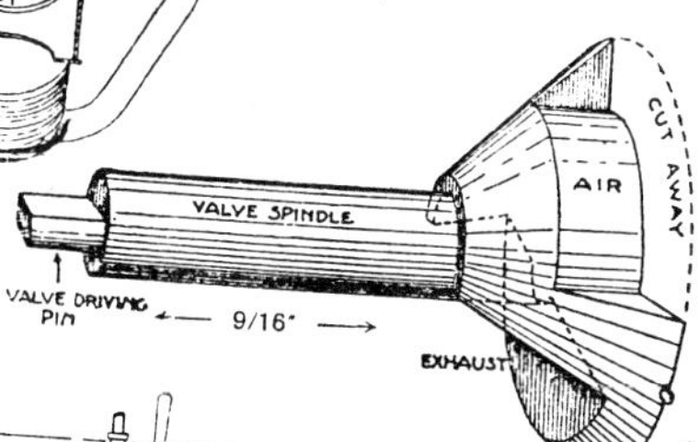

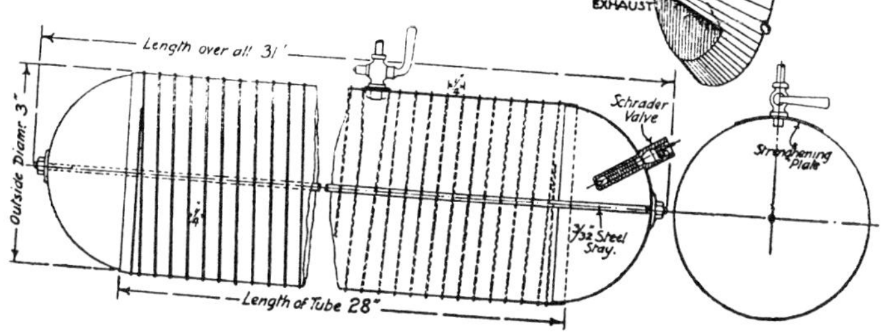

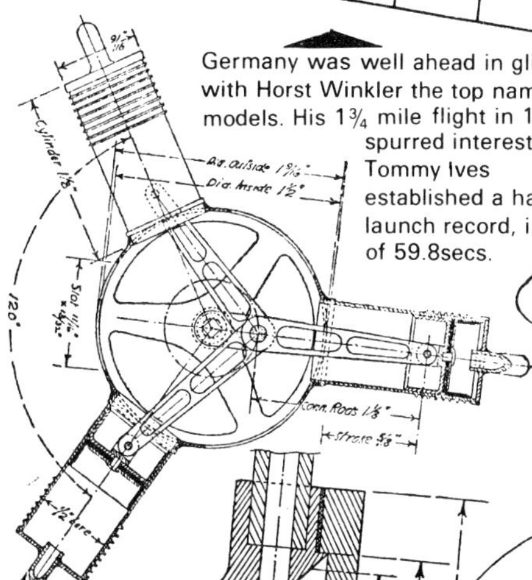

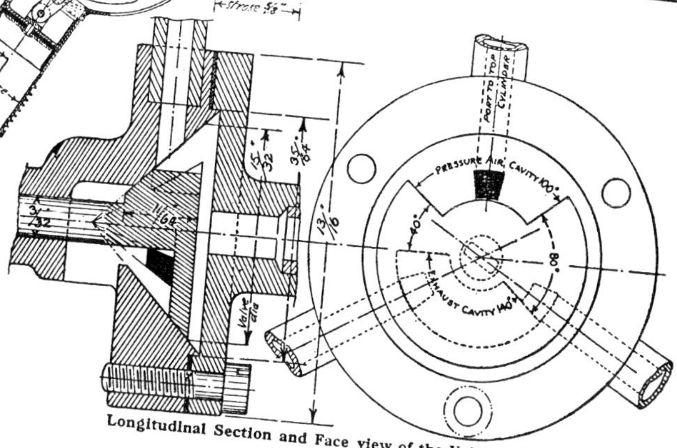

Longitudinal Section and Face view of the Valve and Valve Chest.

1933

Top scorer in the 1932 Wakefield which was subsequently voided was Gordon Light (U.S.A.) whose model (left) recorded 7:57.2. Third was J. E. Pelly Fry (3:22.2) who proxy-flew Light's model to third place in 1933. It was the subject of a degree of disparagement from some proponents of silk-covered, hardwood, geared heavyweights typified by the remarkably ugly 'Ad Astra' at right.

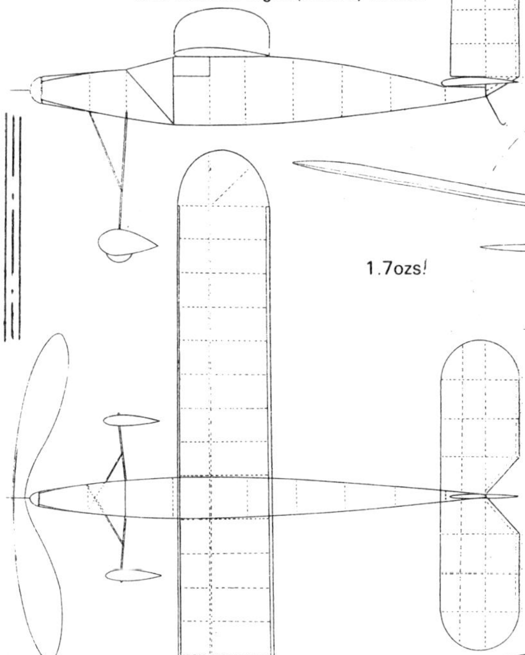

1.7ozs!

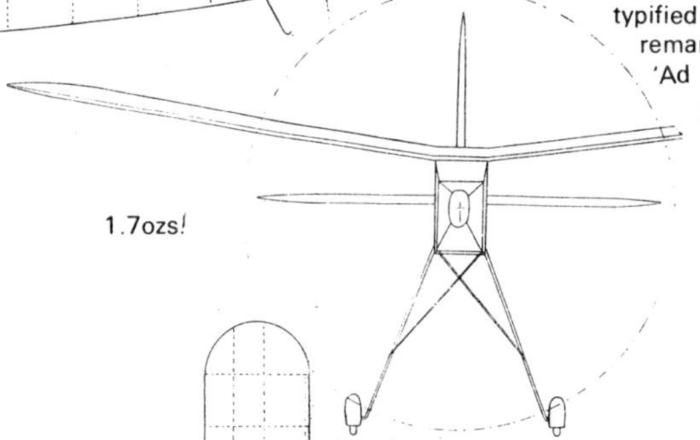

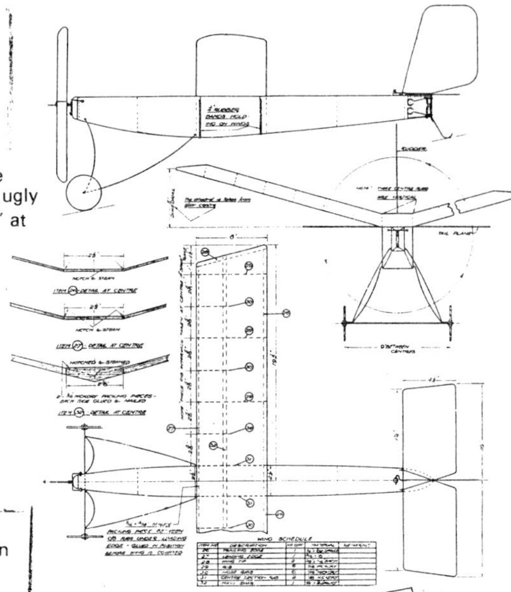

The 'Ibis' design has recently been resuscitated by *Airborne* magazine in Australia for a one-design vintage event in honour of its designer, Tim Hervey, a major influence on model flying down under.

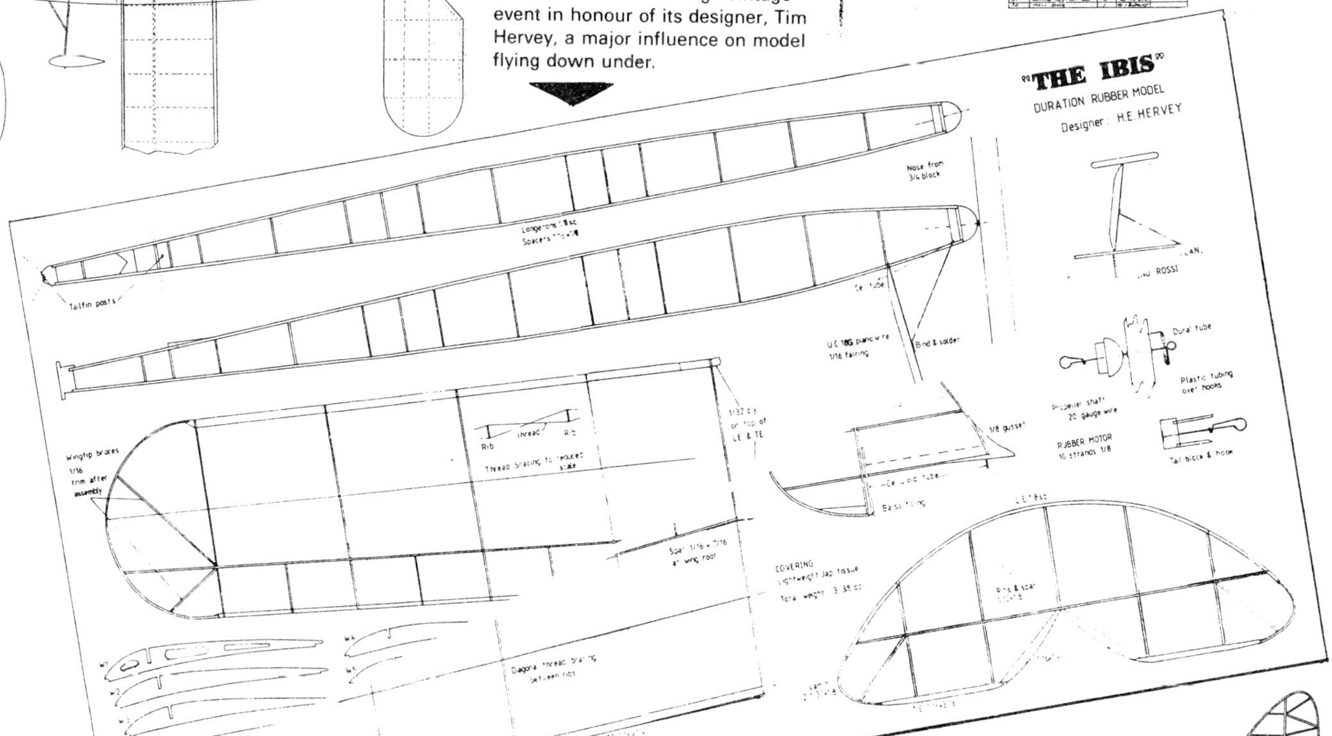

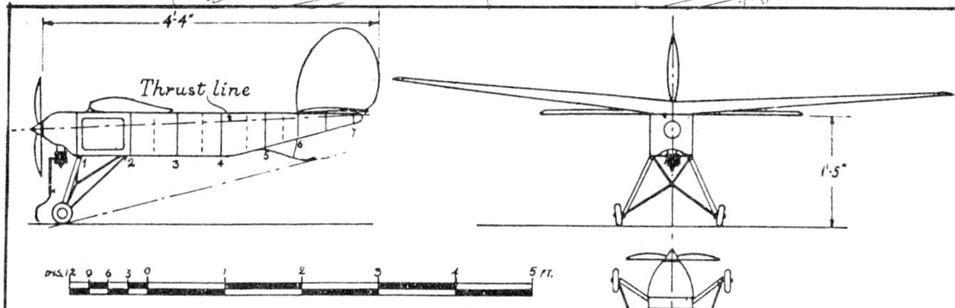

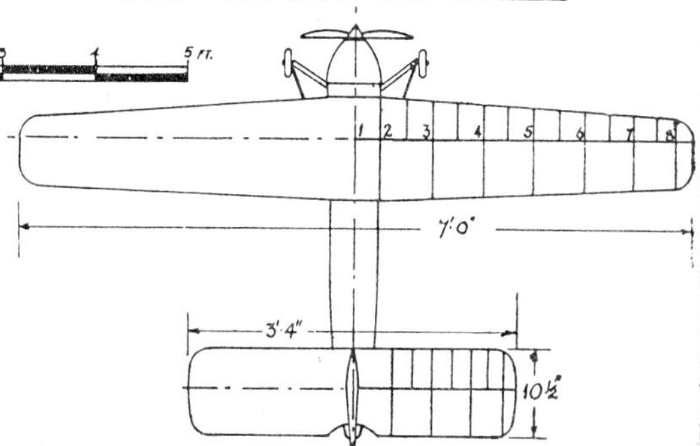

The Bowden 'Bee' was the second model from the Bowden/Westbury team, using an Atom Minor (14.2cc) driving a 15in. cast elektron prop at 3,500r.p.m. and weighing all-up 6½lbs. Even the designer wrote '. . . an ugly but simple and robust model . . .' It raised the record to 86secs.

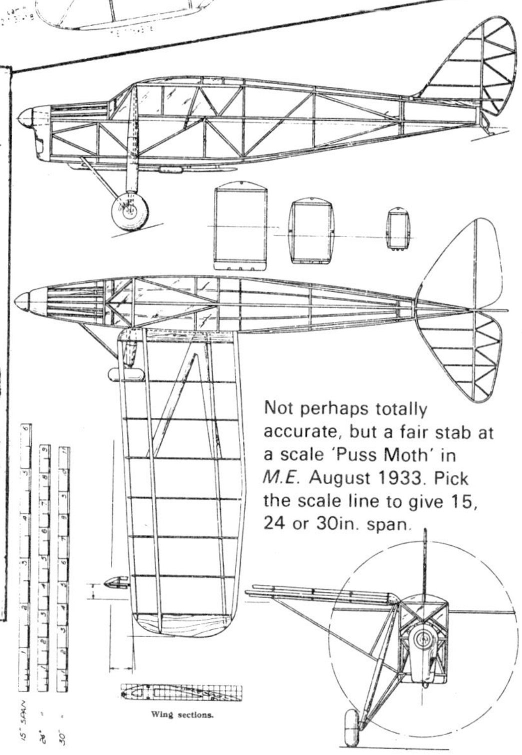

Not perhaps totally accurate, but a fair stab at a scale 'Puss Moth' in *M.E.* August 1933. Pick the scale line to give 15, 24 or 30in. span.

The year 1933 was to trigger the first major change in model flying by the arrival of practical, commercially possible petrol (gasoline) engines. Power flight had been demonstrated in both Britain and U.S.A in 1932 but the competition potential only began to be evident in 1933. For the next dozen years there were marked differences between published British and American power models, the Americans producing graceful, handsome, simple and flyable models.

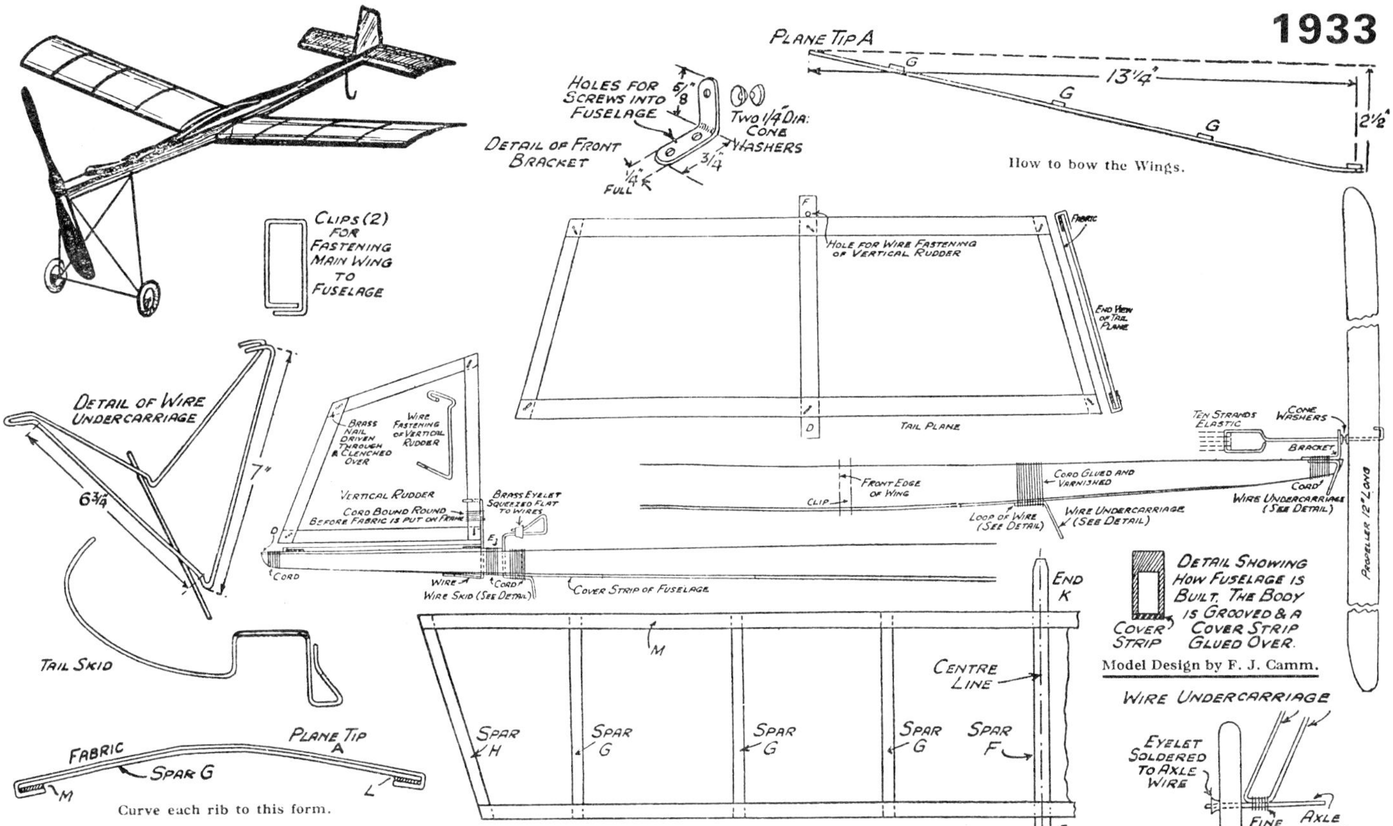

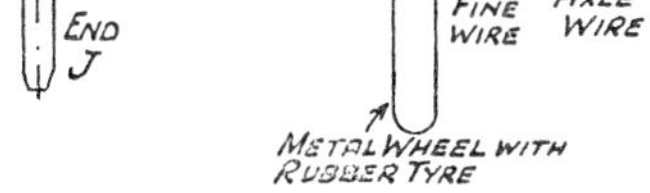

Many mature readers will have started their modelling interest with a spar or stick model in the late 1920s or the 30s. Probably the best-known ready-to-fly range was that by Warneford — R. N. Bullock was responsible for those models — but as has been shown, quite a few designs were published for scratch building. This is one by F. J. Camm, published by Newnes, for beginners, but failing to include dimensions!

An autogiro model was featured in *American Boy* in April 1932 and a formula for microfilm in March 1932.

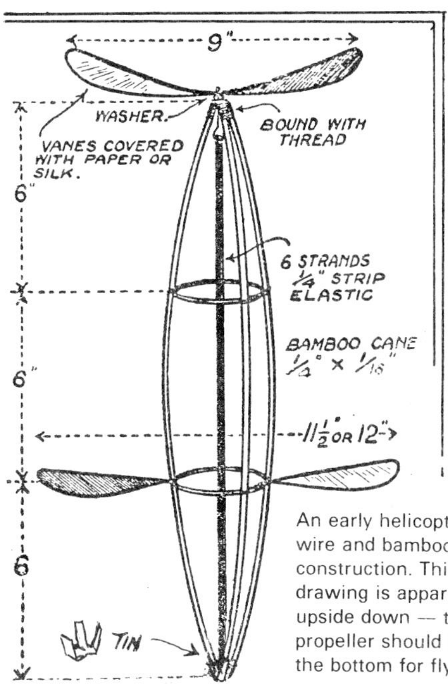

An early helicopter of wire and bamboo construction. This drawing is apparently upside down — the propeller should be at the bottom for flying.

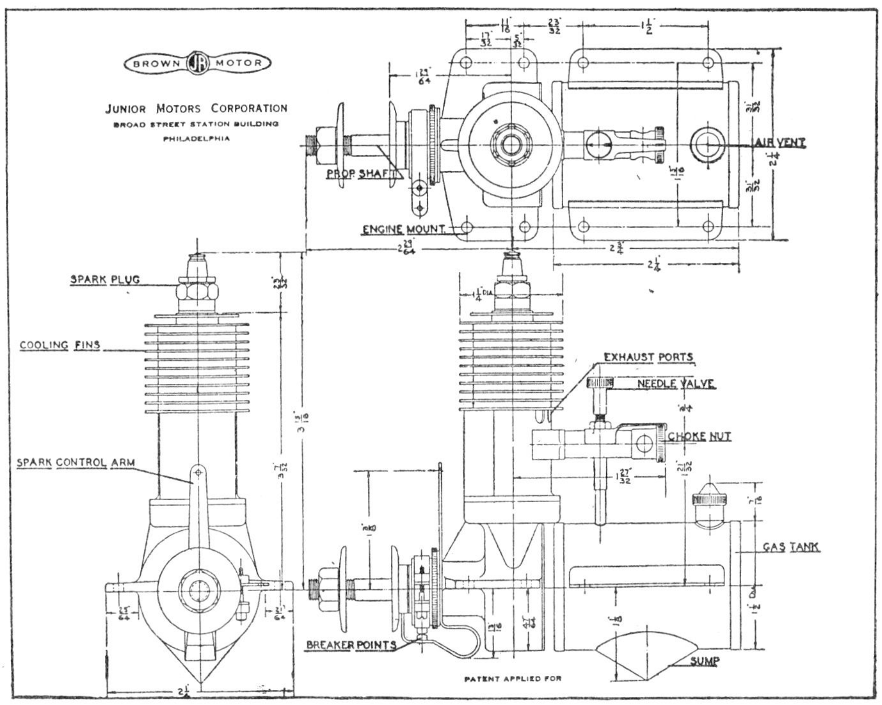

At the U.S. Nationals, New York 1933 Maxwell H. Bassett (Philadelphia), 18 years old, entered two models powered by petrol engines built by Bill Brown, winning three events with 14:55, 22:22.5 and 28:18. Rules subsequently became more specific!

Bill Brown's first successful power flight with Maxwell Bassett was 7.8secs. in May 1932 (later increased to 2:42.6) when Bill was 21 and just graduating from high school to college. He left college in 1933 to work with his father, producing 5,000 Brown Juniors per year; the company effectively ceased in 1938 when stockholders and the Browns disagreed over the need to introduce a smaller motor.

BROWN JR MOTOR

Gas Engine for Model Planes and Boats

Boys! This motor is guaranteed to run. Price $15.75 complete with spark plug, spark coil, condenser and 2 oz. fuel tank. Holds all world's records for gas powered planes. Wide range of speed, 1000 to 6000 R.P.M. with positive control.

Bare Motor 6 1-2 oz. Ready to run with Gas, Two Batteries and Prop. 20 1-2 oz. Approx. 1-6 H. P. SEND ORDER TO

JUNIOR MOTORS CORPORATION

Broad Street Station Building Philadelphia, Pa.

Brown engines were advertised in 1934 at $15.75 with tank, coil and condenser but it was 1936 before gas model kits began to be advertised.

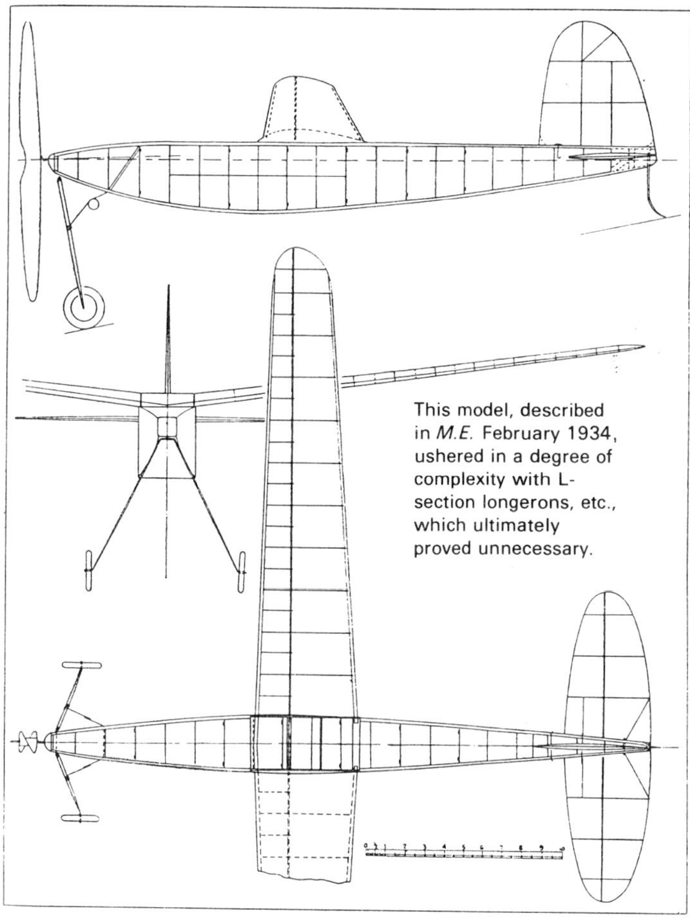

This model, described in *M.E.* February 1934, ushered in a degree of complexity with L-section longerons, etc., which ultimately proved unnecessary.

Mass launches were popular in most clubs. Apart from the thrill of a sky full of models — always impressive to spectators and the Press — there can be no argument over last one down!

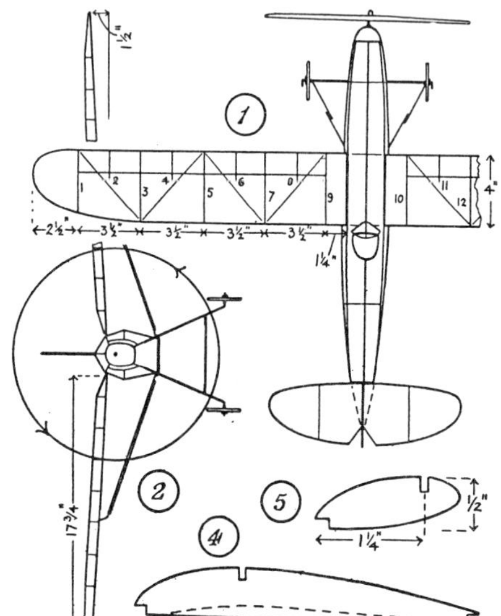

Winner of the 1933 Wakefield was J. W. Kenworthy (GB) with a splendid average of 321secs. His 'Conqueror' model was 48½in. span and flying weight was 2.94ozs., 0.9oz. of which was rubber, driving a 14 × 15½in. pitch airscrew.

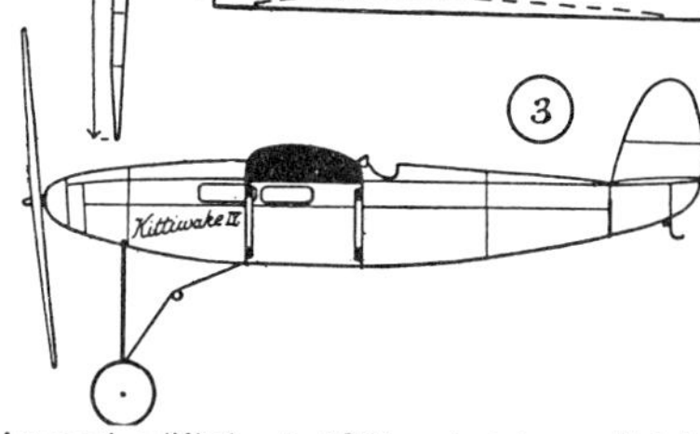

Attractive 'Kittiwake' 38in. strut-braced high-wing model by M. R. Knight, October 1934, relied on accurate wing and strut fittings for correct alignment.

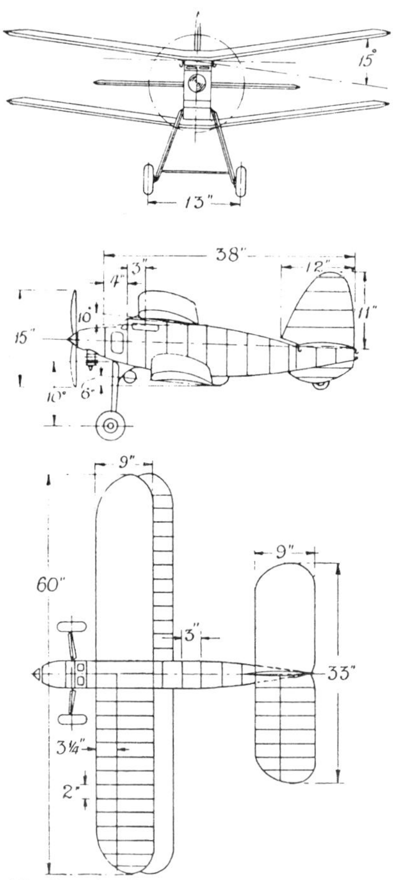

The Bowden 'Baby', at 60in. span considered a very small power model in 1934. Bowden probably wrote more articles on power flying in various magazines than anyone although not everyone agreed with all his beliefs.

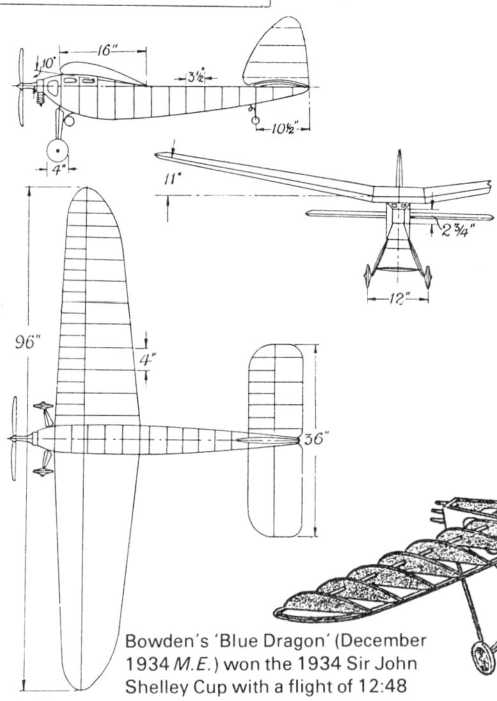

Bowden's 'Blue Dragon' (December 1934 *M.E.*) won the 1934 Sir John Shelley Cup with a flight of 12:48 O.O.S. which was also a new duration record for Britain.

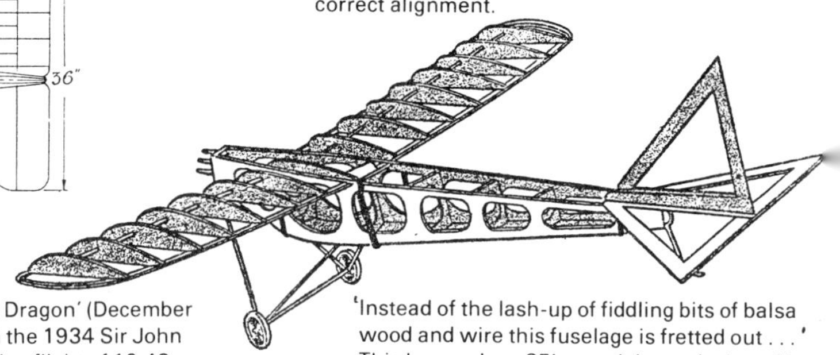

'Instead of the lash-up of fiddling bits of balsa wood and wire this fuselage is fretted out . . .' This horrendous 35in. model was designed in ply, dowel and hardwood strip.

Aircraft enthusiasts will be familiar with the Comper Swift and may have heard of the Comper Streak and the Cranwell (1924 Lympne trials). All were designed by Nick Comper, who as a RAF F/Lt. designed the parasol model on the right for RAF apprentices at Halton to build. Edgar Westbury, a civilian instructor at Halton, provided the 52cc engine, the forerunner of a number of Westbury aircraft motors.

. . . . and was apparently the last rubber model published in the period in which the magazine sponsored the A.M.L.A.

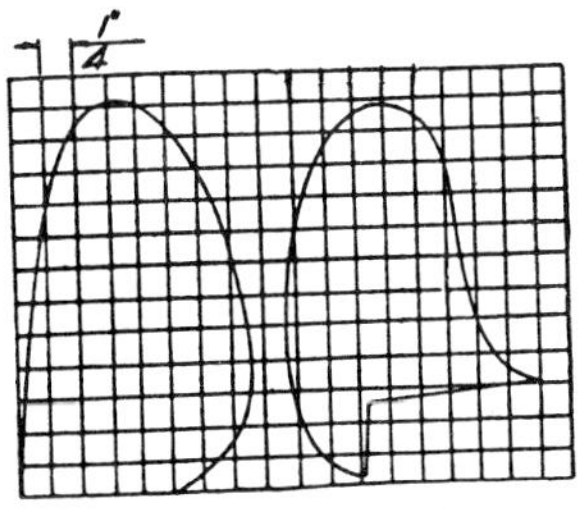

Scale models attracted early interest. This 15in. Fairchild 22 appeared in June 1934 *American Boy*

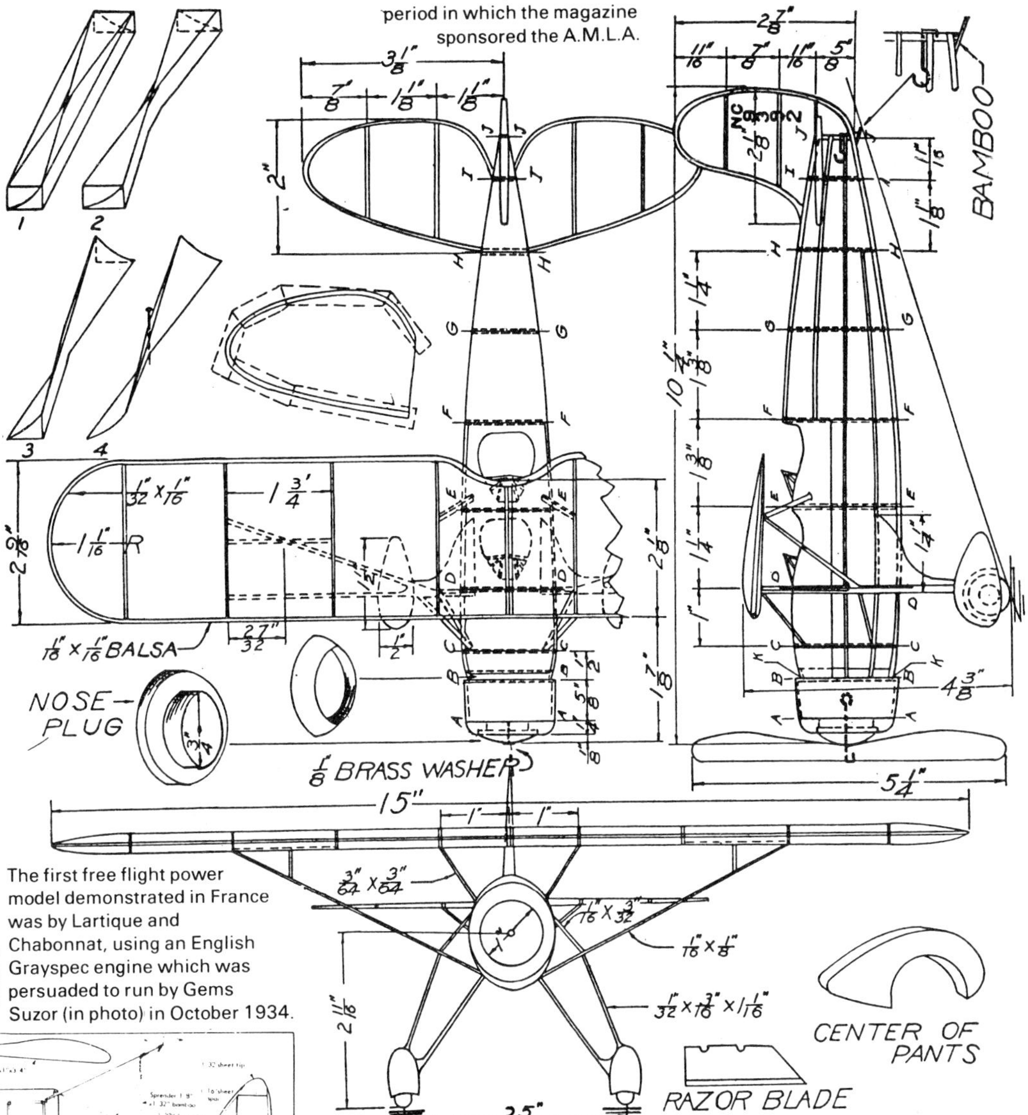

An ad. from a 1933 *Flying Aces* listed six 15in. allegedly scale models.

The first free flight power model demonstrated in France was by Lartique and Chabonnat, using an English Grayspec engine which was persuaded to run by Gems Suzor (in photo) in October 1934.

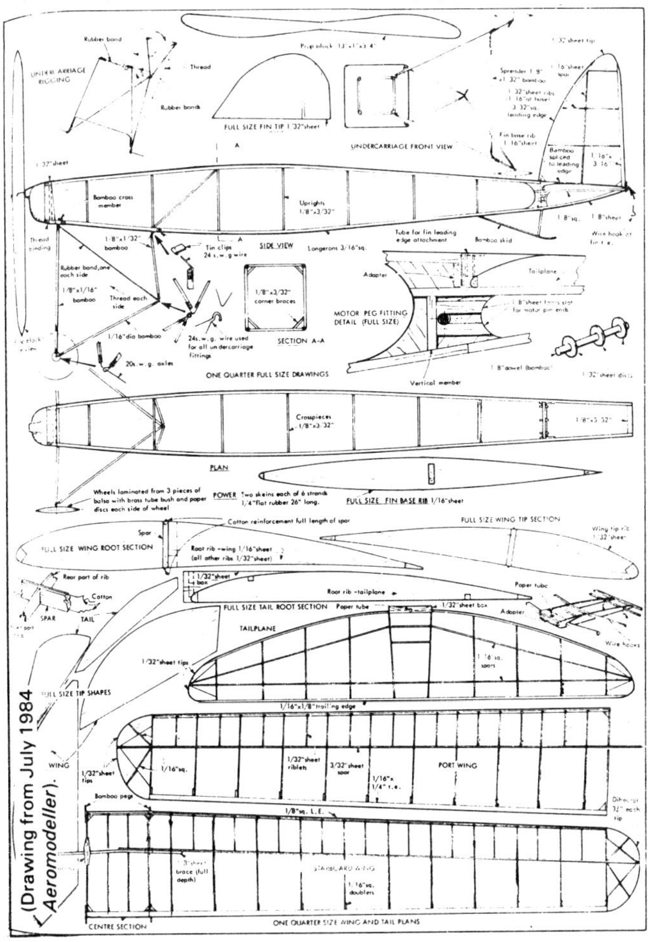

(Drawing from July 1984 *Aeromodeller*).

1934 Wakefield winner was J. B. Allman (GB) averaging 1:51.8 with his 'Grasshopper'. 8ft. towline Darmstadt Konsul below was in August 1934 *American Boy*

The first motors to follow the Brown, advertised in *Flying Aces*, November 1934, a year before the Baby Cyclone.

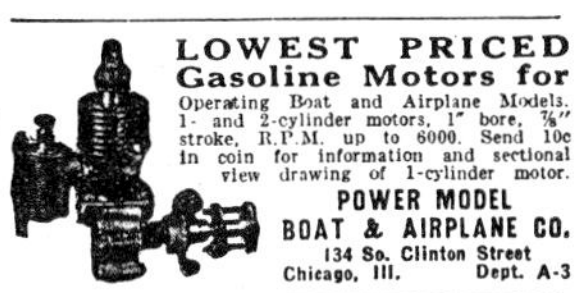

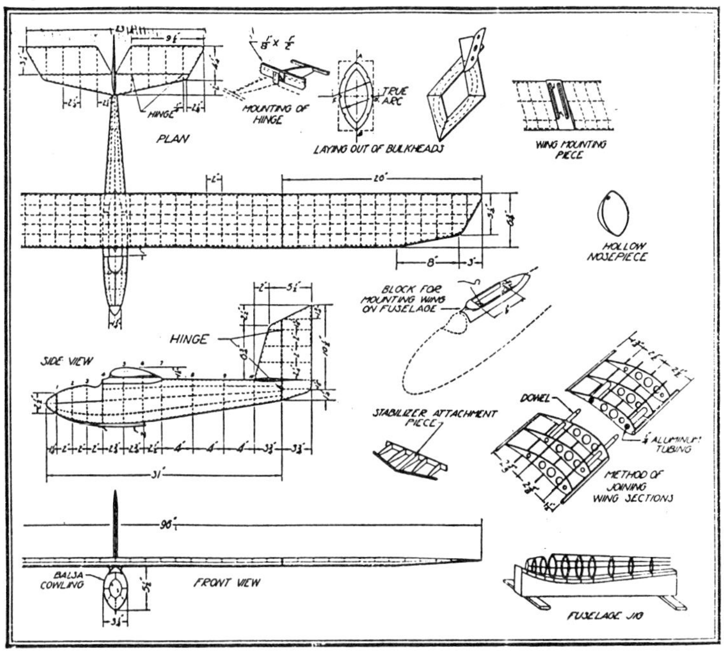

1934-35

Probably the best-known of the early 'gas models' were the KG series designed by Charles Grant and built and flown by Joe Kovel, especially since the KG-3 was published in great detail in *Model Airplane News* (of which Grant was editor) in April/May 1935. Just after publication Kovel set a record of 64:40.0 with KG-2. KG-1 was an 8ft. Brown-powered 7lb. design, KG-2 an enlarged wing version at 10ft. span, with appropriately increased tail areas and a reduction in weight to 6½lbs. KG-3 (part of the plans shown here) retained the tail but returned to an 8ft. wing.

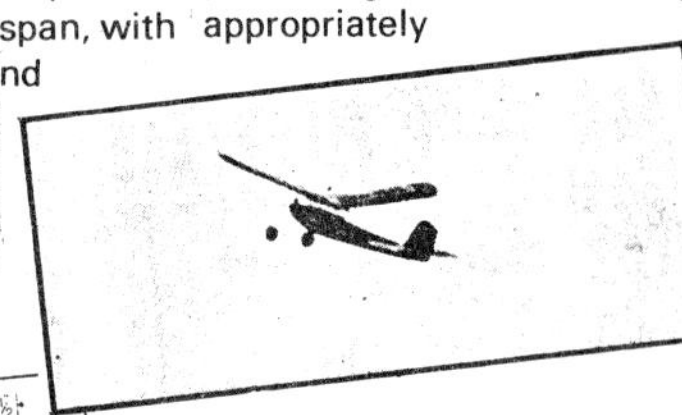

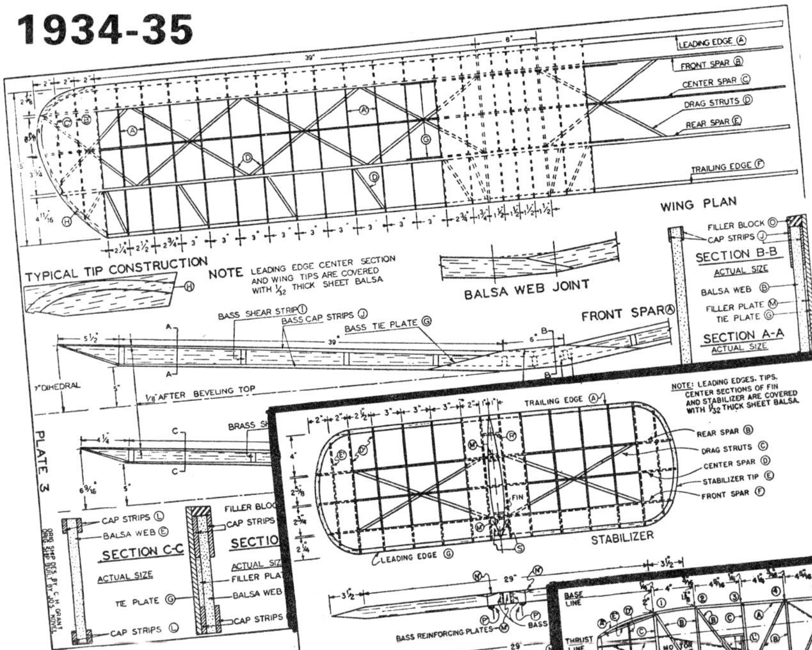

—American flight times seemed very high, but one reason was that ¼oz. of petrol (gasoline) was allowed for every 1lb. model weight. A fifteen minute 1,200ft. climb was not unusual.

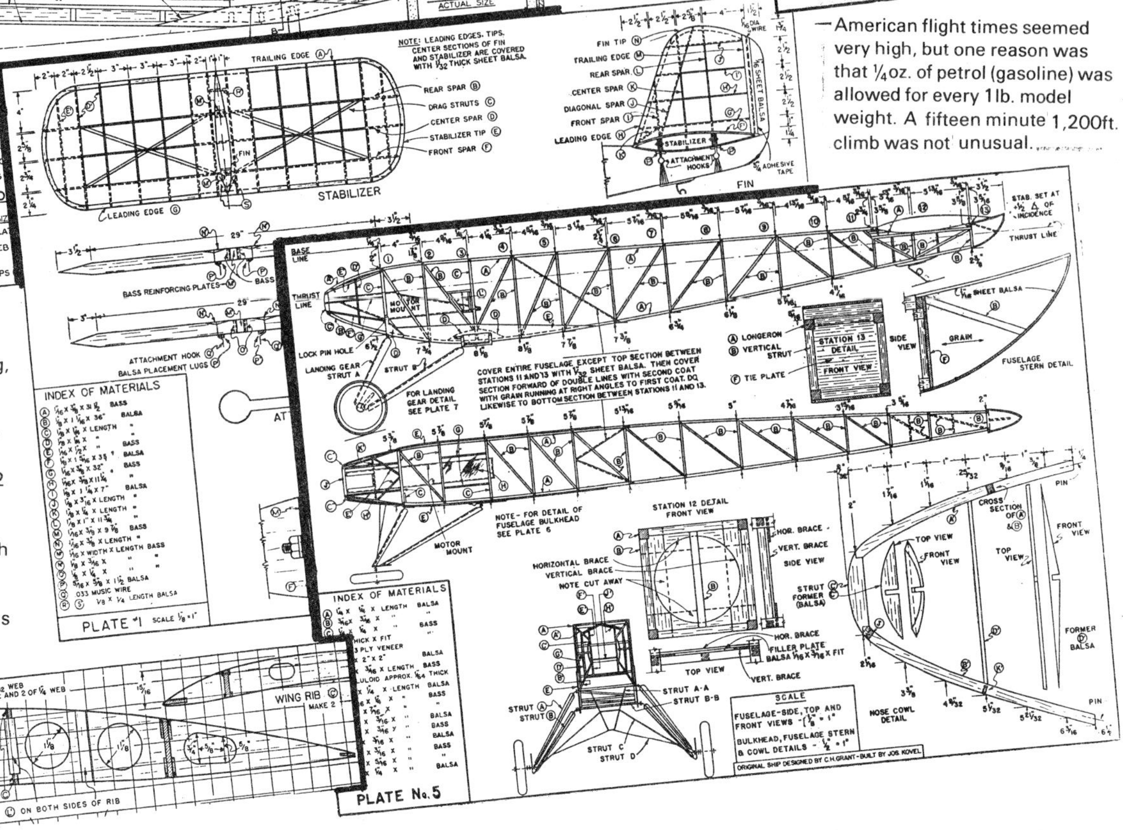

By 1935 emphasis in the U.S had switched to outdoor flying, although armories, hangars and airship sheds were still used for flying. There were in fact 57 categories of indoor record (31 claimed) against 42 outdoors (17 claimed) under the umbrella of the National Aeronautic Association, which represented the U.S. on the F.A.I. and ran national championships, etc. There was also a general tidying-up of terminology.

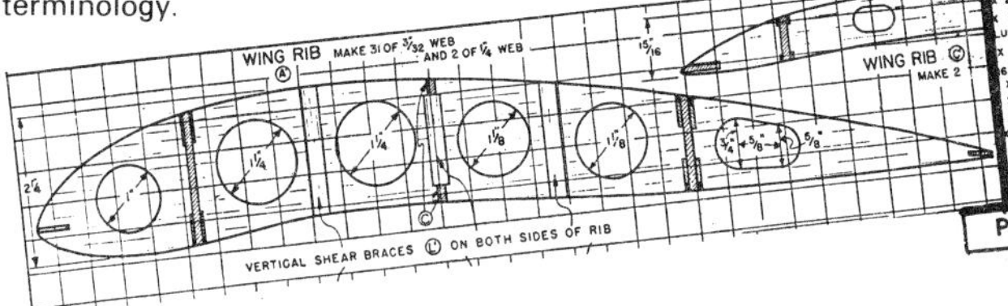

Imp Tornado two- and four-cylinder engines ran on gas generated by a mixture of calcium carbide, water and solid carbon dioxide (dry ice).

The American model trade started to hit its stride in 1935. Among advertisers that year who were to become familiar names were Aircraft (Chicago), Berkeley, Lawrence W. Brown, Bunch, Cleveland, Comet, Douglas, G.H.Q., Guillow, Hawk (solids), Heathe, Ideal, International Models, Jasco, Lindberg, M&M, Megow, Scientific, and Selley. Petrol engines were advertised by Peerless (July), Forster (September) and Aircraft Industries (December), the last offering the Baby Cyclone which became the next best known to the Brown Junior. Engines in 1936 included the Ferguson Eagle, Falcon twin and Condor four-cyl. (February), the Elf (not advertised but mentioned in March) Tlush Super Ace (May), G.H.Q. Loutrel (June), Bunch Gwin Aero (June), Bunch Mighty Midget (September) and Cleveland Tom Thumb (November).

A New Type of Model Fuel-driven Engine

A MINIATURE aircraft engine that uses a patented fuel in stick form and drives a radial engine and propeller at 2,500 r.p.m. has been perfected by a Los Angeles inventor who is marketing first a true scale reproduction of a Lockheed Vega, having 24-in. span.

Doing away entirely with rubber band and clockwork motors the new model plane takes off and flies away within two seconds after an ignited stick of the fuel is placed in a sealed chamber in the fuselage. A quarter-turn valve under the tail surfaces gives access to the fuel chamber which automatically feeds its power to the five-cylinder radial engine.

The engine and fuel chamber and propeller complete weigh only 8 oz. The engine is $2\frac{1}{4}$ in. in diameter, has a $\frac{3}{4}$-in. bore, $\frac{1}{4}$-in. stroke, and develops approximately 1/25 h.p at 2,500 r.p.m.

There is no carburettor nor ignition; the wingspan is 24 in. Other models of Army, Navy and racing planes will soon be ready. The inventor is Fred S. McFarland and his ships are being made by the Mototoy Mfg. Co., whose address readers may have upon application.

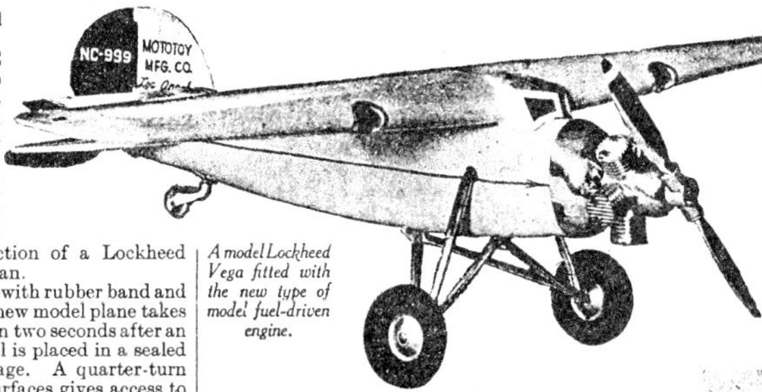

A model Lockheed Vega fitted with the new type of model fuel-driven engine.

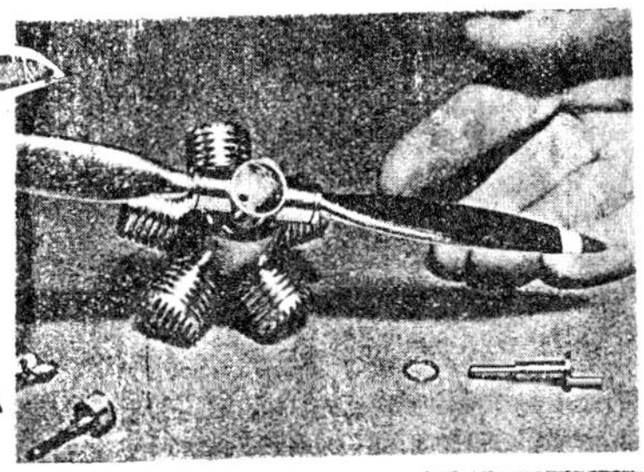

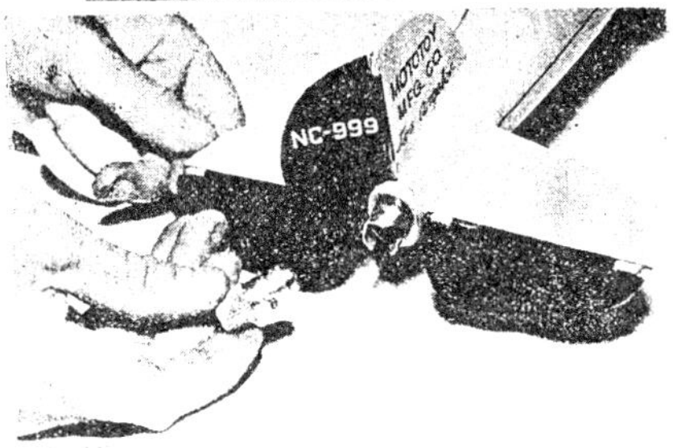

Anyone ever heard of this engine? Write-up and photos appeared in June 1934 P.M. The inference is that a burning stick or pellet of fuel generated gas with pressure enough to run the five-cylinder motor, which has no visible ports or valves. However, 8oz. motor weight in a 24in. model with no dihedral may give a clue, . .

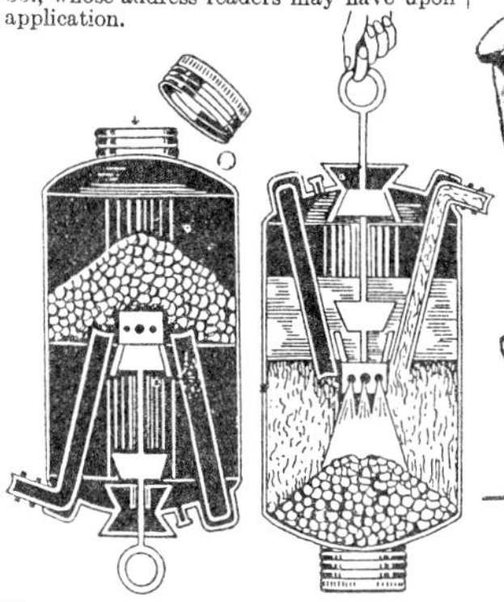

The container is first filled with carbide, as shown on the left, and the latter is then covered with dry ice. When inverted, as on the right, and water is allowed to drip on to the ice and carbide, a non-inflammable gas is given off at high pressure.

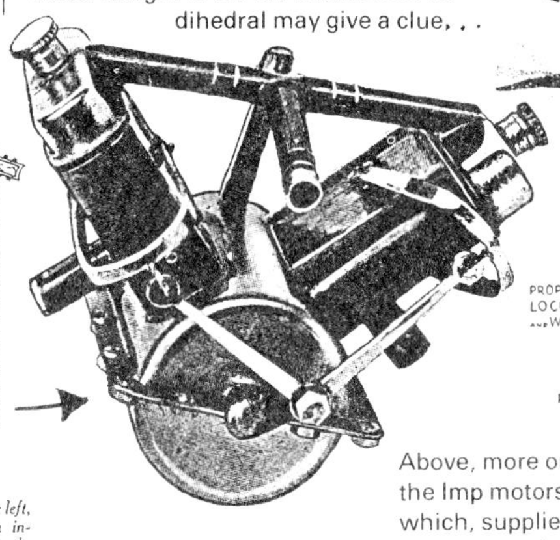

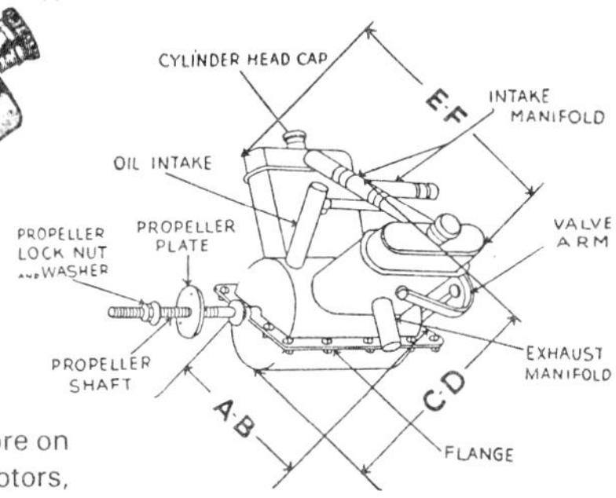

Above, more on the Imp motors, which, supplied direct, cost about £1.50 for the four-pot and £1.25 for the twin. The twin was $2\frac{1}{2}$oz. weight and developed '$\frac{1}{8}$th horsepower' at 3500r.p.m., the four was $3\frac{1}{2}$oz. and about 30 per cent more powerful. The generator was 8oz., $7\frac{1}{2}$in. high by over 3in. dia.

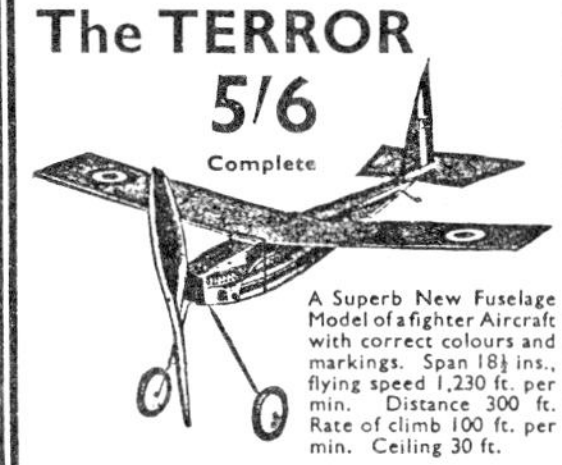

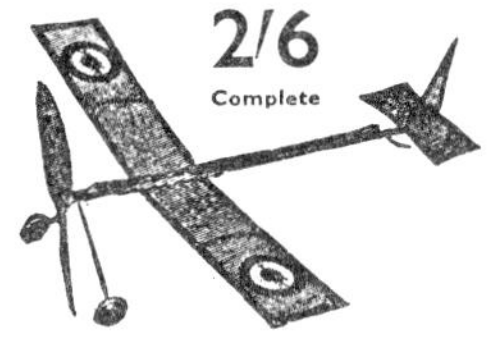

Various advertisements from 1934/5 P.M., M.A.N. and *Flying Aces* show variety of models offered. National Model had around 50 kits from 25c to \$2.50, for example.

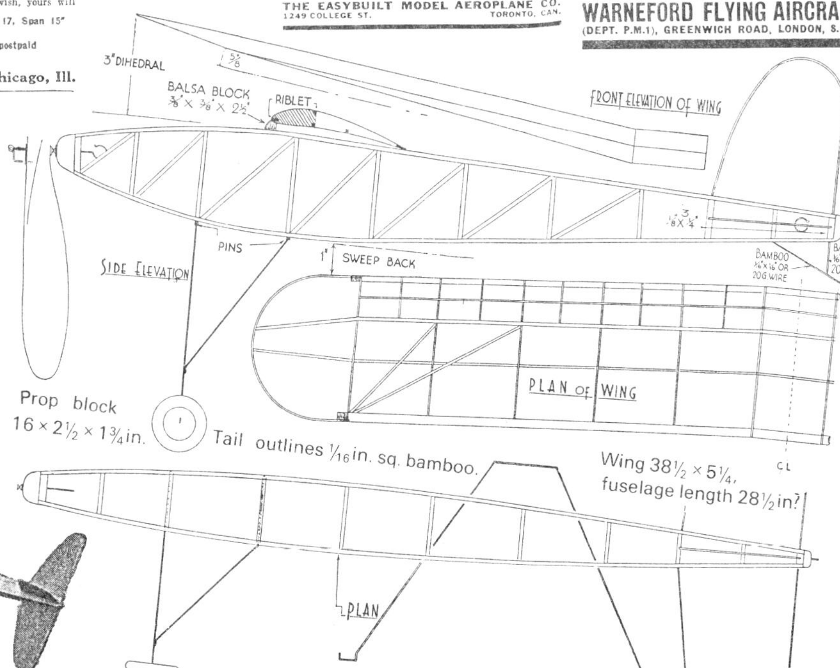

Prop block $16 \times 2\frac{1}{2} \times 1\frac{3}{4}$in.

Tail outlines $\frac{1}{16}$in. sq. bamboo.

Wing $38\frac{1}{2} \times 5\frac{1}{4}$, fuselage length $28\frac{1}{2}$in?

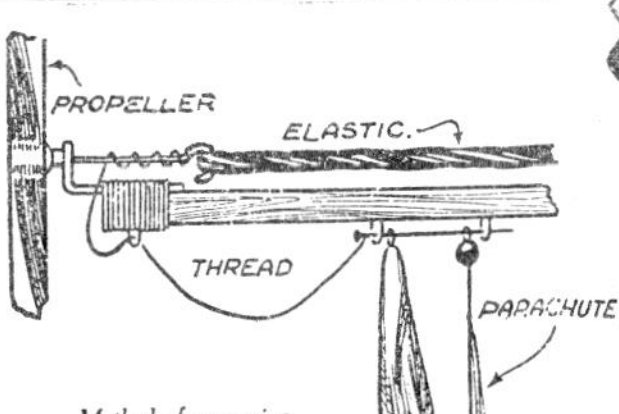

Method of arranging automatic parachute-dropping device.

What price the photo caption? ▶ *Showing the finished model in the hands of a model aeroplane enthusiast.*

A criticism of *Practical Mechanics* was its omission of names and, often, dimensions. This 'Milton Special' was, it appears, the first Australian Wakefield entry and was flown proxy by S. R. Crow to 5th place in the 1935 event.

1935

Extraordinary weather conditions prevailed in 1935 for about 200 contestants at the Eastern States meeting (May 25) and again for about 300 at the Nationals (June 27-29). At the first, the gas event was for up to 7lb. models, ⅛oz. per lb. and Joe Kovel won with his record 64:40. Second was R. Long, 47:20, 3rd L. Vartanian, 45:50, 4th Frank Ehling, 35:06 and 5th B. Noble, 27:28. There was a twin pusher comp. (Class C, 100-150sq. in.) — 1. John Haggerty 18:58, 2. R. Gable 12:20, 3. M. Malley 11:32. In rubber fuselage (Class D, 150-300sq. in.) it was 1. William Ying 41:19, 2. J. Throckmorton 20-30, 3. J. Ginnetti 17:22, 4. Henry Struck 14:44, while in open duration top four were 16:08, 15:04, 14:15 and 14:04.

At the Nats., at St. Louis, the Texaco (gas, under 21) was won by Leo Weiss, 64:12, 2. B. Marchi 41:55, 3. Maxwell Bassett 36:49, 4. Joe Kovel 31:05, 5. Frank Tlush, 17:56. The Mulvihill (rubber stick, under 21) was won by Dick Korda with 24:40.8, the Stout (fuselage R.O.G., under 21) by K Ernst, 20:05, the Moffett Trials (fuselage R.O.G., open) by Robt. Cahill, 33:00, runner-up Vernon Boehle, 28:30.

These phenomenal outdoor times indicate ideal thermally conditions with low drift, but they are also the result of the American practice of allowing timekeepers to follow a model by any means (even lightplanes!)

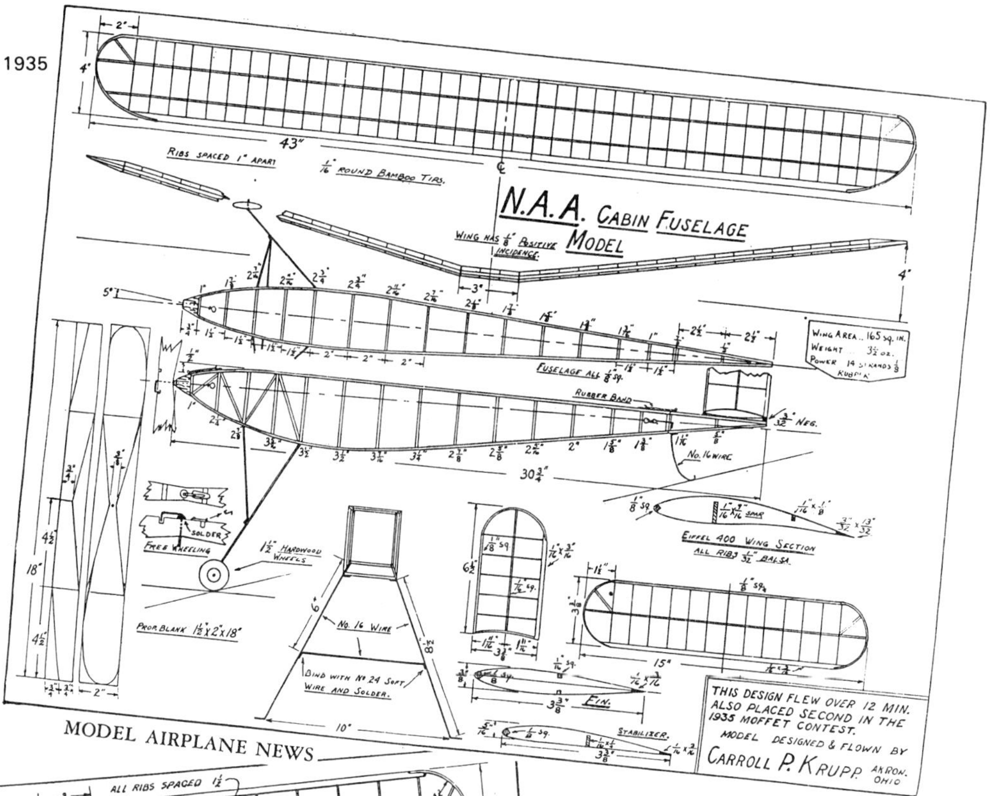

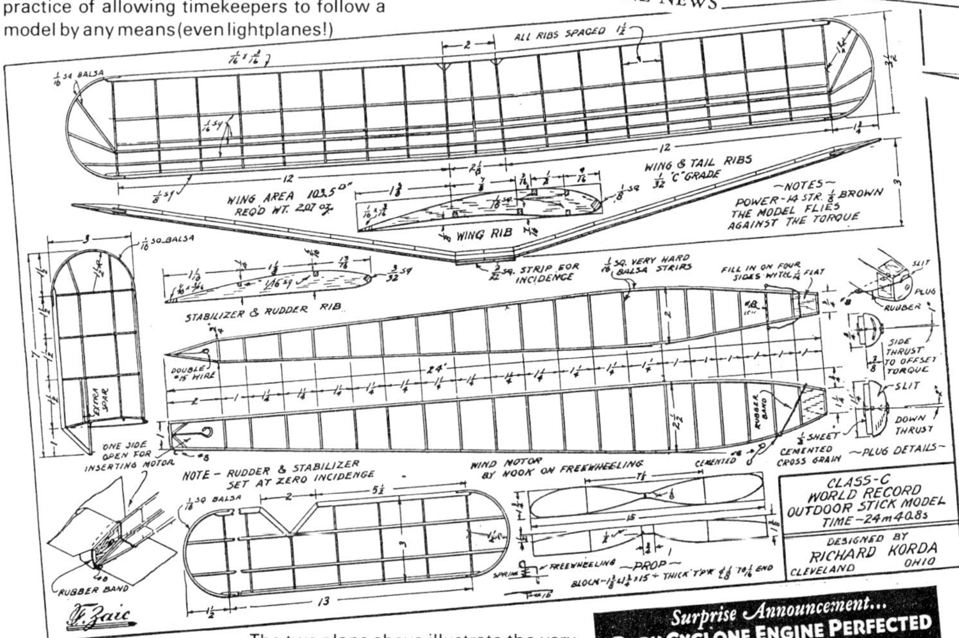

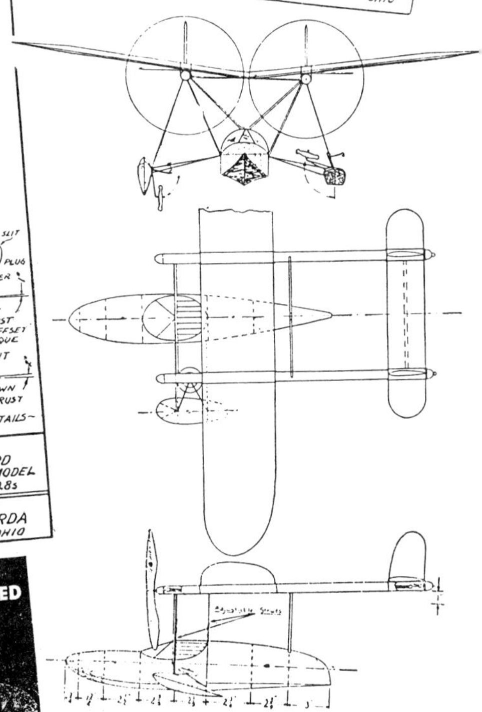

The two plans above illustrate the very slight differences between American fuselage and stick models, really only amounting to fuselage cross-sectional area and whether an undercarriage was fitted. Some builders used two fuselages to qualify for either category. Classes were based on wing area. American organisation at this time was in the hands of the National Aeronautic Association. Chuck gliders (below) were first taken seriously in the U.S.A.

Amphibians seemed to attract 1930s modellers, possibly influenced by full-size interest. This 32in. design by Gordon Light and Alan Booton was a fast climber (M.A.N. Sept. 35). Below, never managed to get our model from this plan trimmed!

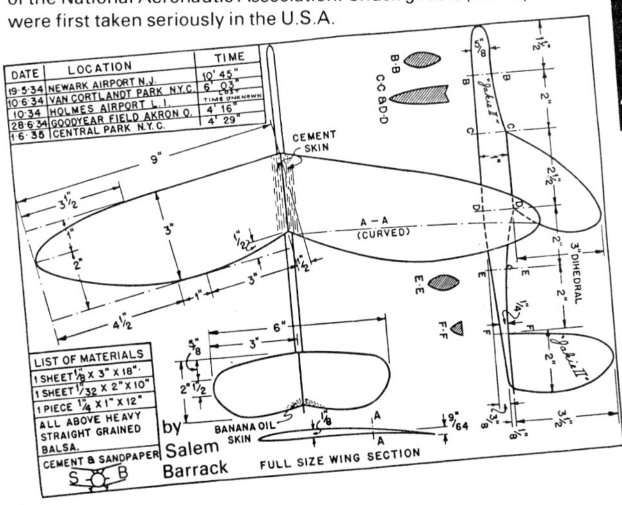

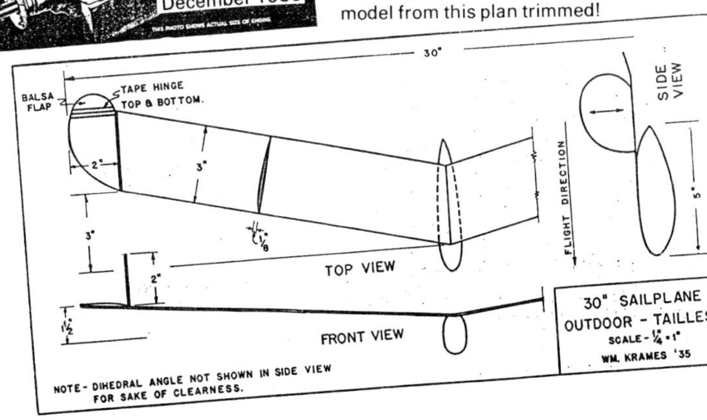

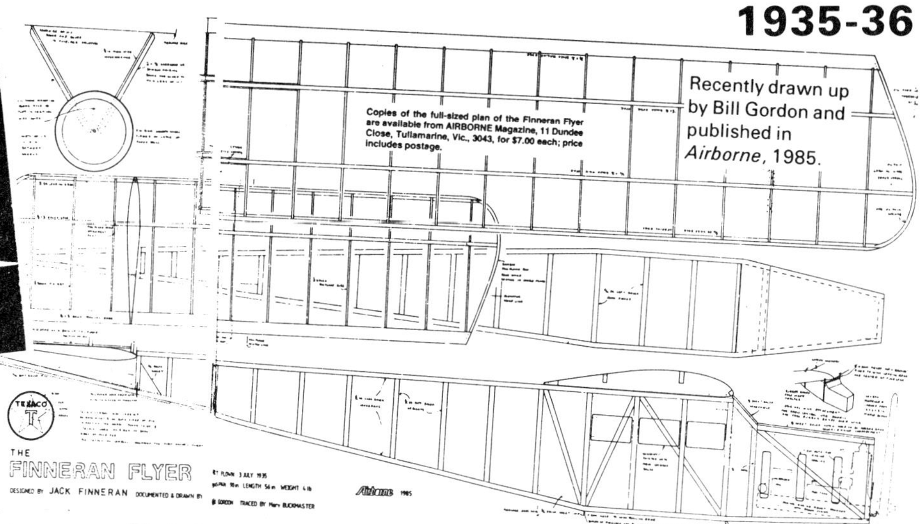

Recently drawn up by Bill Gordon and published in *Airborne*, 1985.

The first power model to fly in Australia is believed to be the 'Finneran Flyer' (3 July 1935), span 90in., powered by an engine provided by the Texaco Company

The Brooklyn-based Selley kit manufacturer launched a new method of construction in June 1935 *Flying Aces* The new system involved parts moulded in a fabric impregnated with a filler and made construction much easier and quicker.

Photo from July 1936 *Model Aeroplane Constructor* of the 40in. *Ibis* (see page 22) which was timed 24:52 O.O.S. by the Victoria club, Australia.

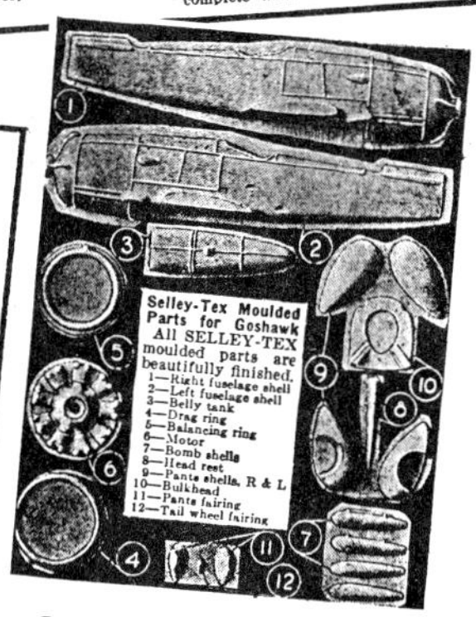

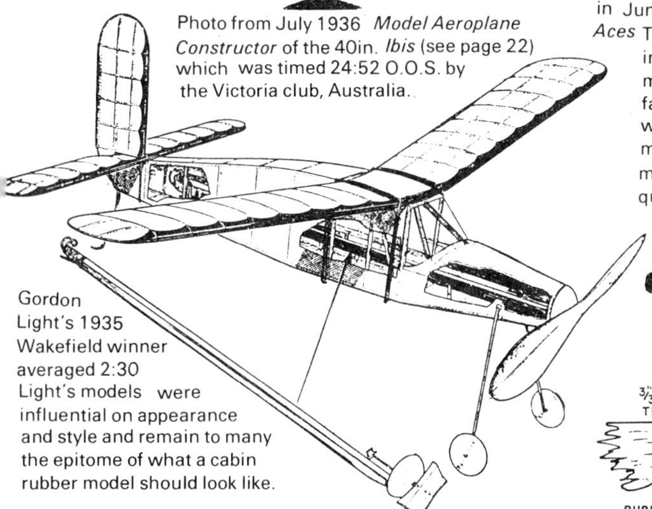

Gordon Light's 1935 Wakefield winner averaged 2:30 Light's models were influential on appearance and style and remain to many the epitome of what a cabin rubber model should look like.

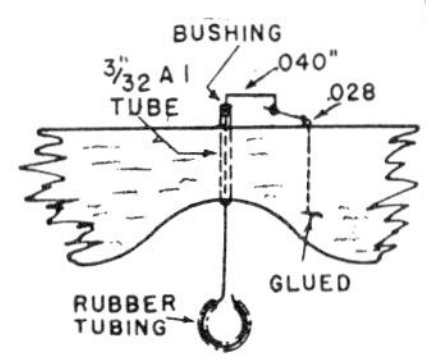

German tailless slope-soarer by Klaus Schmidtberg) which recorded a h.l. flight of 37:20 at the Wasserkuppe gliding centre early in 1936.

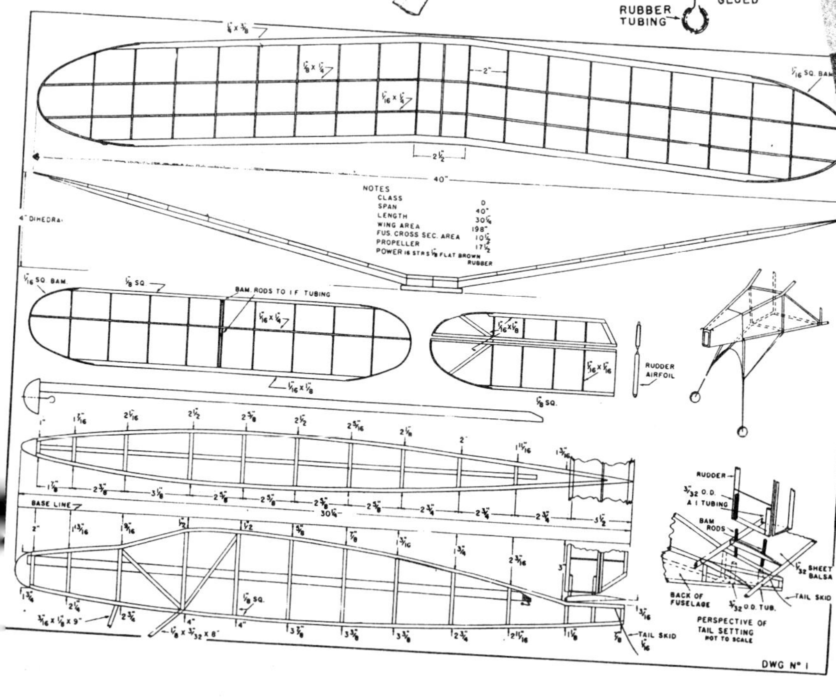

William Ying's model which flew 41:19 O.O.S. at the 1935 Eastern States meeting shows some Gordon Light influence. The model was lost when the timer's car was stopped by railway lines.

Cut ribs had been marketed since about 1928, but most kits settled for printed sheet.

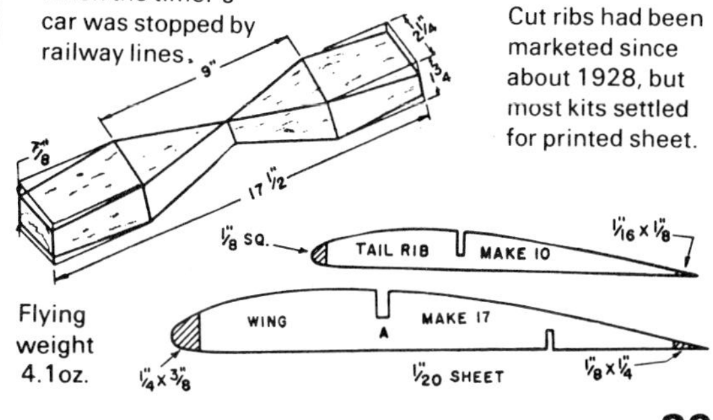

Flying weight 4.1oz.

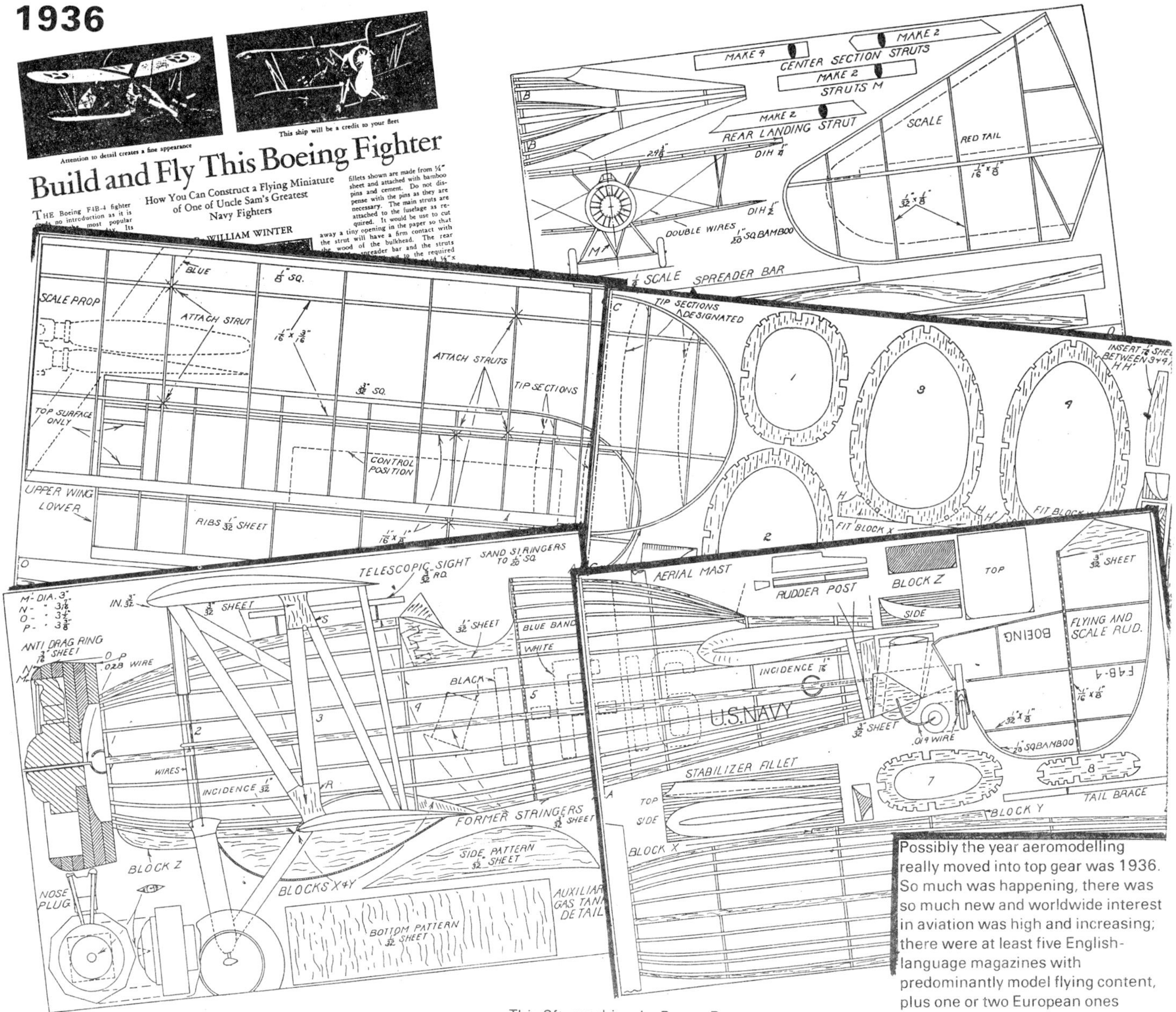

Possibly the year aeromodelling really moved into top gear was 1936. So much was happening, there was so much new and worldwide interest in aviation was high and increasing; there were at least five English-language magazines with predominantly model flying content, plus one or two European ones starting up. A stream of full-size plans had begun, a typical example being this Boeing F4B-4 by Bill Winter in February 1936 M.A.N. Bill was then, and has remained, one of the most prolific American designers.

This 8ft. machine by Bunny Ross won the 1936 Sir John Shelley Cup Like an enlarged rubber model!

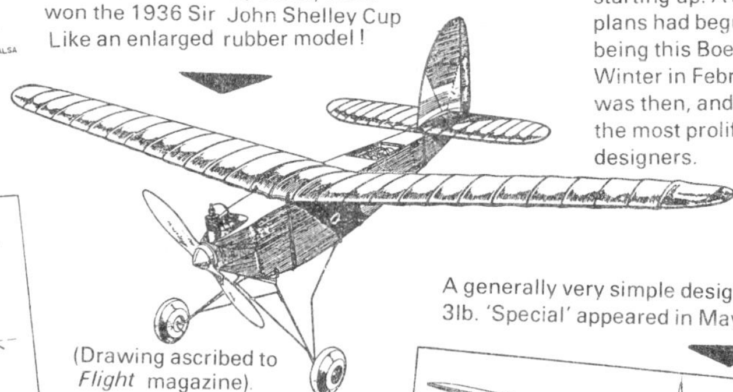

(Drawing ascribed to *Flight* magazine).

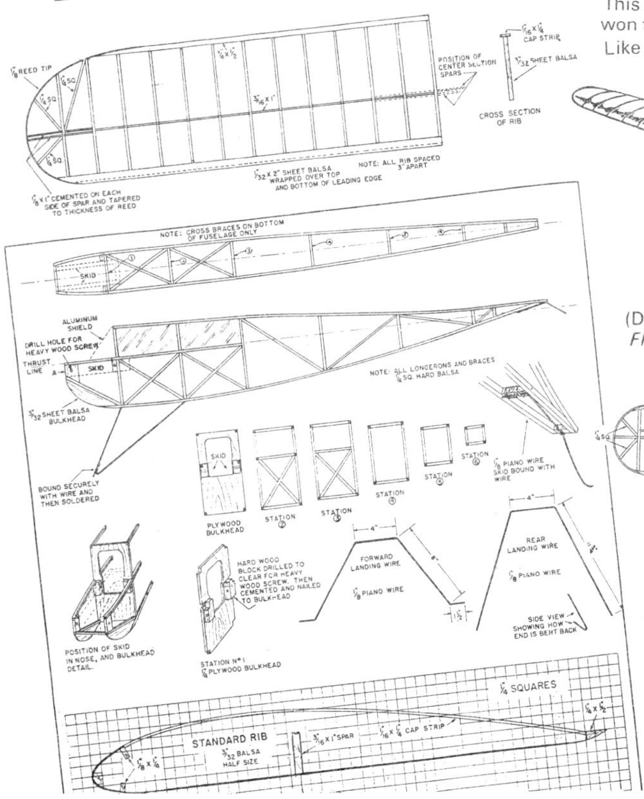

A generally very simple design, Allen Turner's 7ft., 3lb. 'Special' appeared in May 1936 M.A.N.

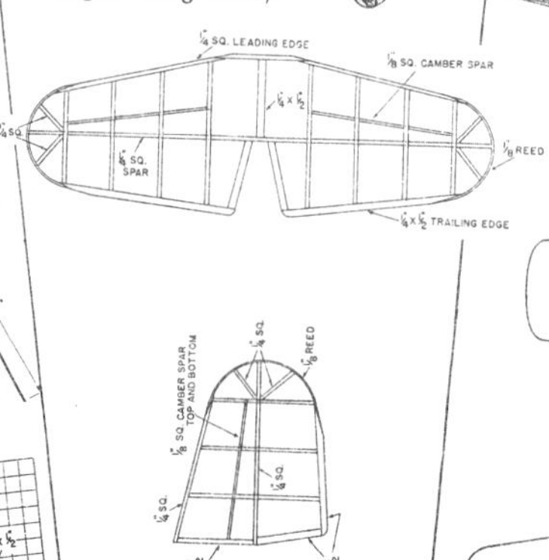

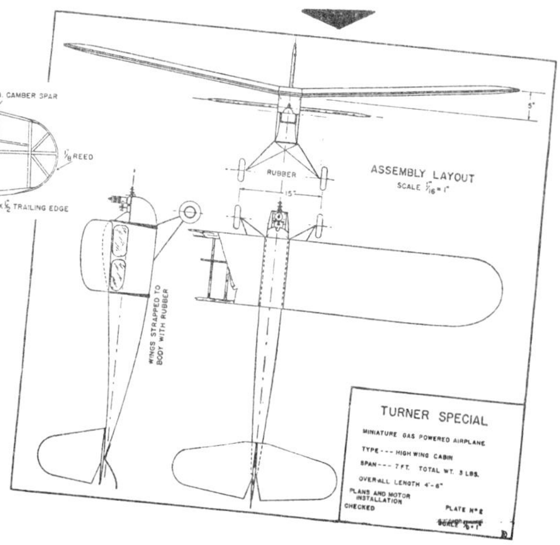

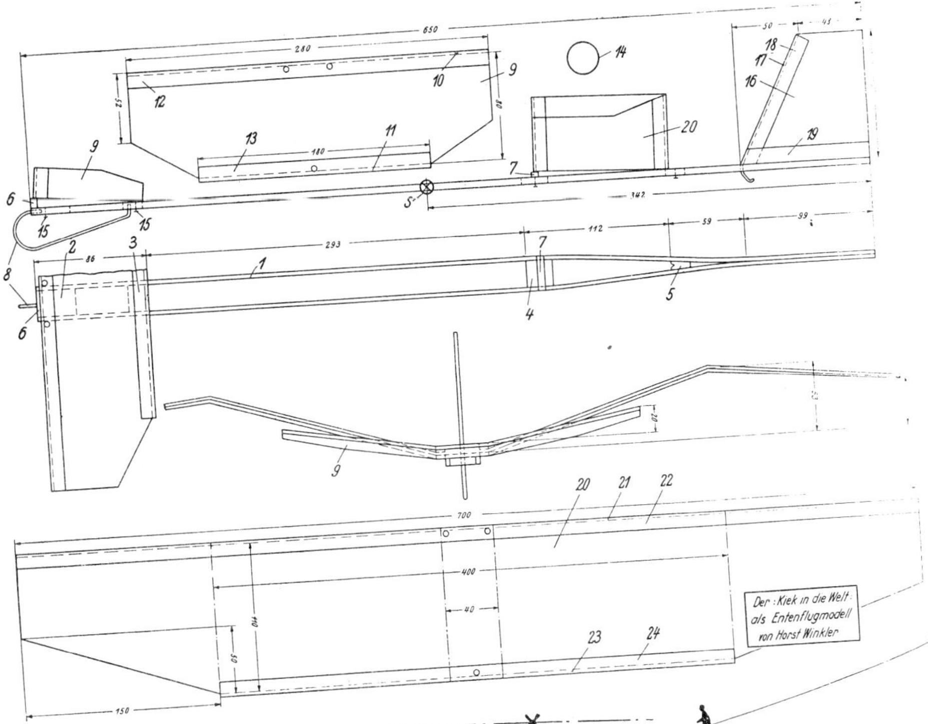

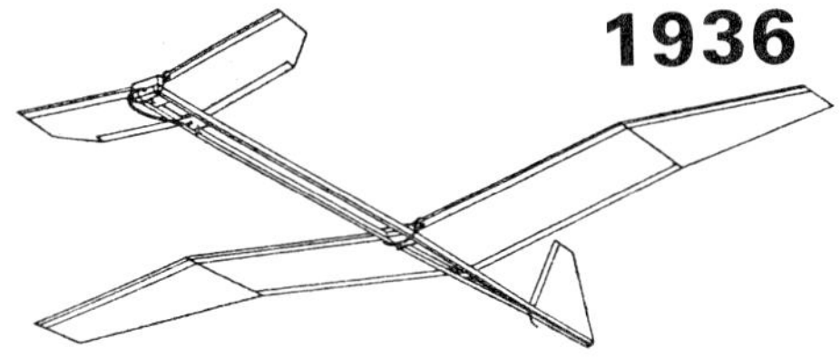

The magazine *Modellflug* started in April 1936 in Germany and this largely paper 27½in. canard glider by Horst Winkler was in the first issue. Winkler was editor of the new publication but had been an active modeller, especially in gliders, for many years and had written a widely-read manual. By 1931 he had established the principle of the 25p.c. catapult or hi-start launch, hook position at 60° to c.g., pulley launching and much else.

Stückzahl	Bezeichnung	Teil-Nr.	Werkstoff	Rohabmessungen
2	Rumpfleisten ...	1	Kiefer....	3 × 5 × 650 mm
1	Füllklotz	2	〃	5 × 20 × 30 mm
1	〃	3	〃	5 × 15 × 20 mm
1	〃	4	〃	5 × 20 × 20 mm
1	〃	5	〃	5 × 5 × 15 mm
1	Aufleimer	6	〃	5 × 7 × 26 mm
1	〃	7	〃	3 × 5 × 26 mm
1	Landekufe	8	Eisendraht	∅ 1,8 mm, lg. 160 mm
1	Höhenleitwerksfläche	9	Wellpappe	80 × 280 mm
1	Versteifung.....	10	Eisendraht	∅ 1,8 mm, lg. 280 mm
1	〃	11	〃	∅ 1,8 mm, lg. 180 mm
1	Klebestreifen ...	12	Papier ...	25 × 280 mm
1	〃 ...	13	〃 ...	25 × 180 mm
4	Gummiringe ...	14	Gummi ..	
4	Drahtstifte	15	—	8—10 mm lg.
1	Seitenleitwerksfläche	16	Wellpappe	95 × 95 mm
1	Versteifung	17	Eisendraht	∅ 1,8 mm, lg. 125 mm
1	Klebestreifen ...	18	Papier ...	25 × 115 mm
2	〃 ...	19	〃 ...	25 × 100 mm
1	Tragflügel	20	Wellpappe	110 × 700 mm
1	Versteifung.....	21	Eisendraht	∅ 1,8 mm, lg. 700 mm
1	Klebestreifen ...	22	Papier ...	25 × 700 mm
1	Versteifung.....	23	Eisendraht	∅ 1,8 mm, lg. 400 mm
1	Klebestreifen ...	24	Papier ...	25 × 400 mm
	Leim		—	
	Bindezwirn		Heftzwirn	
	Trimmgewicht ..		Blei oder Staniol	

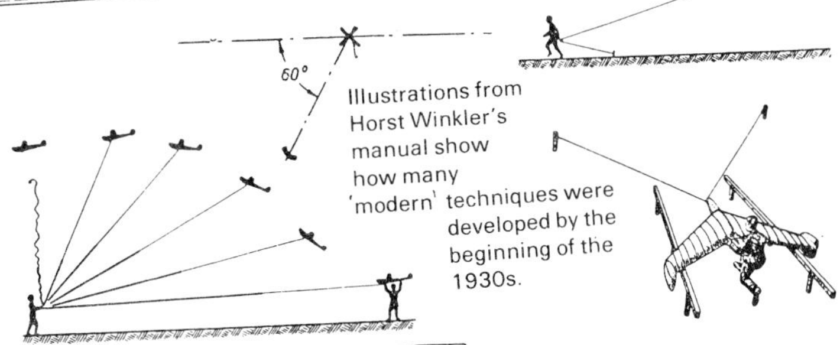

Illustrations from Horst Winkler's manual show how many 'modern' techniques were developed by the beginning of the 1930s.

Once a little 'gas' experience had been gained some American thoughts turned to scale models. The 'Bullpup' was a popular prototype and several kits appeared. Note that this plan was drawn by Bill Effinger, later to design the 'Buccaneer' and a wide range of other Berkeley kits.

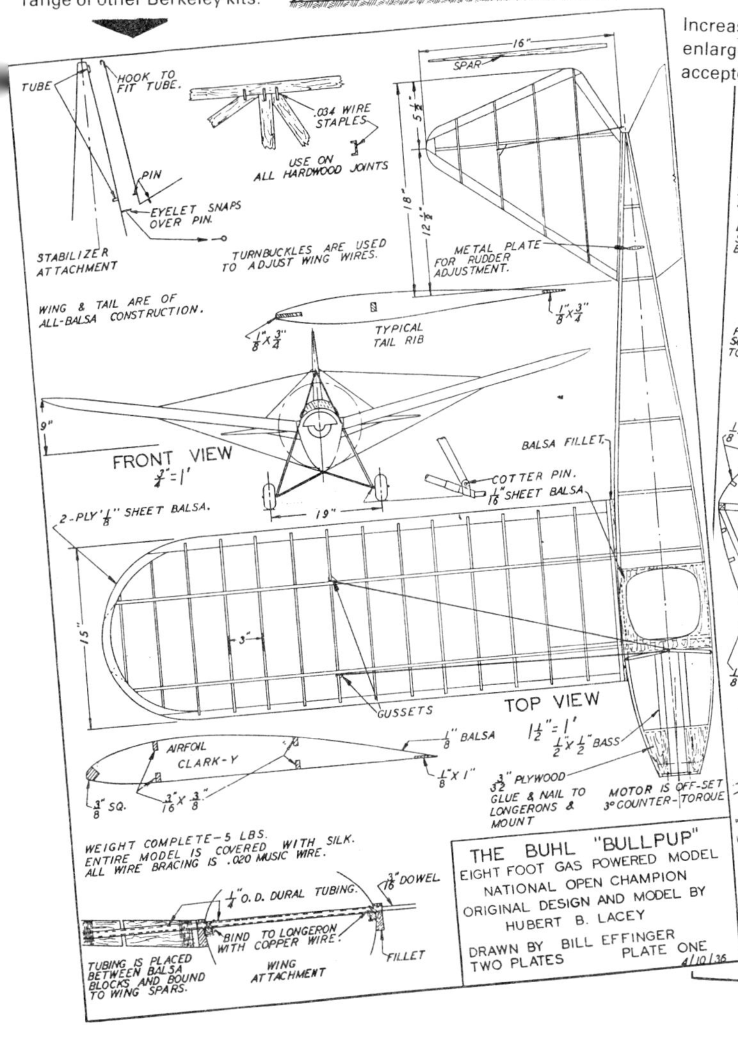

Increased dihedral and enlarged tail surfaces were accepted for many years.

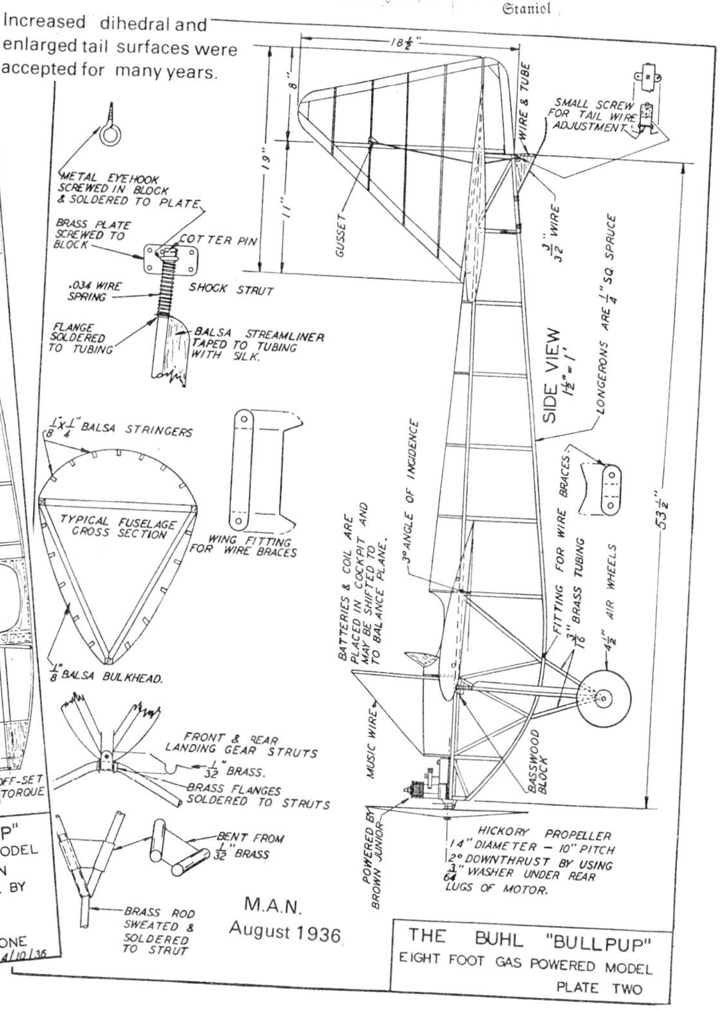

M.A.N.
August 1936

1936

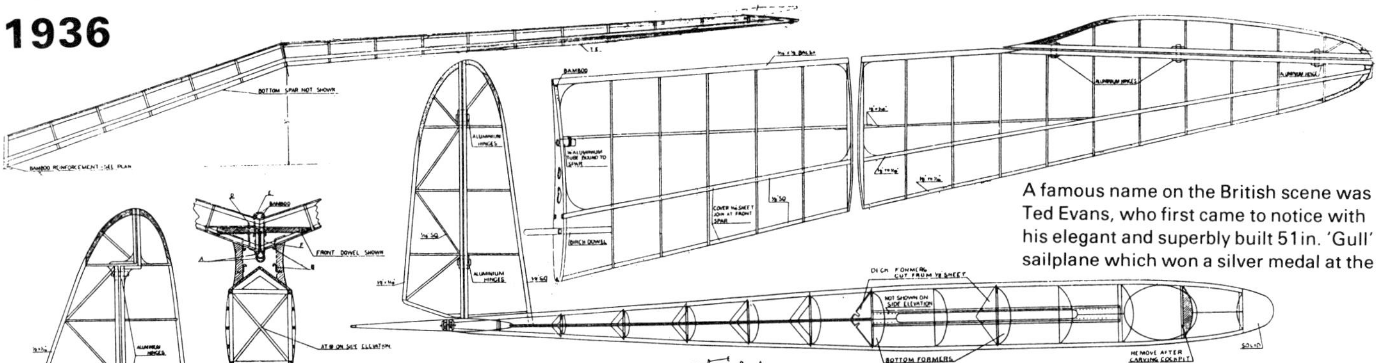
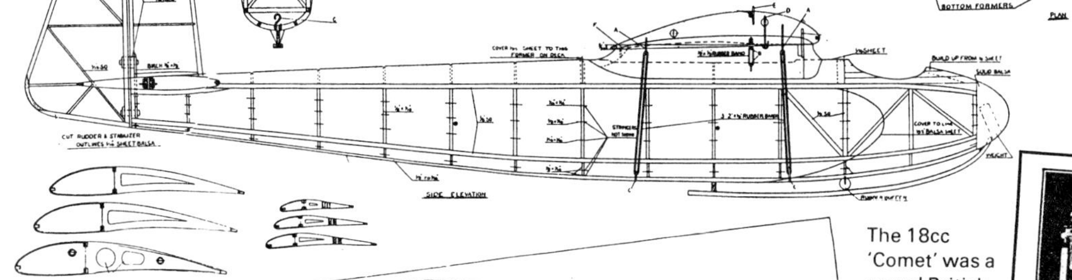

A famous name on the British scene was Ted Evans, who first came to notice with his elegant and superbly built 51in. 'Gull' sailplane which won a silver medal at the 1936 Model Engineer Exhibition and was later detailed in *M.E.* (November 1936) and *The Model Aeroplane Constructor* (May 1937). Ted was to become one of the outstanding Wakefield figures.

The 18cc 'Comet' was a sound British engineof 1936

And three rubber kits for 4/- (20p) including carriage?

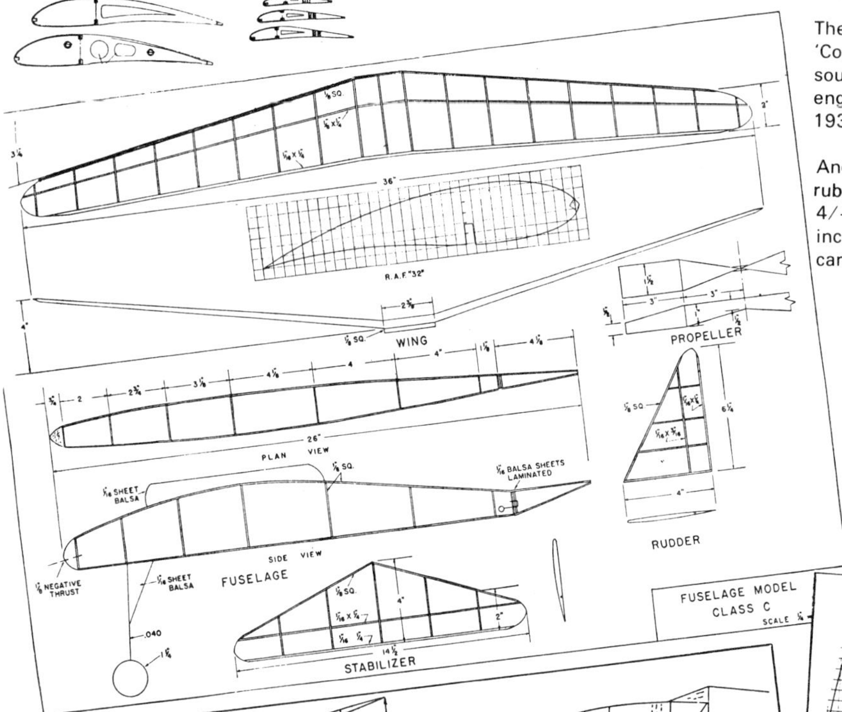

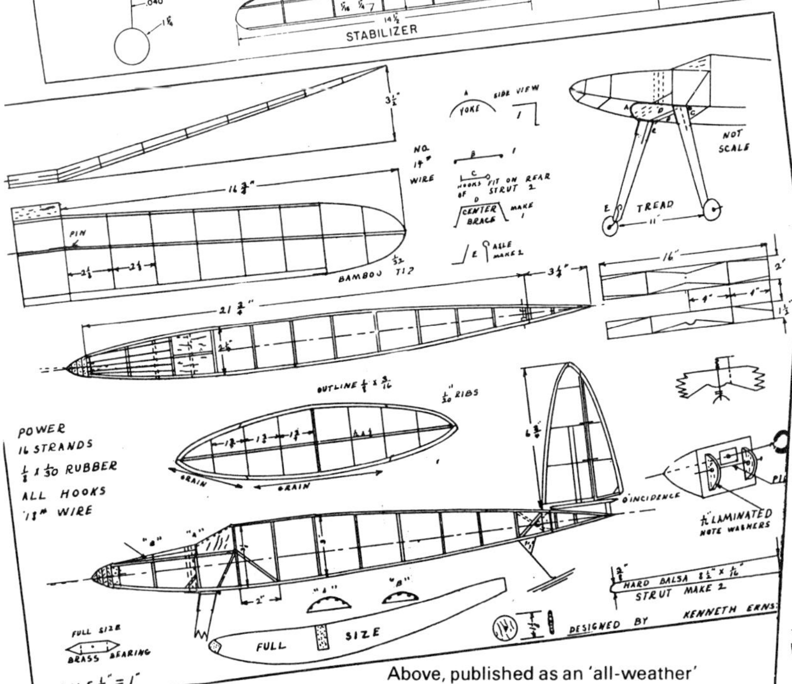

Above, published as an 'all-weather' contest model' in April '36, this Alan Orthof design looks a shade light, but the fuselage is $\frac{1}{32}$in. sheet covered. The wing section is claimed to be RAF32 and power is '12 strands of brown rubber'; it would be safe to presume $\frac{1}{8}$in. Immediately above, Kenneth Ernst's 1935 Stout Trophy winner. Right, Julius Unrath drew up the 'Sportster' (M.A.N. July '36), slightly unusual in that the model had only just appeared in kit form for the G.H.Q.- Loutrel engine a month before.

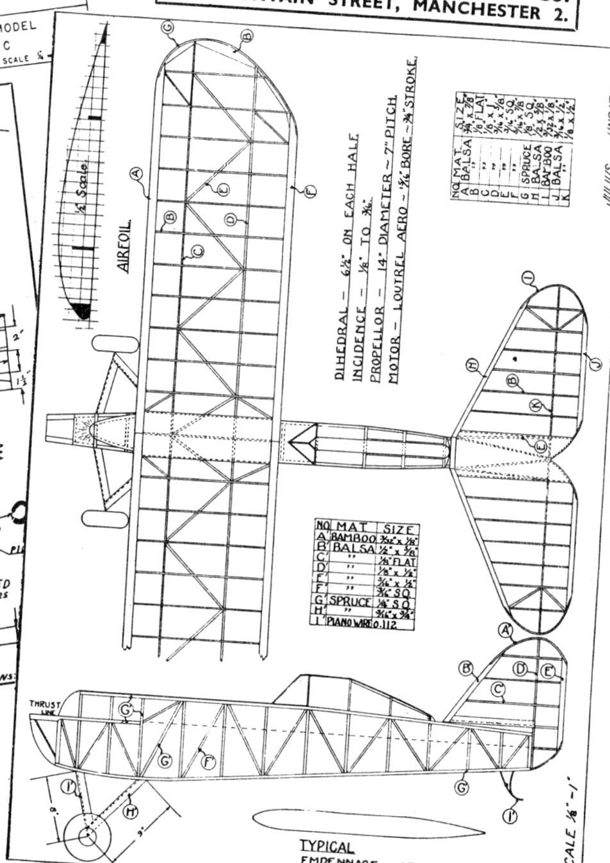

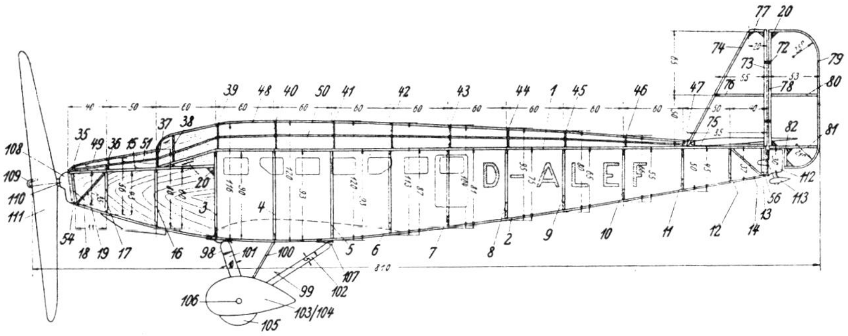

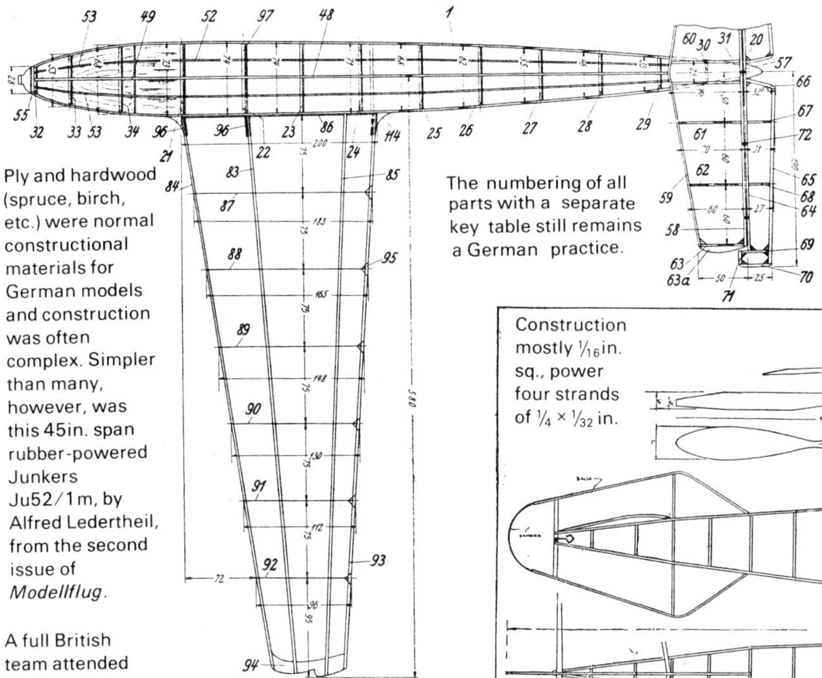

Ply and hardwood (spruce, birch, etc.) were normal constructional materials for German models and construction was often complex. Simpler than many, however, was this 45in. span rubber-powered Junkers Ju52/1m, by Alfred Ledertheil, from the second issue of *Modellflug*.

The numbering of all parts with a separate key table still remains a German practice.

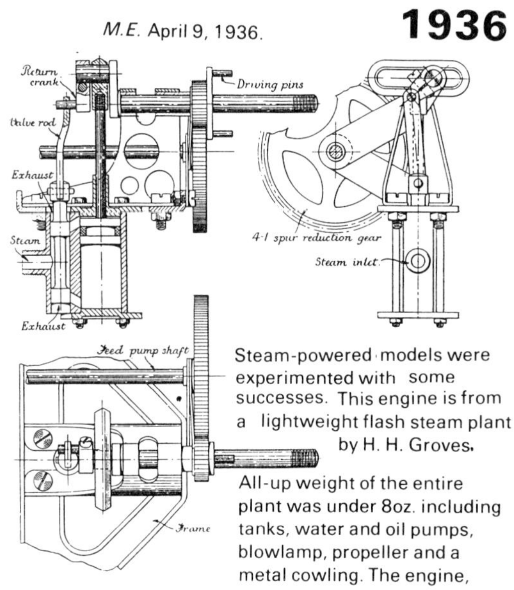

Steam-powered models were experimented with some successes. This engine is from a lightweight flash steam plant by H. H. Groves.

All-up weight of the entire plant was under 8oz. including tanks, water and oil pumps, blowlamp, propeller and a metal cowling. The engine, $\frac{1}{4}$in. bore and $\frac{7}{16}$in. stroke, had piston valve steam admission.

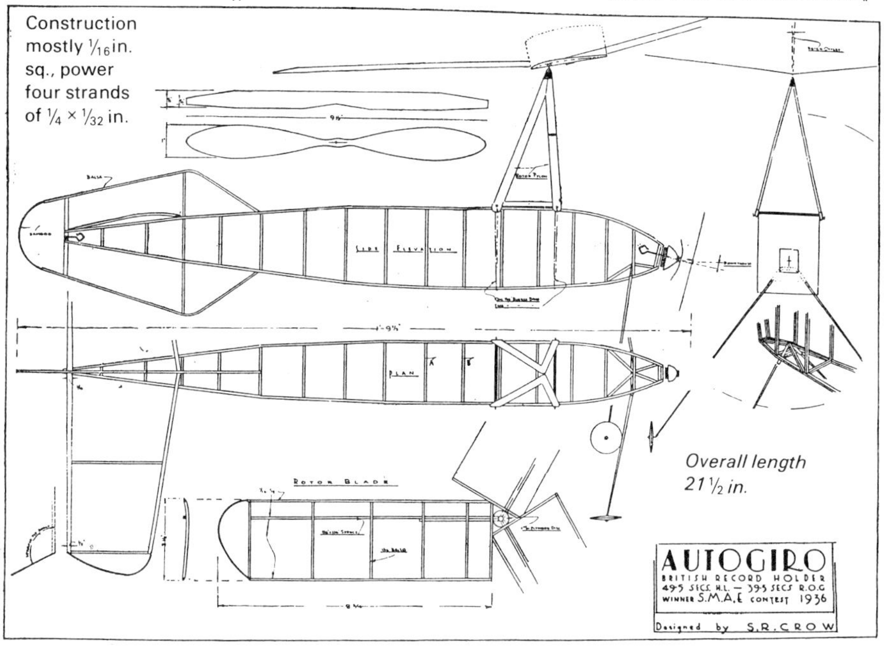

The S.M.A.E. ran autogiro events in the mid-30s and this was the 1936 winner as well as holder of h.l. and r.o.g. records for its designer, S. R. Crow. (P.M., December 1936).

A full British team attended the American Nationals at Detroit in 1936, where the major event was the Wakefield Trophy. This was won by A. A. Judge (4:99) with other GB flyers Bob Copland 3rd, J. B. Allman 5th and Dennis Fairlie 7th, spaced by Americans Roy Wriston, Dick Everett and Gordon Light and with André Vincre (France) 8th.

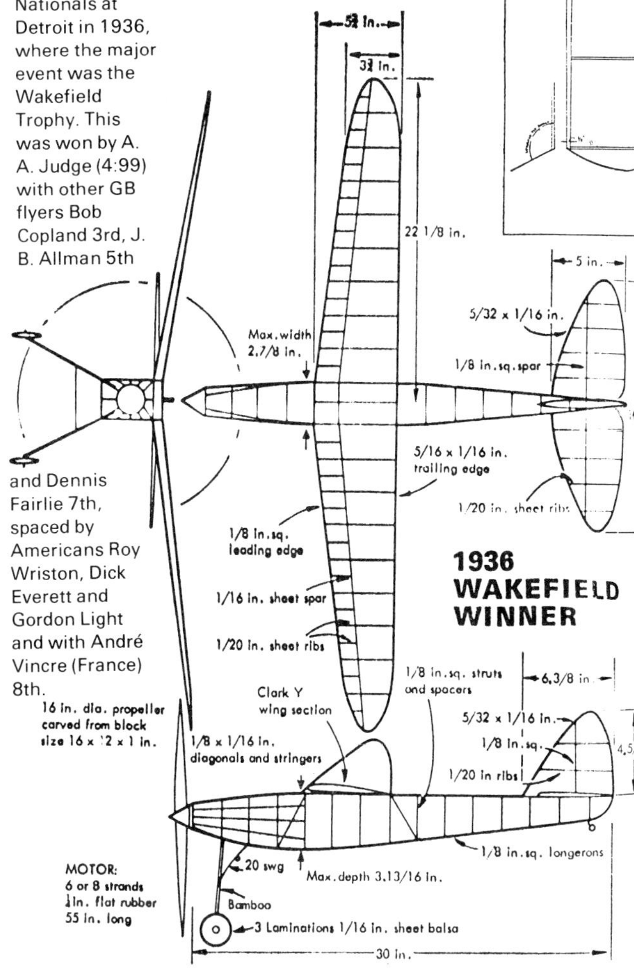

November, 1936 NEWNES PRACTICAL MECHANICS 101

SUPER-DURATION RUBBER MOTOR FOR MODEL AIRCRAFT

By D. A. Russell, A.I.Mech.E.

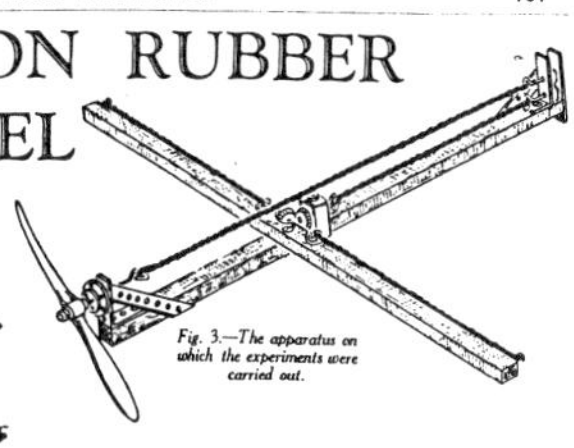

Fig. 3.—The apparatus on which the experiments were carried out.

This Ingenious Rubber Motor will Add Considerably to the Duration of the Flight of A Model Aeroplane

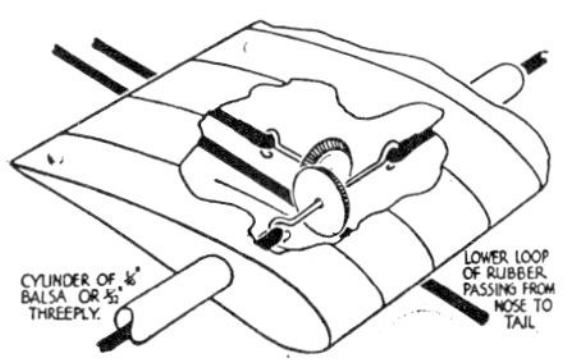

A familiar name, before he had become owner of *Aeromodeller* but a suggestion which does not appear to hold water. The idea was to increase the length of rubber by including spanwise motors

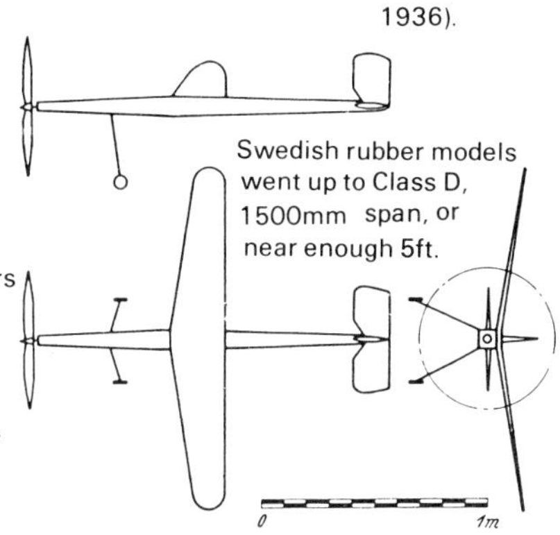

Swedish rubber models went up to Class D, 1500mm span, or near enough 5ft.

Swedish model of 1936 by A. Palmgren, Goteborg. Span 148cm (58in.), length 157cm (62in.) with 95g of rubber turning a 60cm ($23\frac{1}{2}$in.) prop. Total flying weight just under 8ozs.

1936

A good scale prototype was the Arado Ar.76 and this 37½in. model was in *Modellflug* in December 1936.

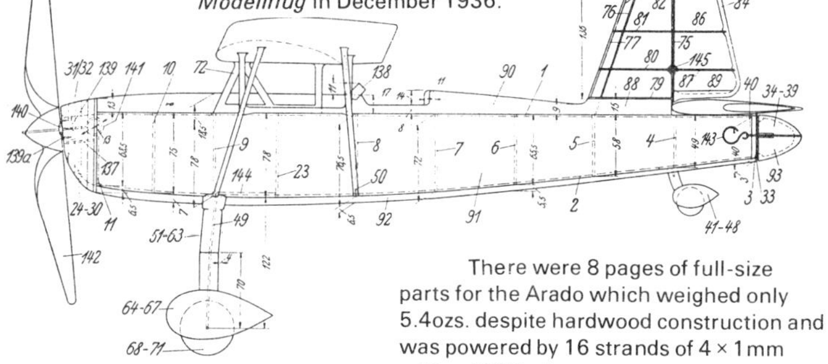

There were 8 pages of full-size parts for the Arado which weighed only 5.4ozs. despite hardwood construction and was powered by 16 strands of 4 × 1mm rubber 23½in. long.

Gems Suzor again, with a somewhat odd-looking petrol model which won the Coupe de France in 1936.

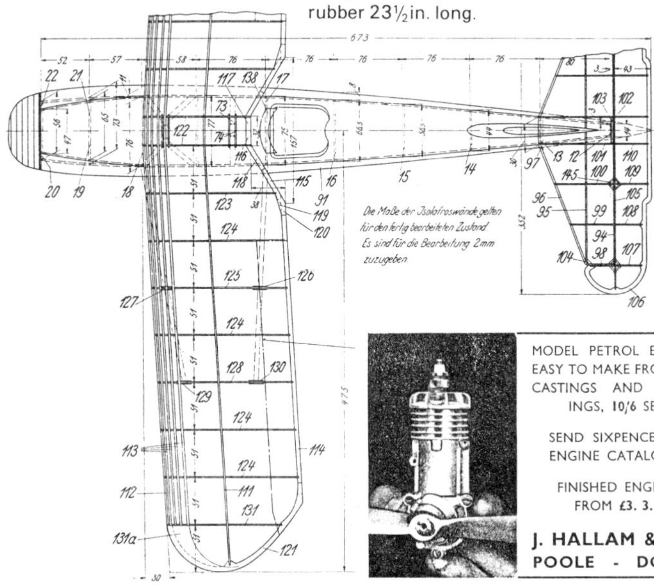

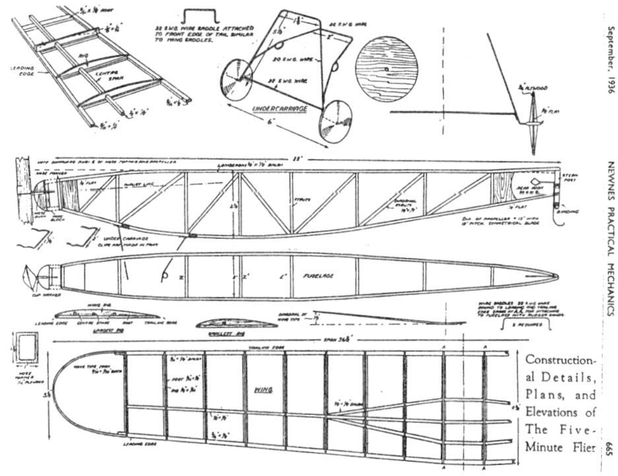

Construction-al Details, Plans, and Elevations of The Five-Minute Flier

Hallam engines covered a wide range and were available for several years.

There were some strange occurrences. The 'Lincol' appeared in P.M. in 1936 as 'The Five Minute Flier' but clearly a slightly different drawing, acknowledged to Model Aircraft Supplies Ltd.

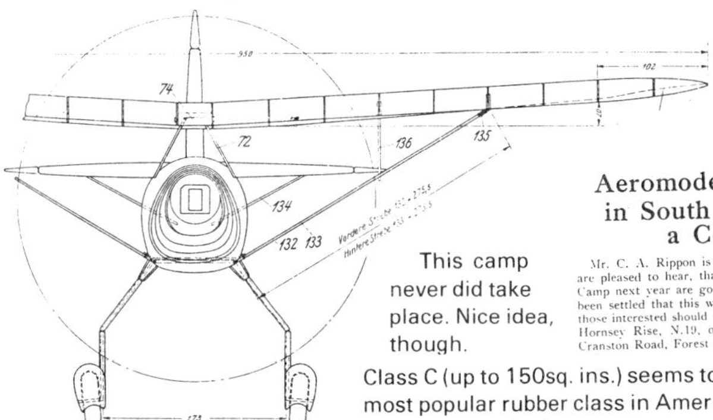

Aeromodellers Camp in South of England a Certainty

Mr. C. A. Rippon is pleased to inform us, and we are pleased to hear, that plans for an Aeromodellers' Camp next year are going forward. It has definitely been settled that this will be held at Herne Bay. All those interested should write to Mr. C. A. Rippon, 2a, Hornsey Rise, N.19, or to Mr. H. Chatterley, 117, Cranston Road, Forest Hill, S.E.

This camp never did take place. Nice idea, though.

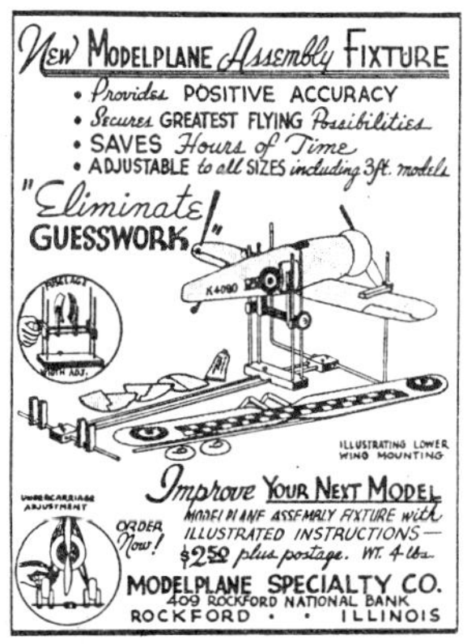

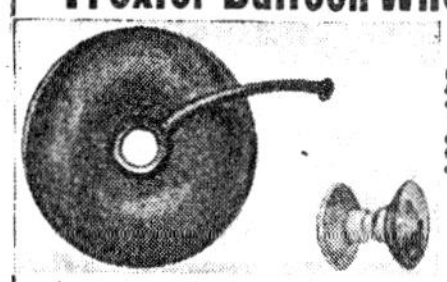

Class C (up to 150sq. ins.) seems to have been the most popular rubber class in America. Consistency of drawing standard was yet to come.

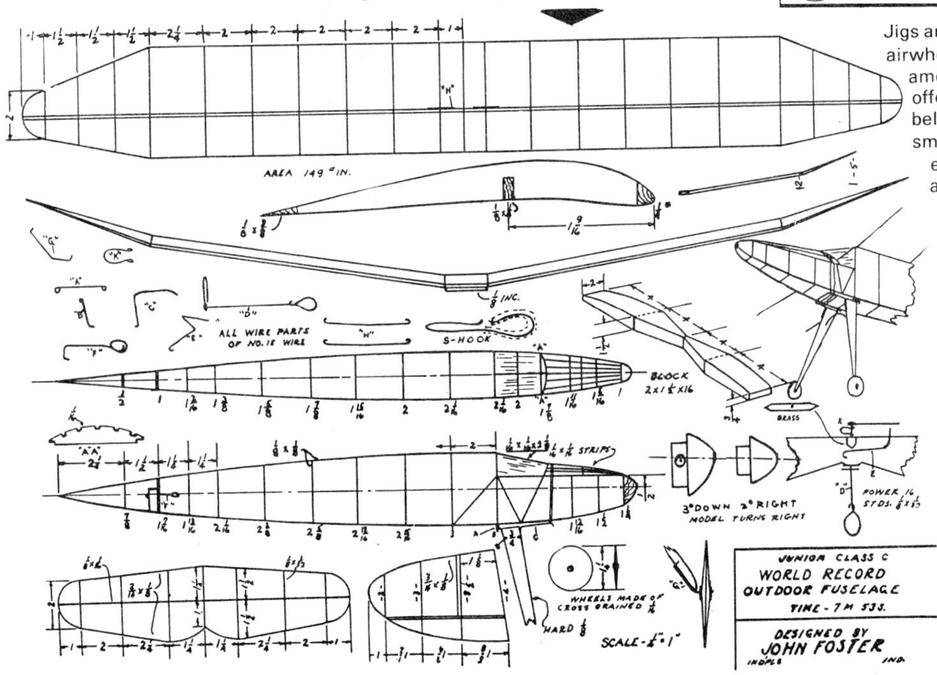

Jigs and ultra-light airwheels were among 1936 U.S. offerings, while, below, a relatively small British engine was advertised.

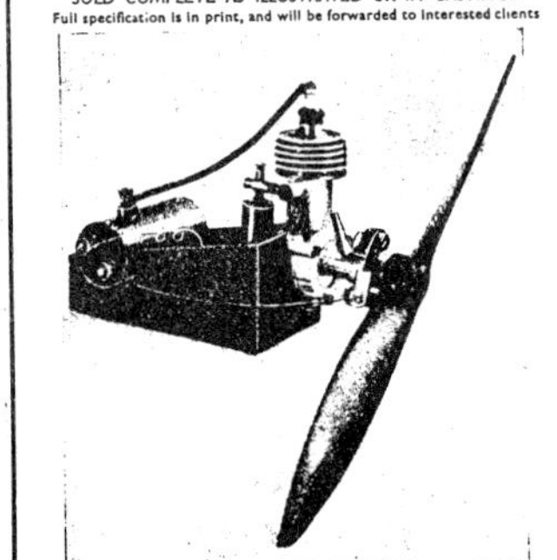

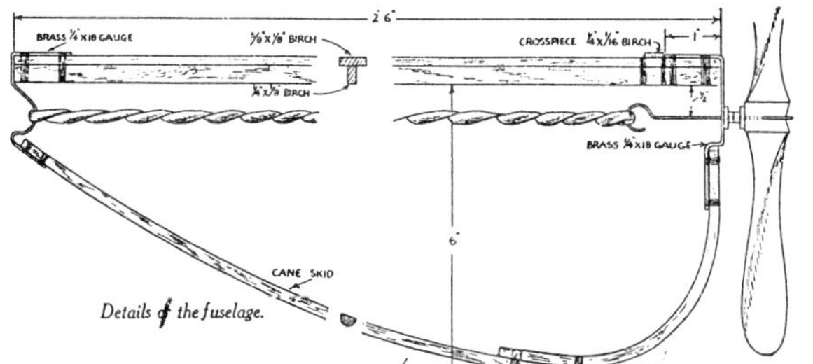

Details of the fuselage.

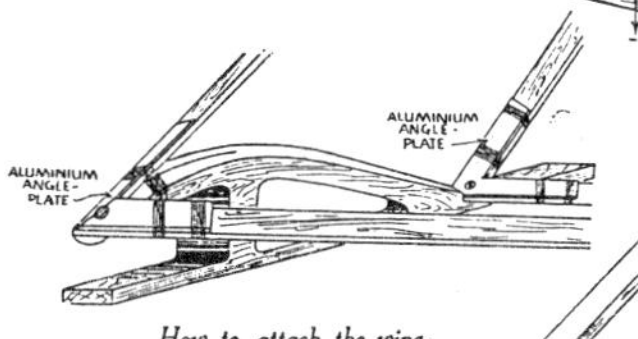

How to attach the wing-spars to the centre section.

For beginners, it says (P.M. Dec. '36). It's a pusher, and we think the gull bit is the t.e., but . . . !

A MODEL PTERODACTYL

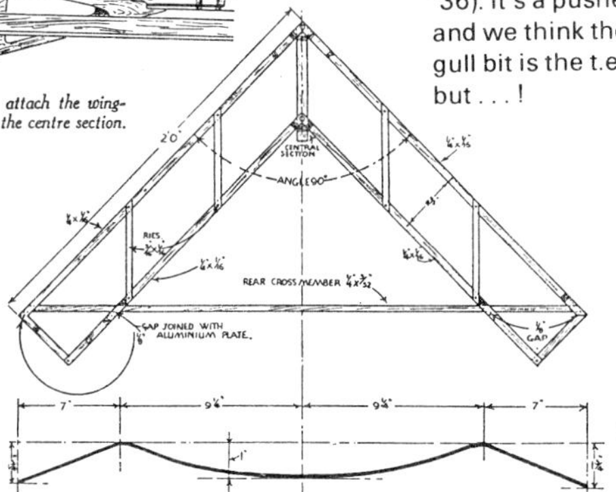

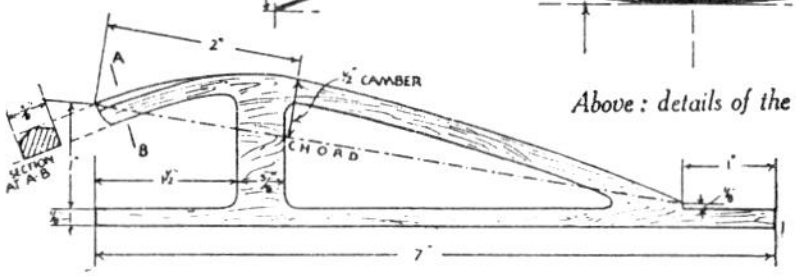

Above: details of the mainplane. Left: the centre rib.

(See page 39).

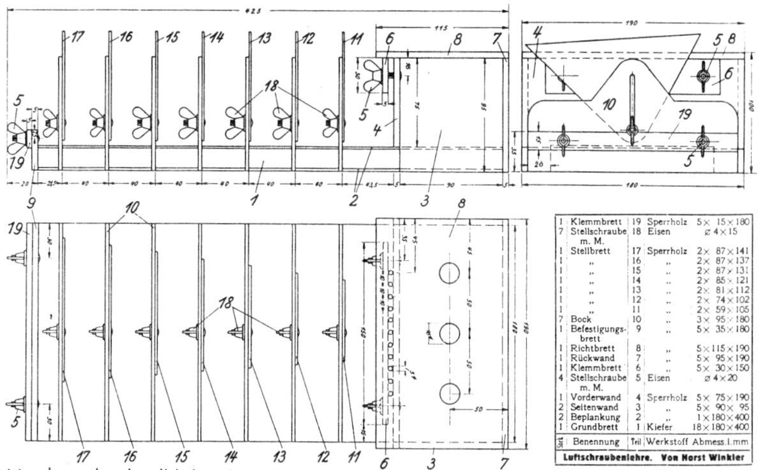

1	Klemmbrett	19	Sperrholz	5× 15×180
7	Stellschraube m. M.	18	Eisen	⌀4×15
1	Stellbrett	17	Sperrholz	2× 87×141
1	,,	16	,,	2× 87×137
1	,,	15	,,	2× 87×131
1	,,	14	,,	2× 85×121
1	,,	13	,,	2× 81×112
1	,,	12	,,	2× 74×102
1	,,	11	,,	2× 59×105
7	Bock	10	,,	3× 95×180
1	Befestigungs-brett	9	,,	5× 35×180
1	Richtbrett	8	,,	5×115×190
1	Rückwand	7	,,	5× 95×190
1	Klemmbrett	6	,,	5× 30×150
4	Stellschraube m. M.	5	Eisen	⌀4×20
1	Vorderwand	4	Sperrholz	5× 75×190
2	Seitenwand	3	,,	5× 90× 95
2	Beplankung	2	,,	1×180×400
1	Grundbrett	1	Kiefer	18×180×400
Stck.	Benennung	Teil	Werkstoff	Abmess. i. mm

Luftschraubenlehre. Von Horst Winkler

Just to show that he didn't only make gliders, Horst Winkler devised this propeller pitch gauge. Each wing-nutted piece is a ply triangle which can be adjusted for height and angle.

A Model Pterodactyl for Beginners
A few years ago I published a design suitable for beginners for a model Pterodactyl. As the issues are now out of print and I have received many requests for this design, I repeat it herewith.

The first U.S. 'gas kit' we could find was an unnamed 7ft. design from Cyclone Aircraft, Brooklyn, at $5 in January 1936. In February, Modelcraft advertised a Corben Super Ace and in March plans for the Berkeley Buccaneer and Cavalier were offered; kits followed in July.

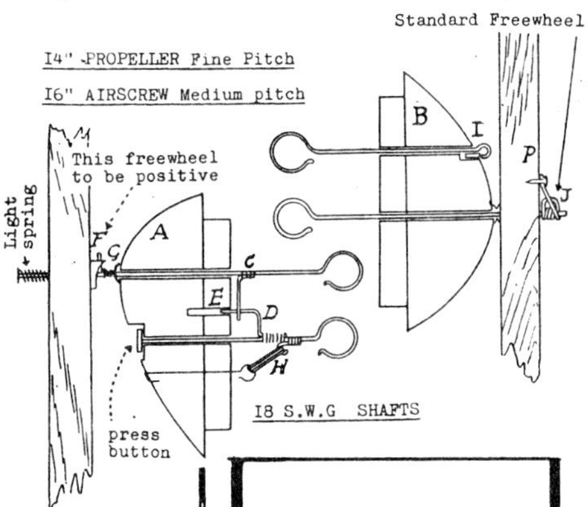

By C. A. RIPPON, S.M.A.E., N.H.M.F.C.

Reprinted in *Model Aeroplane Constructor* from *Flight*, 1912, this heading introduced Rip's ideas on extending the running time of rubber motors by releasing a second motor on a tractor prop when the first, pusher, motor ran out.

In 1936, the S.M.A.E. was able to secure the use of the Albert Hall for indoor flying and fortnightly meetings were held. Bob Copland set the first record of 6:36 with this model, included because it is strong and simple. Span was 22½in. (1 × 8 and 2 × 7¼) and chord 3¼in. Note plan taken from bound volume, hence distortion at centre. Power 6 × 16in. strands 1/32in. sq.

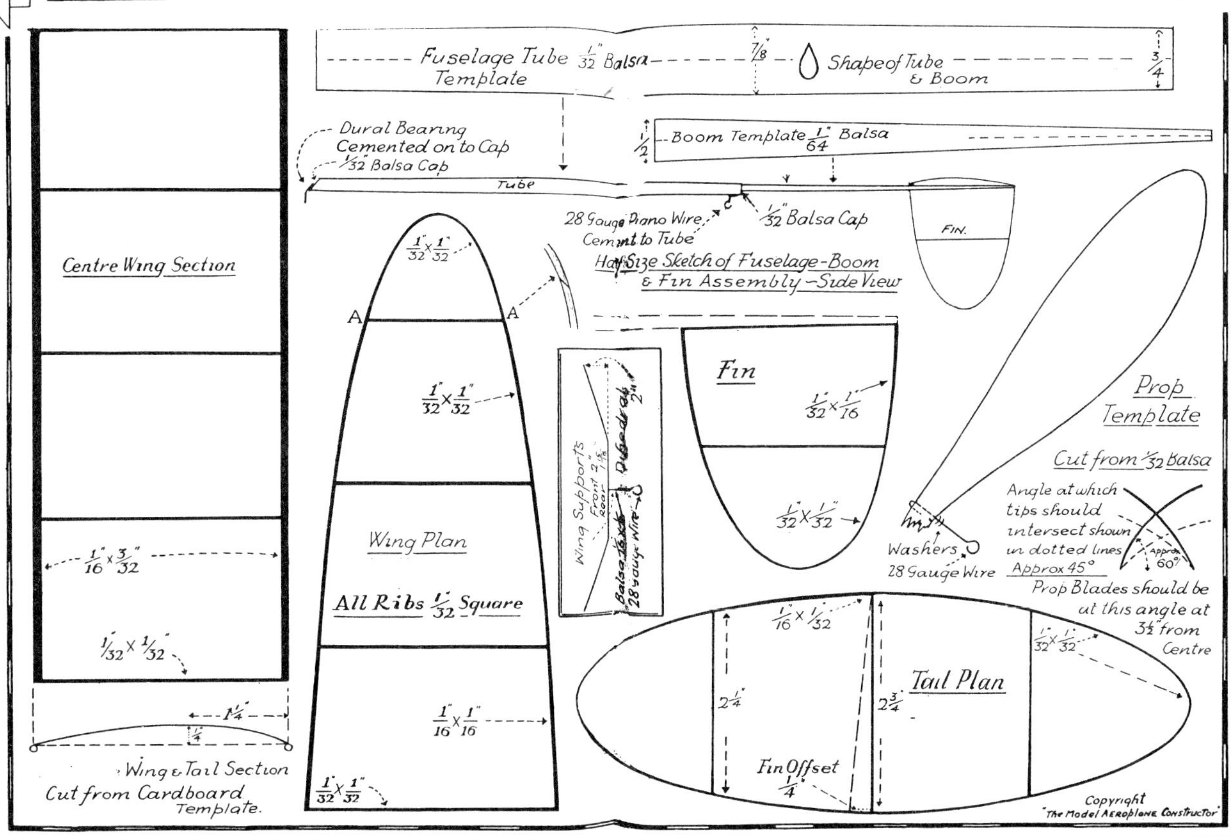

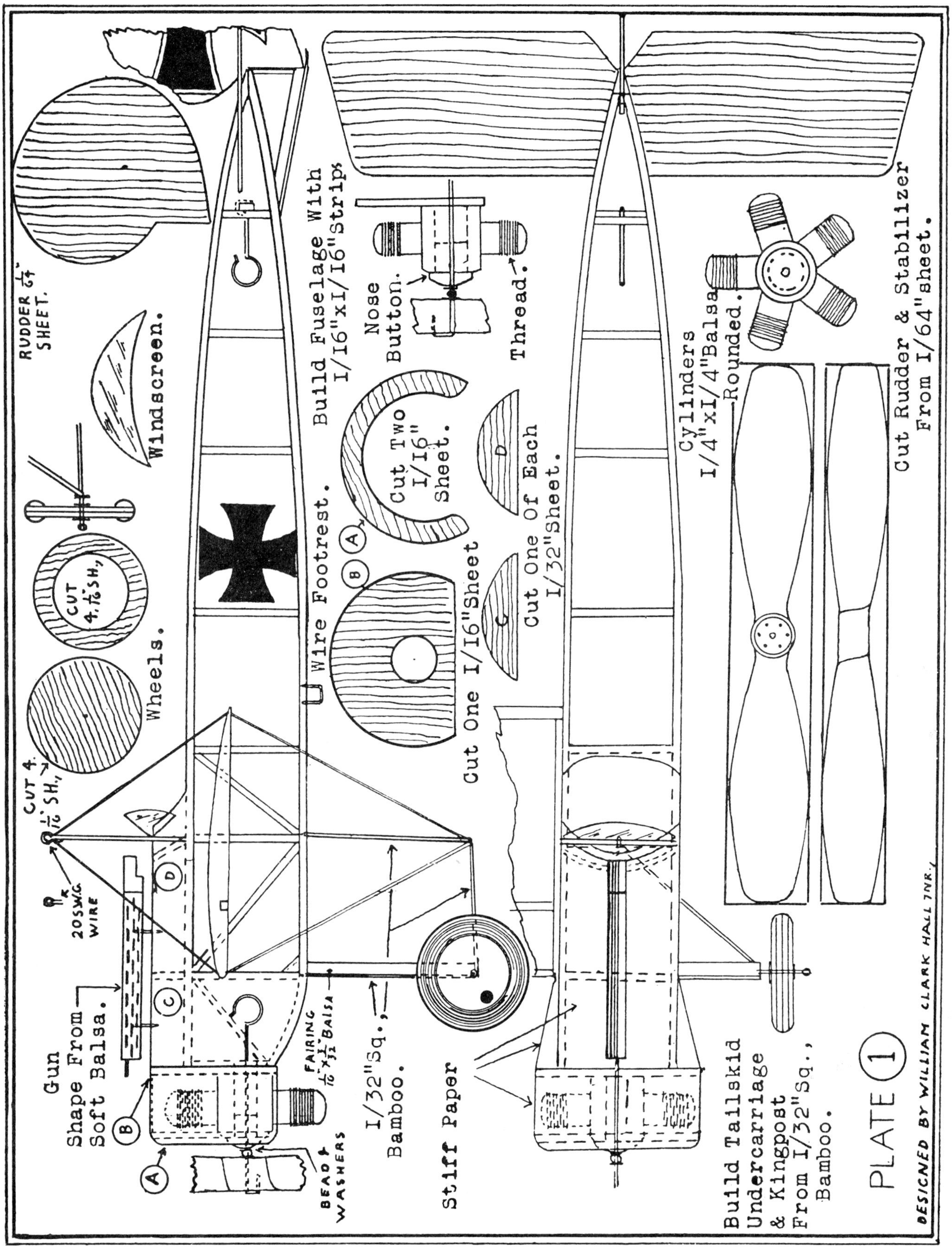

Full-size plans from which to build were popular inclusions in most model aircraft magazines and in its early days William Clarke Hall provided several designs for *Aeromodeller*. This (approximately) 12½in. DI predated modern Peanut scale (13in.) models by around 40 years. Similar size models by Herb Weiss and others were featured in American magazines of the period.

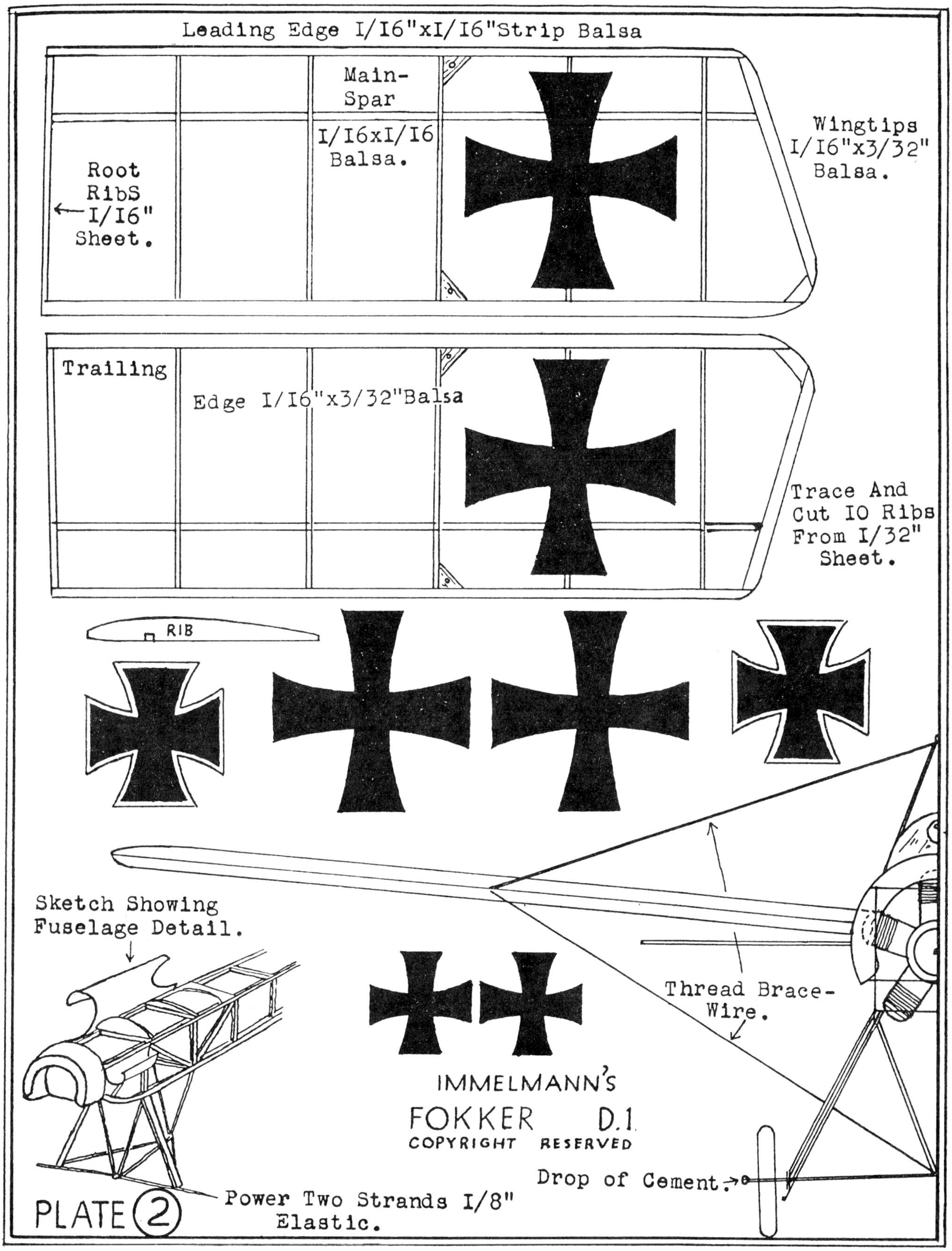

Begun in November 1935, *Aeromodeller* was initially produced by the Barron-Dean Publishing Company but passed to the ownership of D. A. Russell in mid-1937 and to Argus Press Ltd. in 1947. A companion book to this, titled 'Fifty Years of Aeromodeller', was published in 1986 to celebrate the Golden Jubilee of the magazine at the end of 1985.

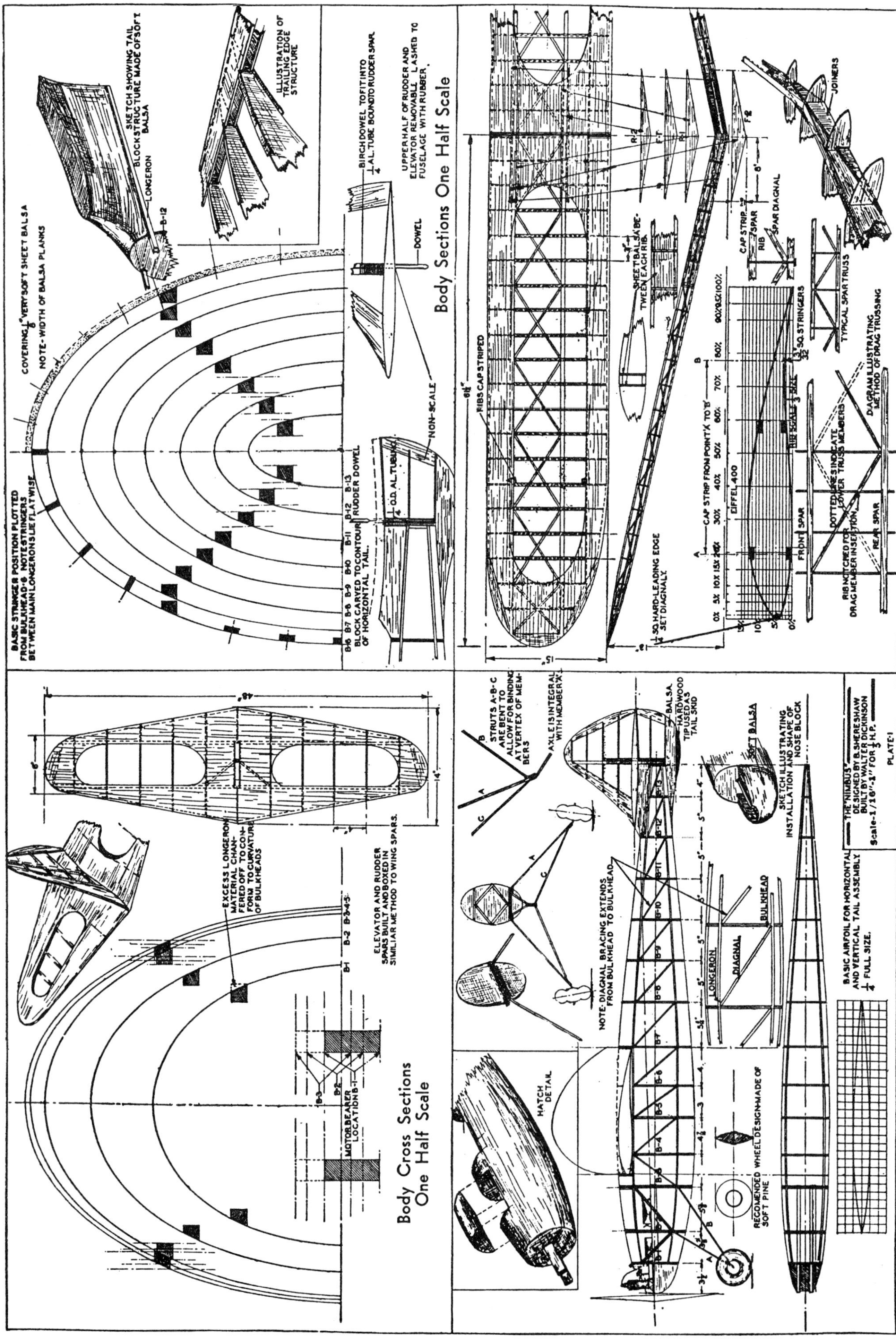
BASIC STRINGERS POSITION PLOTTED FROM BULKHEAD-6. NOTE: STRINGERS BETWEEN MAIN LONGERONS LIE FLATWISE
COVERING 1/8" VERY SOFT SHEET BALSA
NOTE - WIDTH OF BALSA PLANKS
SKETCH SHOWING TAIL BLOCK STRUCTURE MADE OF SOFT BALSA
LONGERON
B-12
ILLUSTRATION OF TRAILING EDGE STRUCTURE
BIRCH DOWEL TO FIT INTO 1/4 AL. TUBE BOUND TO RUDDER SPAR.
UPPER HALF OF RUDDER AND ELEVATOR REMOVABLE LASHED TO FUSELAGE WITH RUBBER.
DOWEL
Body Sections One Half Scale
NON-SCALE
1/4 O.D. AL. TUBING
B-6 B7 B-8 B-9 B-10 B-11 B-12 B-13
RUDDER DOWEL
BLOCK CARVED TO CONTOUR OF HORIZONTAL TAIL.
RIBS CAP STRIPED
SHEET BALSA BETWEEN EACH RIB.
1/8 SQ. HARD-LEADING EDGE SET DIAGNALY
CAP STRIP FROM POINT 'A' TO 'B'
EIFFEL 400
1/32 SQ. STRINGERS
RIB SCALE 1/2 SIZE
CAP STRIP
SPAR
RIB
SPAR DIAGNAL
JOINERS
TYPICAL SPAR TRUSS
DIAGRAM ILLUSTRATING METHOD OF DRAG TRUSSING
DOTTED LINES INDICATE LOWER TRUSS MEMBERS
RIB NOTCHED FOR DRAG MEMBER IN SECTION
FRONT SPAR
REAR SPAR
EXCESS LONGERON MATERIAL CHAMFERED OFF TO CONFORM TO CURVATURE OF BULKHEADS
B-1 B-2 B-3-4-5
ELEVATOR AND RUDDER SPARS BUILT AND BOXED IN SIMILIAR METHOD TO WING SPARS.
MOTOR BEARER LOCATION B-1
Body Cross Sections One Half Scale
STRUTS A-B-C ARE BENT TO ALLOW FOR BINDING AT VERTEX OF MEMBERS
AXLE IS INTEGRAL WITH MEMBER 'A'
BALSA
HARDWOOD TIP USED AS TAIL SKID
SOFT BALSA
SKETCH ILLUSTRATING INSTALLATION AND SHAPE OF NOSE BLOCK
NOTE: DIAGNAL BRACING EXTENDS FROM BULKHEAD TO BULKHEAD
LONGERON
DIAGNAL
BULKHEAD
HATCH DETAIL
RECOMENDED WHEEL DESIGN-MADE OF SOFT PINE
BASIC AIRFOIL FOR HORIZONTAL AND VERTICAL TAIL ASSEMBLY 1/4 FULL SIZE.
"THE NIMBUS" DESIGNED BY B. SHERESHAW BUILT BY WALTER DICKINSON Scale-1/16"=1" FOR 1/5 H.P.
PLATE 1

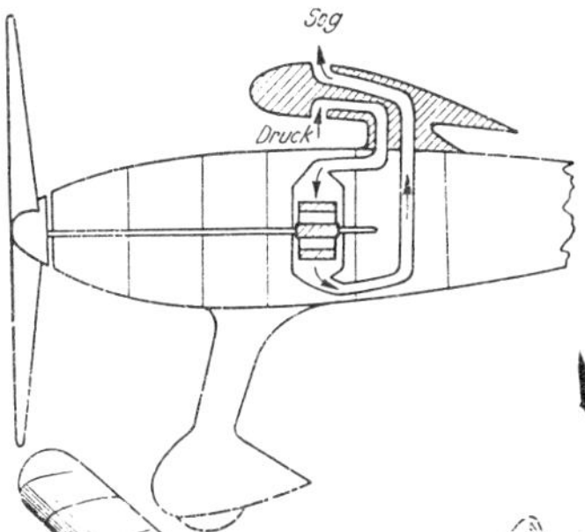

This notion in *Modellflug*, April 1937, sparked off enormous controversy. Werner Funfe suggested that the pressure differential between below and above wing airflow could be tapped to drive a turbine and made remarkable claims for an experimental model — last seen by an aircraft crew at some 26,000ft. An April Fool leg-pull? Well, 'funf' can mean 'connive at'!

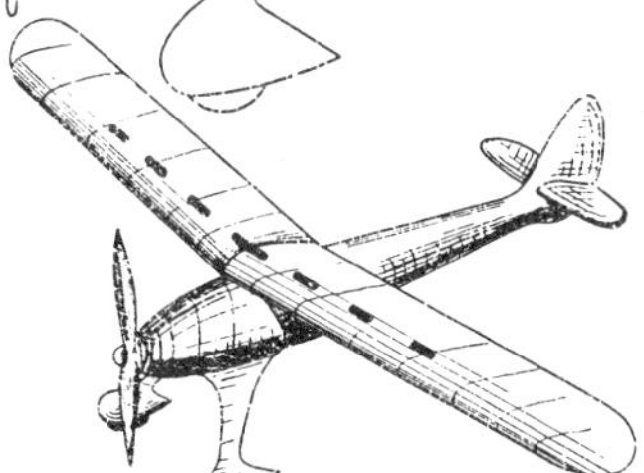

Alfred van Wymersch flew this model in the Belgian team in 1937.

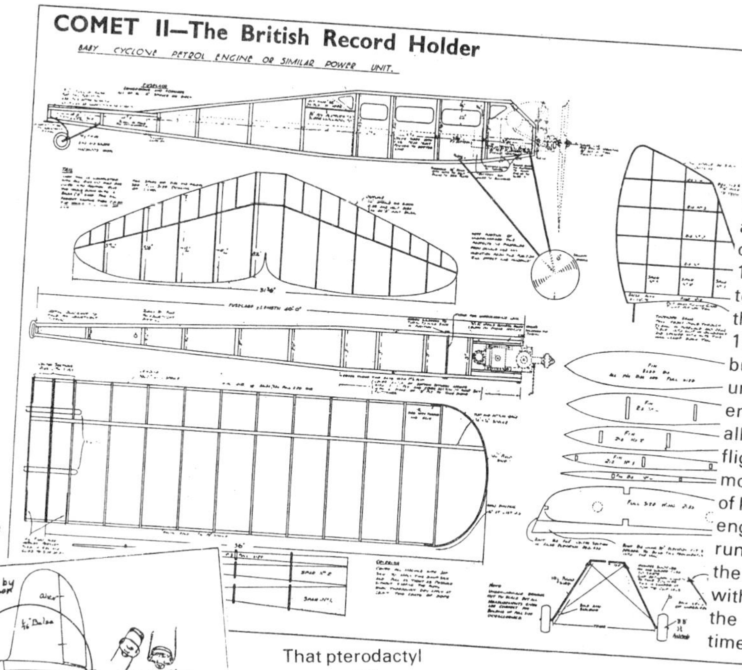

The 'Comet' design was sturdy even if not aerodynamically refined. A. T. Fraser achieved an o.o.s. flight of 16:25 in 1937 to displace the Bowden 1932 record, but with unlimited engine run allowed such flights seem more of a test of keeping the engine running and the model within sight of the timekeepers.

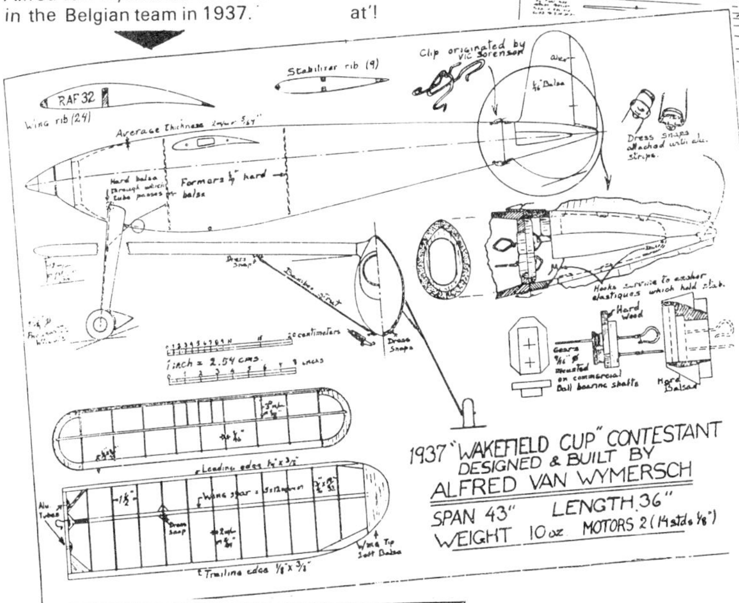

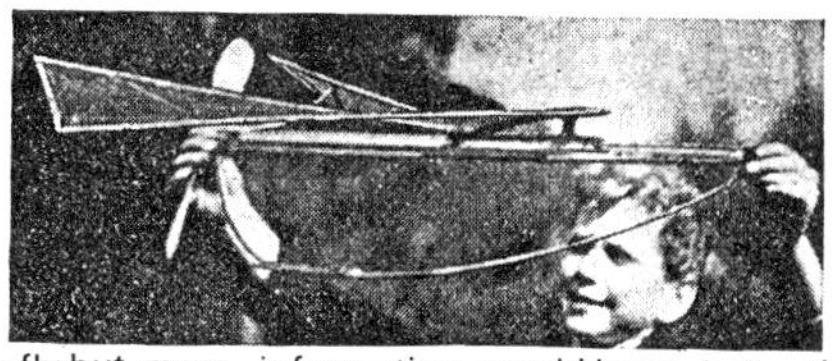

That pterodactyl on page 35 photographed. There is a lot of washout, but no guidance on how it is achieved or held. It could certainly be made to fly but more information would have helped!

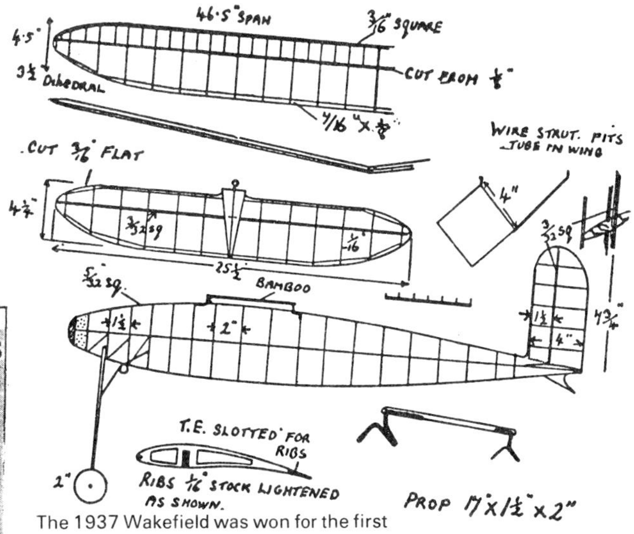

The 1937 Wakefield was won for the first (and so far only) time by a French enthusiast, Emmanuel Fillon, with the diamond-fuselage parasol design above, which averaged 4:13.2 and was retrieved after a 15-mile chase on its second flight.

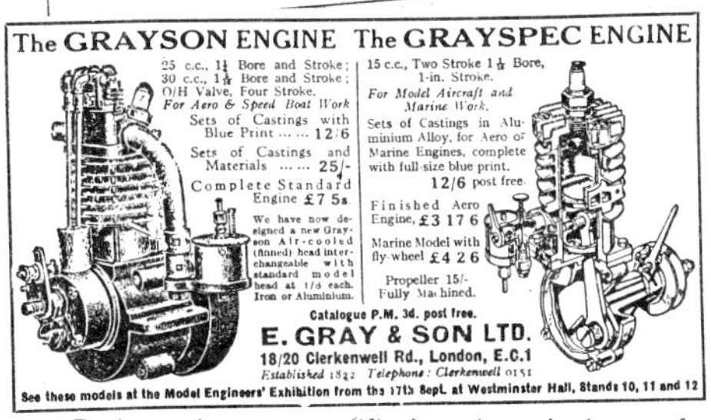

Perhaps the most prolific American designer of early power models Ben Shereshaw produced a seemingly endless flow of attractive and elegant designs, many of which appeared in *Flying Aces*; he later designed motors. The *Model Aeroplane Constructor* photo shows him holding his 10ft. 3in. 'Nimbus' (one of 21 gas models he entered in the 1936 Nationals!) the drawings for which appeared in June 1937 *M.A.N.* (opposite page).

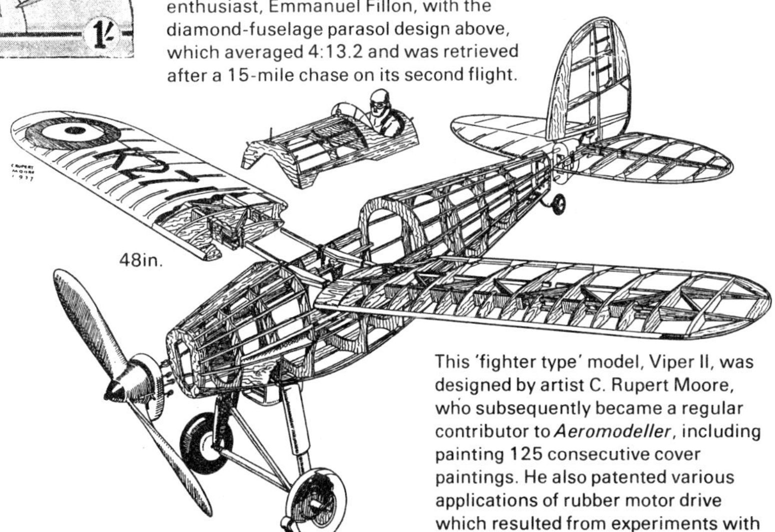

This 'fighter type' model, Viper II, was designed by artist C. Rupert Moore, who subsequently became a regular contributor to *Aeromodeller*, including painting 125 consecutive cover paintings. He also patented various applications of rubber motor drive which resulted from experiments with accurate scale models.

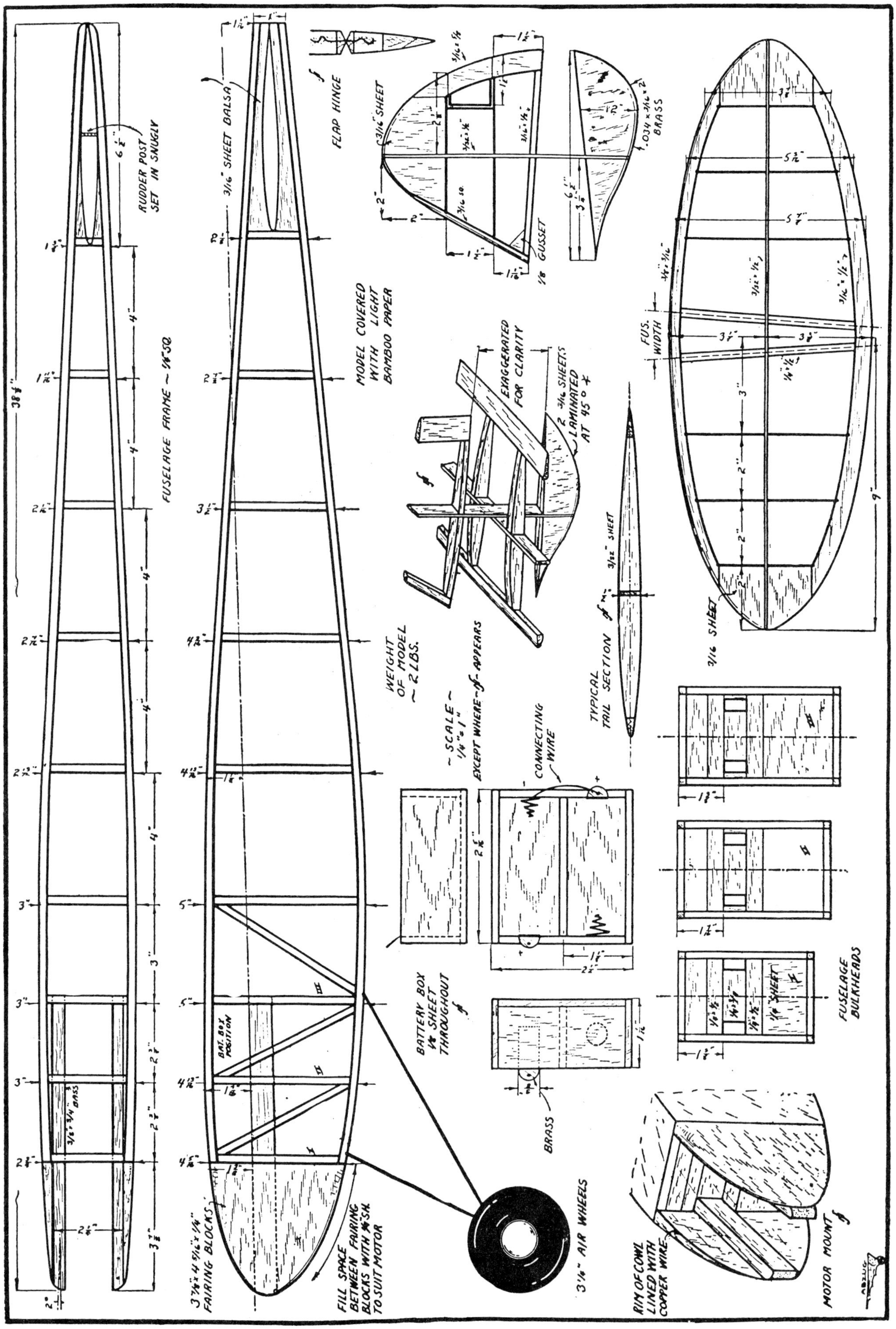

'Flying Midget', 300sq. in., Brown Junior power, by Thracy Petrides and Malcolm Abzug.

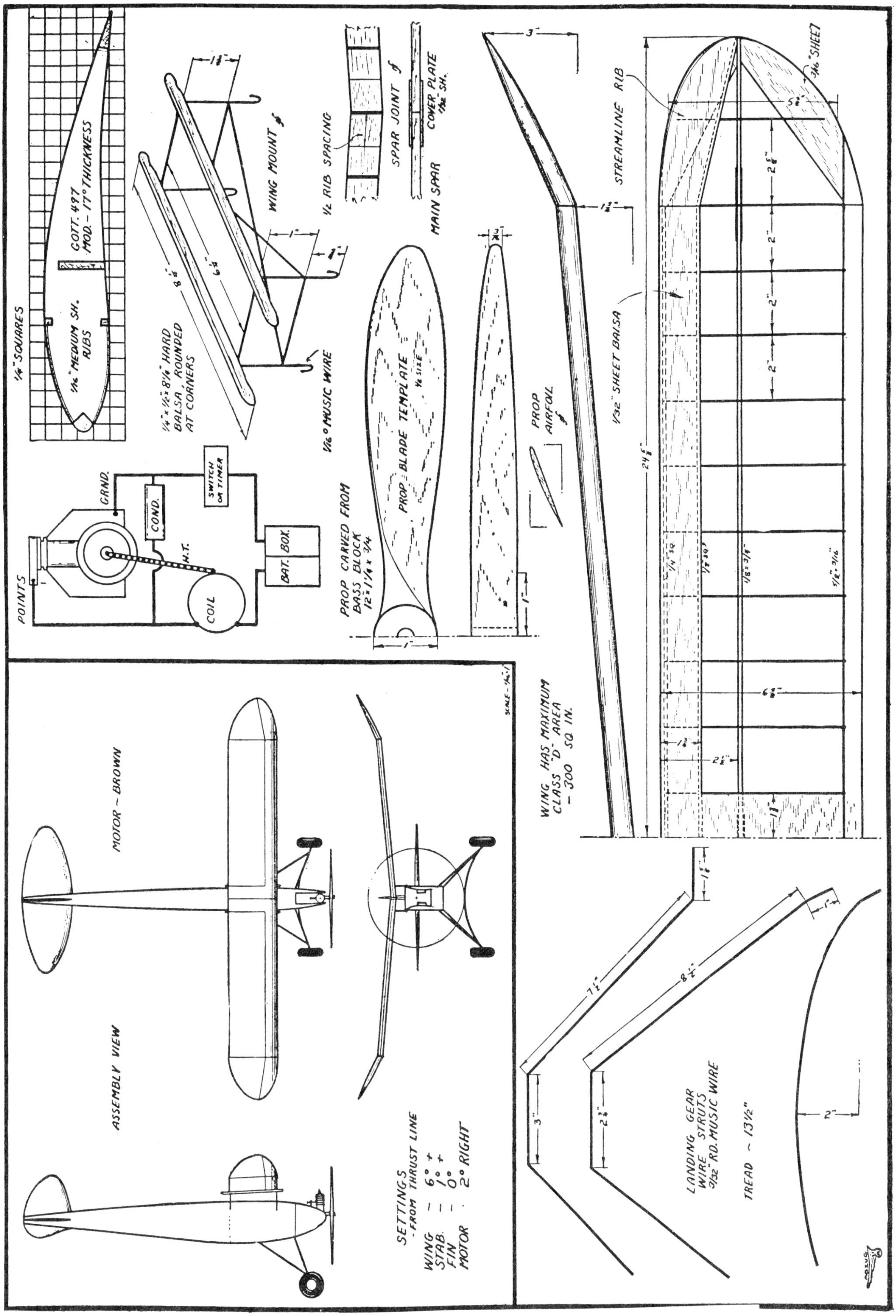

A very simple 49¼in. 'box-car' model requiring the motor to be throttled down! (*M.A.N.* August 1937).

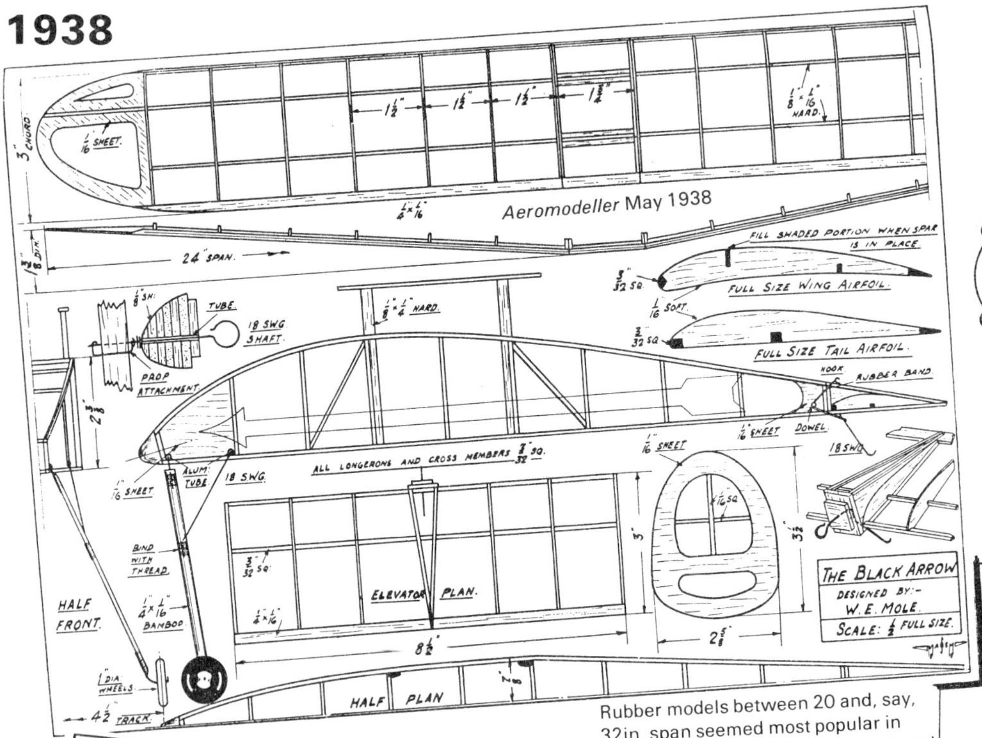

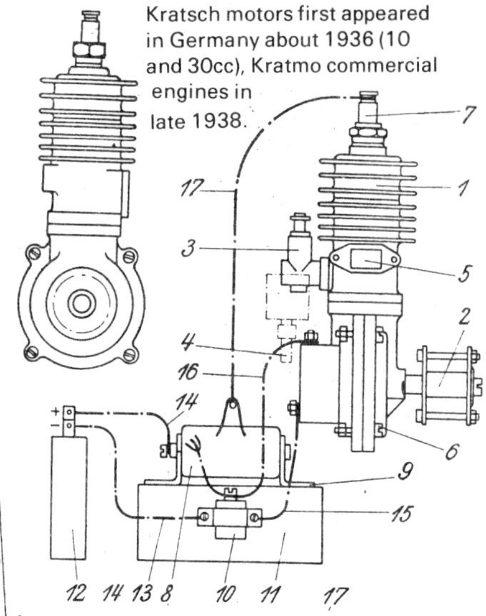

Kratsch motors first appeared in Germany about 1936 (10 and 30cc), Kratmo commercial engines in late 1938.

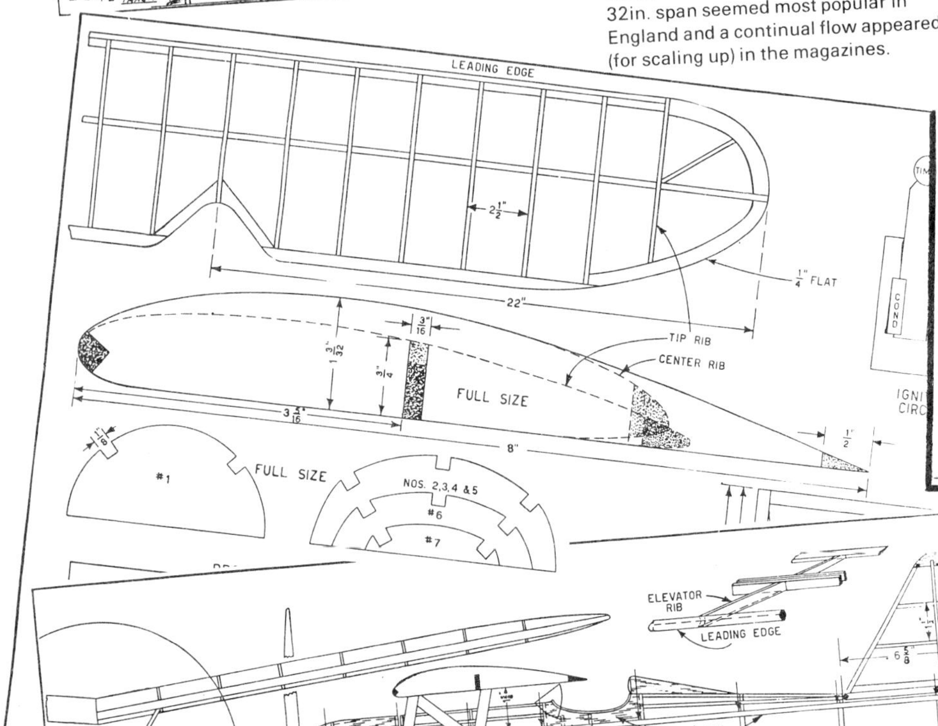

Rubber models between 20 and, say, 32in. span seemed most popular in England and a continual flow appeared (for scaling up) in the magazines.

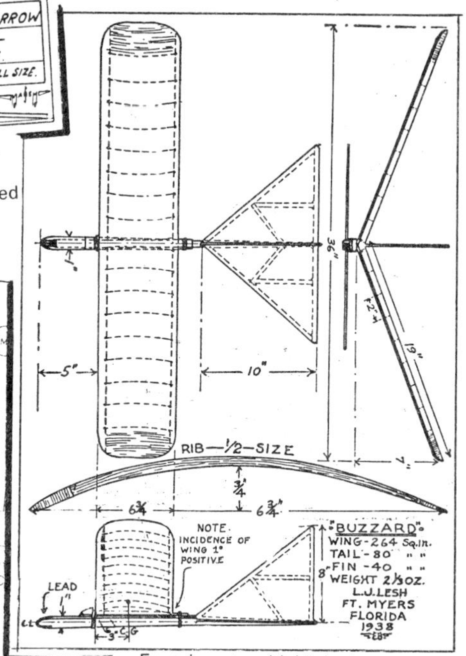

Experiments with bird wings, actual and simulated, were conducted in 1937/8 by L. J. Lesh and reported in *M.A.N.* July '38. The conclusion was that bird sections are good but critical on stability.

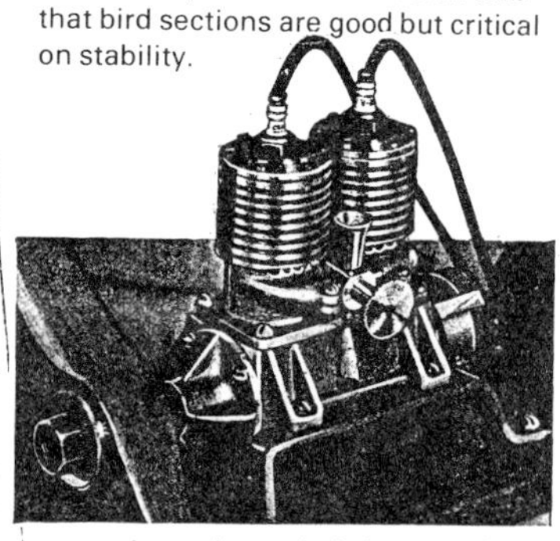

An early vertical alternate-firing twin was advertised at $20 by Southern Model Engines, Nashville, in mid-38. Bore ⅝in., flying weight with batteries 18ozs. American motors proliferated in 1937-8.

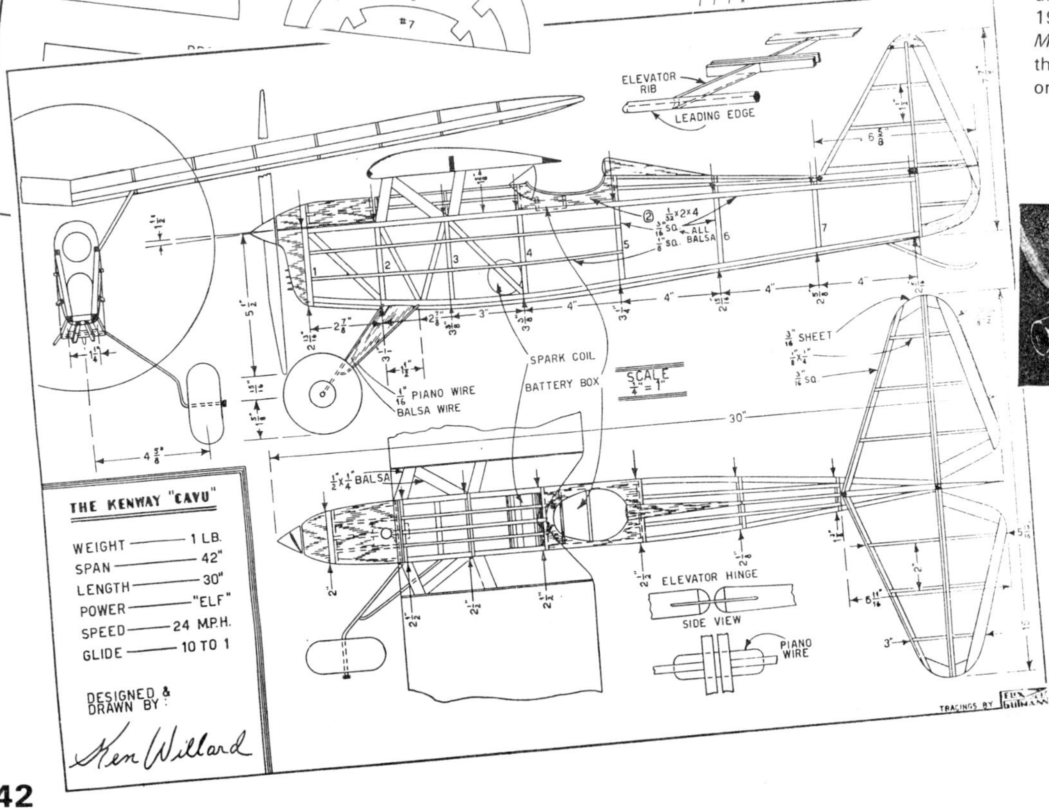

Ken Willard's 'Cavu' (ceiling and visibility unlimited) in April 1938 *M.A.N.* aimed for simplicity with attractive appearance. Aluminium-hinged elevators were because the wing and tail were permanently cemented in place.

A RADIO CONTROL SYSTEM FOR MODEL AIRCRAFT

By LEO A. WEISS

This article appeared in *Aeromodeller*, August 1938. Chester Lanzo is believed to be the first modeller to make a successful flight with R/C.

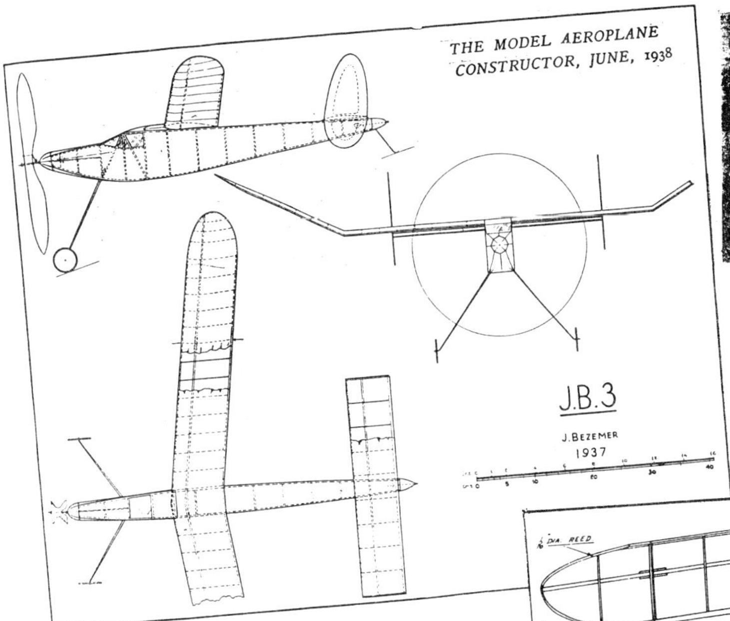

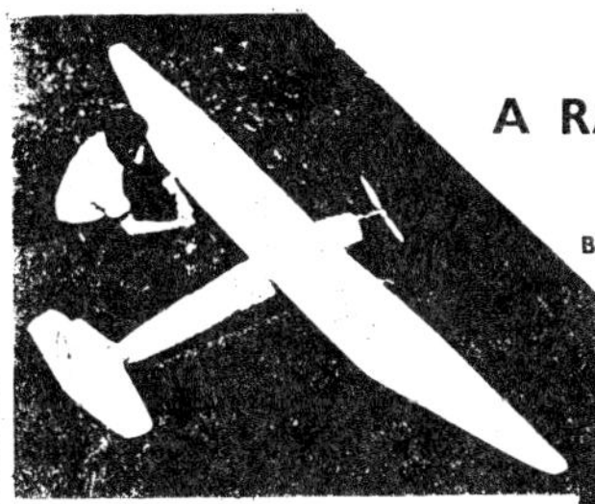

Leo Weiss' radio as described, using audio modulation and tuned reeds, weighed some six pounds. Probably the most successful and consistent results were obtained pre-war by the Good Brothers (U.S.A.)

Simple Dutch design which remained popular for some 25 years through Aeromodeller Plans Service.

Two popular American kit subjects, the Howard DGA8 'Mr. Mulligan' and the Waco Custom Cabin Biplane, from a Scientific advert.

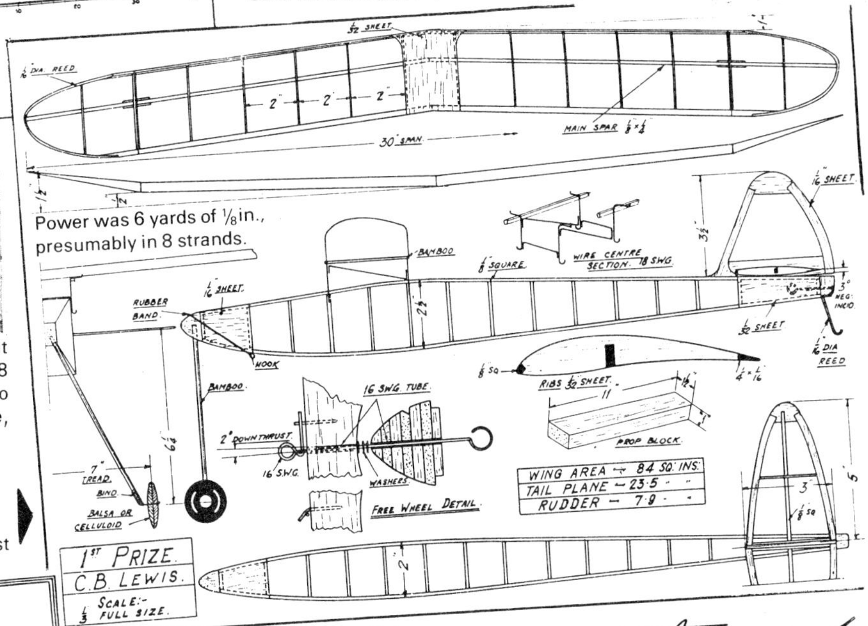

Power was 6 yards of 1/8 in., presumably in 8 strands.

Another simple small rubber design, this one from *Aeromodeller* August 1938.

Leon Shulman's 'Skyscraper' (November 1938) had an appearance that offended many conventional modllers but no-one could argue with the performance

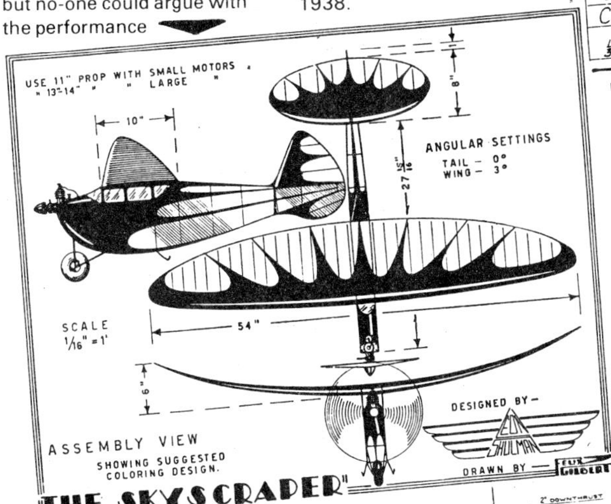

Proposal for tandem airscrews using spanwise rubber motors, from *Modellflug*. Using contrate gears and pinions would certainly simplify dihedral allowance, as in the detail sketch.

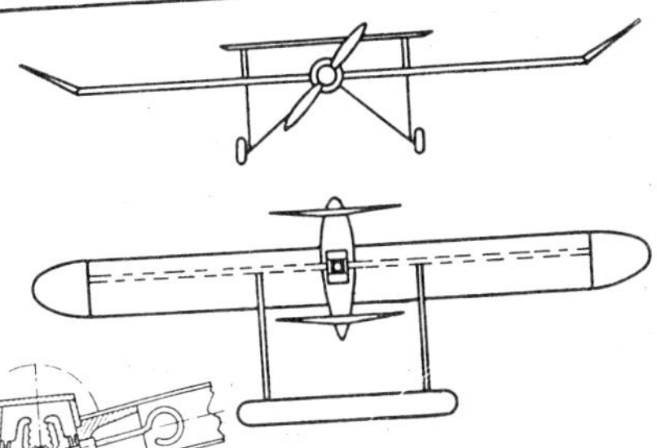

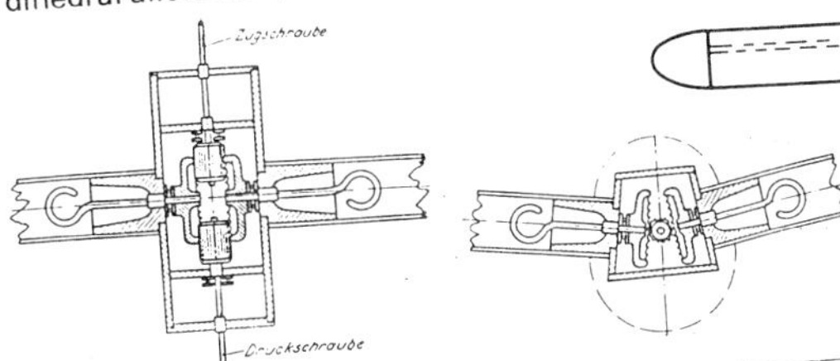

German Kratmo engines were available in 4, 10 and 30cc sizes.

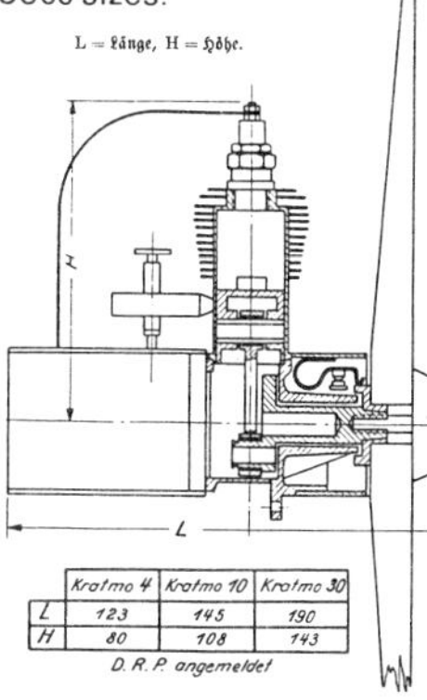

	Kratmo 4	Kratmo 10	Kratmo 30
L	123	145	190
H	80	108	143

A Marshall book, *Models for Flying*, by L. H. Sparey and C. A. Rippon, appeared originally in 1936/7 and was reprinted several times The 'Northern Arrow' was typical of the lightweight (4oz.) models around 36in. span popular for competition flying in the late 1930s and was featured in reprints of the book.

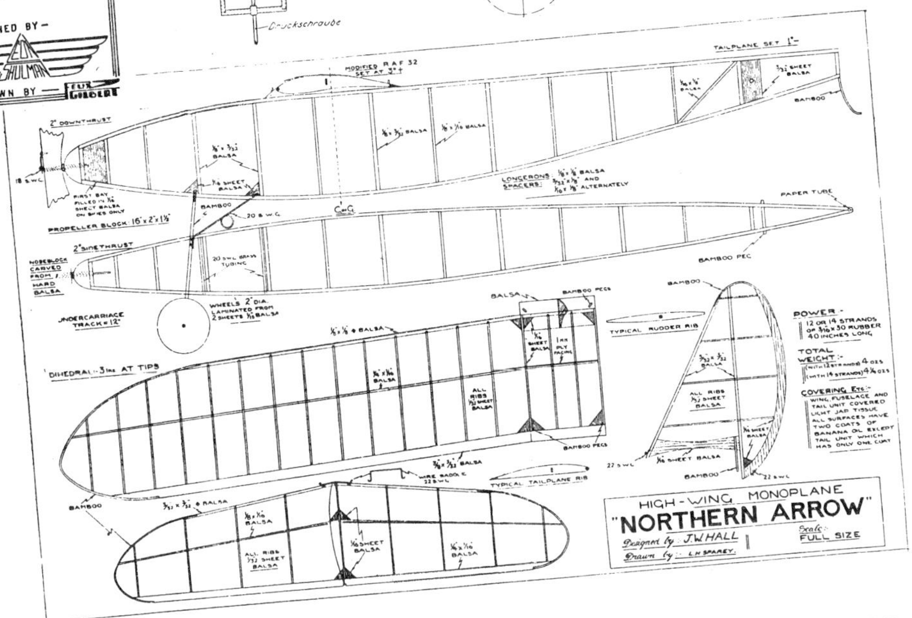

1938

Another Ben Shereshaw design (*Flying Aces*, September 1938). Though perhaps not as stylish as most from this designer, it embodies sound structural practice allied with simplicity.

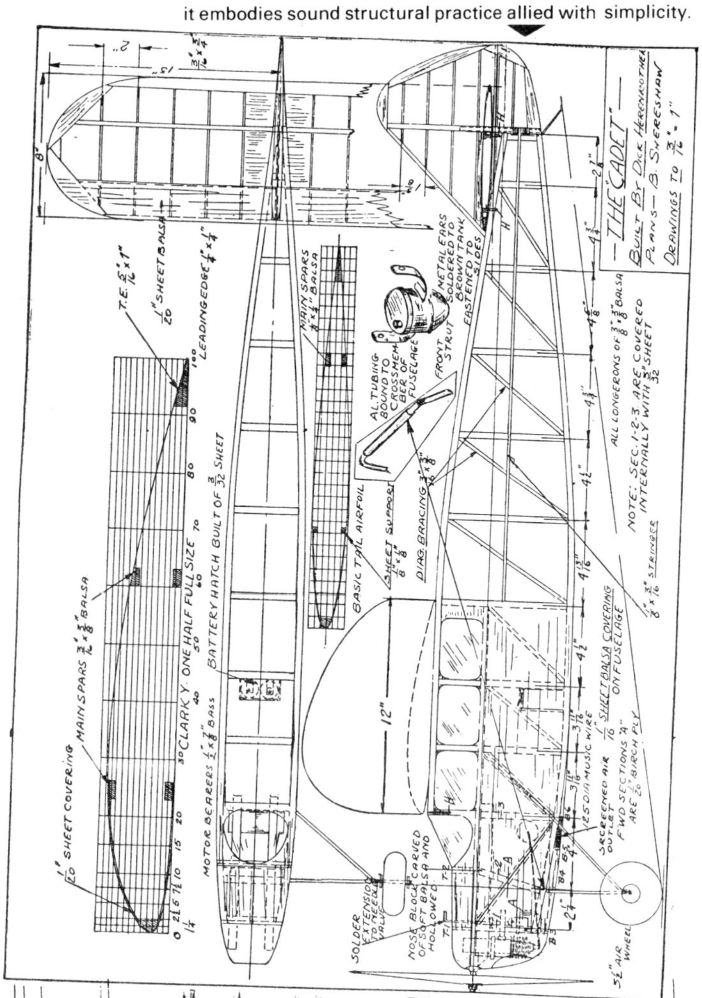

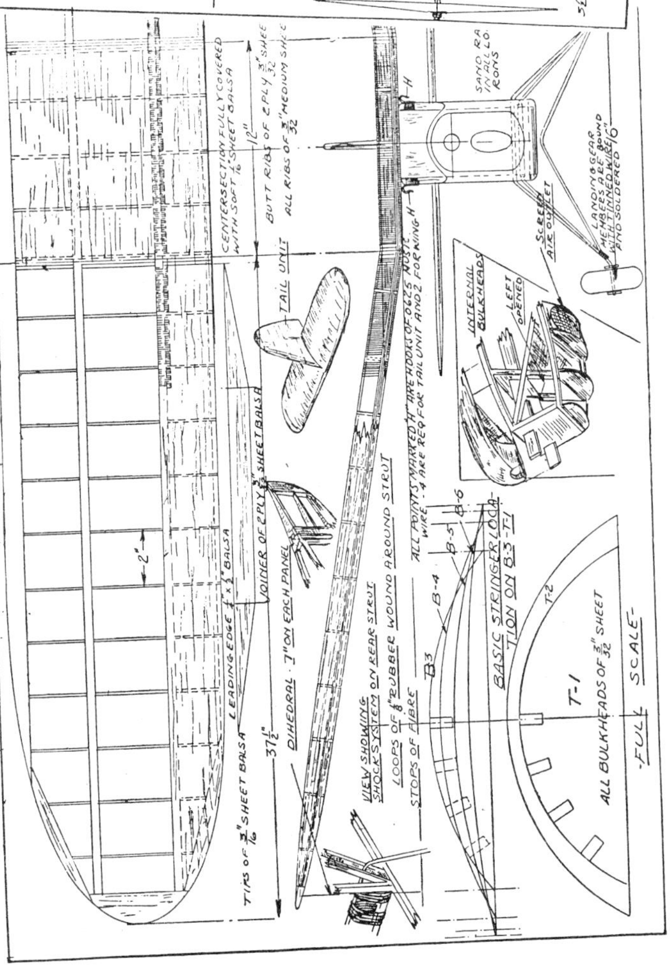

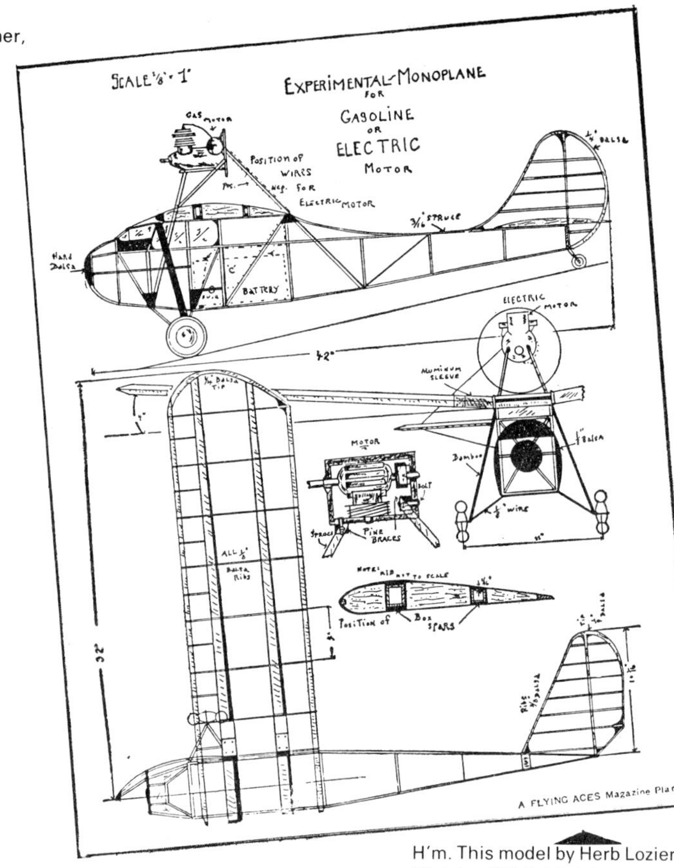

H'm. This model by Herb Lozier (October '38 *F.A.*) claims excellent results using an electric motor from a discarded automobile horn and two C size dry batteries. Chief objection, he says, is expense of batteries . . .

. . . Is it the man from Missouri who demands 'Show me'?

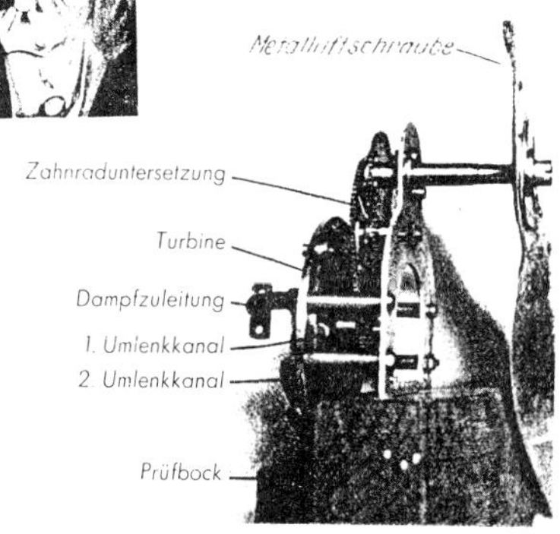

Steam turbine (right) described by Hermann Happel in *Modellflug*, may 1938, and shown installed with generator.

Rubber-powered speed models enjoyed a vogue in the early '30s in Britain and later in America. This single wheel 20in. span design with hollowed block fuselage and 20-30 strands of ⅛in. flat claimed repeated flights of just under 70m.p.h. (112k.p.h.).

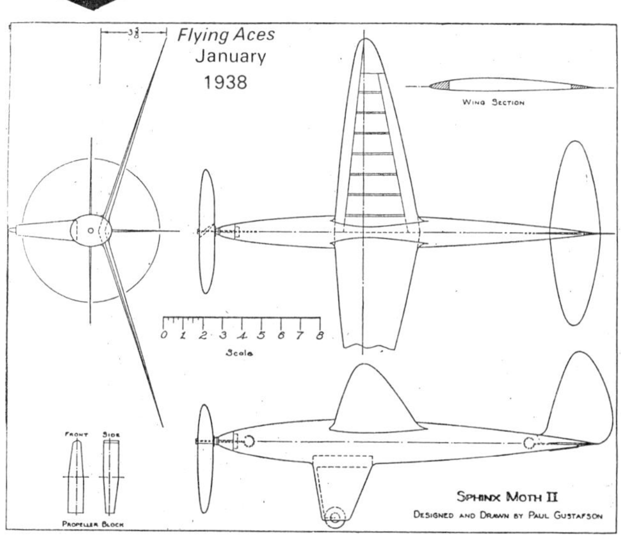

Jim Cahill won the Wakefield for the U.S. with an average of 10:54 in a high-scoring competition between 14 nations.

Ralph Bullock introduced r.t.p. flying in 1936 in *Aeromodeller*; this follow-up appeared in *M.A. Constructor* in August 1938.

AN INTERESTING METHOD OF

INDOOR FLYING

by

Ft. Lt.

R. N. BULLOCK

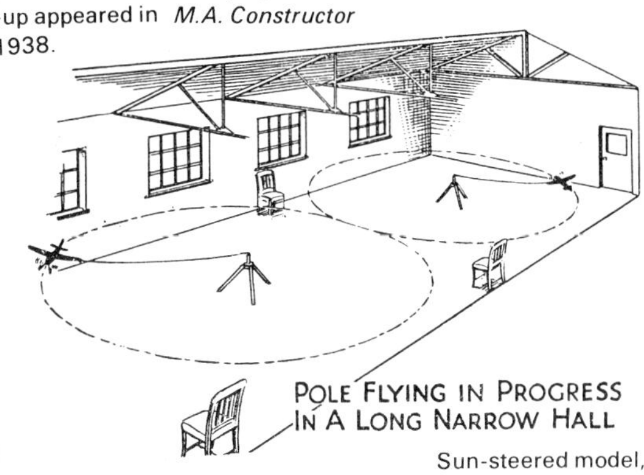

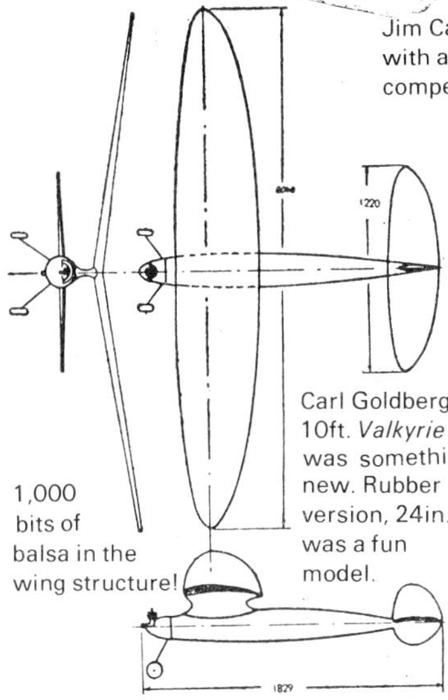

1,000 bits of balsa in the wing structure!

Carl Goldberg's 10ft. *Valkyrie* was something new. Rubber version, 24in. was a fun model.

PLANEFILM
The Magic Covering

Megow's are now exclusive distributors of this marvelous light, tough, elastic covering for model planes.

Non-porous, smooth as glass, PLANEFILM can be applied in a fraction of the time required by old methods. Actually stretches itself drumlike over irregular curves . . . easy to get even tension on surface areas, even distribution of weight and perfect balance so necessary to good performance. PLANEFILM comes attached to backing sheet—it is easy to handle and cut to shape. Stripped from backing, it is soft and pliable, and forms to the many curves necessary in model building.

PLANEFILM eliminates the use of glue, dope, finishes and stretching, is far superior in lasting quality and produces an opaque brilliant, colorful, glassy effect that cannot be approached by old methods.

A forerunner of modern film coverings?

The 44in. 'Eaglet' from Scientific, a very pretty little model.

Downthrust adjusted by motor tension on a rubber block, an idea

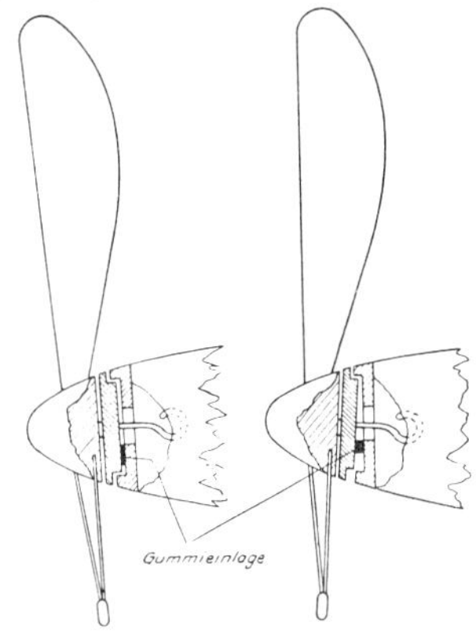

seen at the Wakefield Trophy competition in Paris.

Sun-steered model, the sun shining through a window in the fuselage on to a neutral area. However, if the model turns the light will fall on a photo-electric cell which will apply rudder to return the model to its original heading. This is an abbreviated interpretation of the system devised by Helmut Sinn. (*Modellflug*, July 1938).

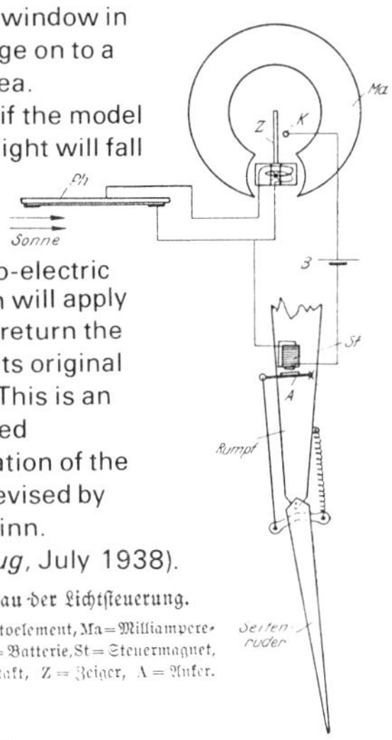

Aufbau der Lichtsteuerung.

Ph = Photoelement, Ma = Milliamperemeter, B = Batterie, St = Steuermagnet, K = Kontakt, Z = Zeiger, A = Anker.

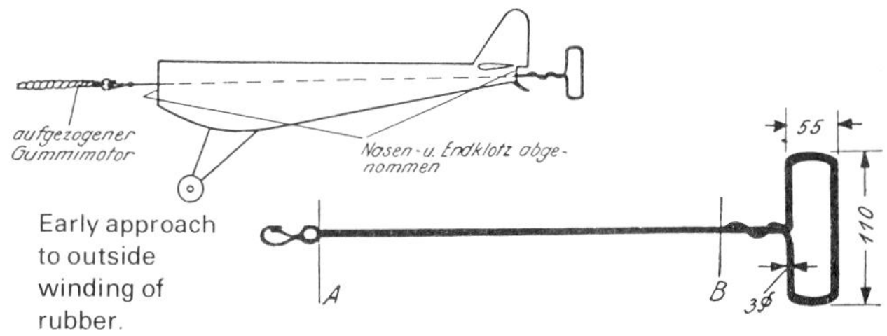

Early approach to outside winding of rubber.

German Thaler motor advertised in 1938 'Preis RM 1.80'.

"HE TOLD ME TO WASH OUT THE WING TIPS"

Cartoonist Freddie began in *Model Aeroplane Constructor*

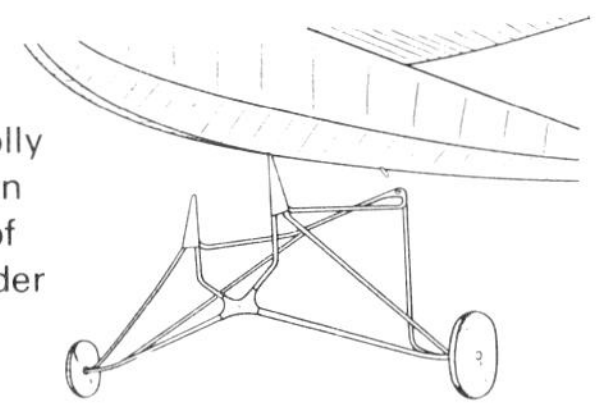

Drop-off dolly for one-man operation of towline glider launching.

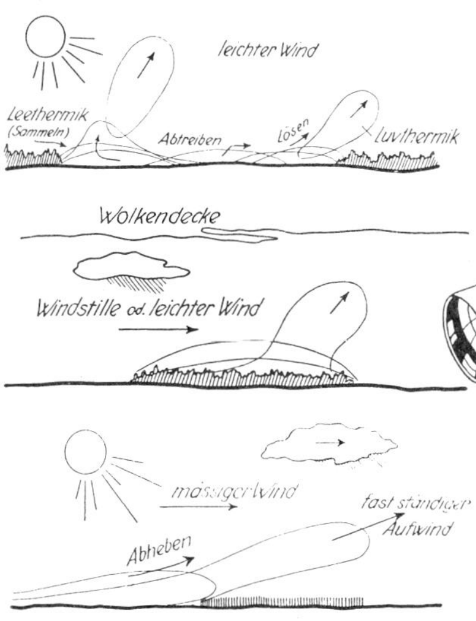

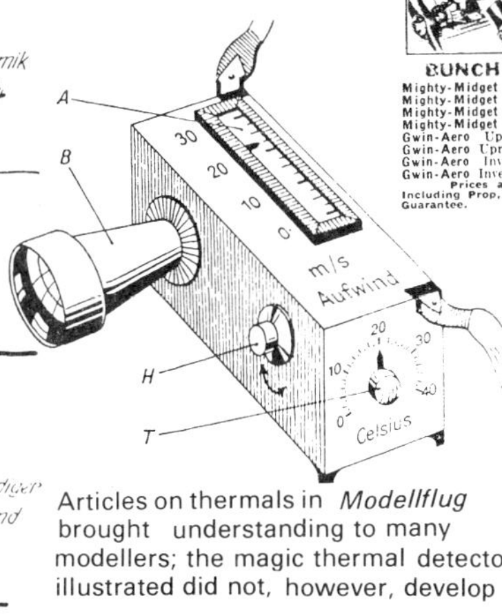

Articles on thermals in *Modellflug* brought understanding to many modellers; the magic thermal detector illustrated did not, however, develop

BUNCH MOTORS

Mighty-Midget Uprt. Asbld.	$9.50
Mighty-Midget Upright Kit	7.85
Mighty-Midget Inv. Asbld.	9.75
Mighty-Midget Inverted Kit	7.85
Gwin-Aero Upright Asbld.	12.00
Gwin-Aero Upright Kit	9.85
Gwin-Aero Inverted Asbld.	12.50
Gwin-Aero Inverted Kit	9.85

Prices are postpaid Including Prop, 70 Oil and Double Guarantee.

ROBOT TIMER

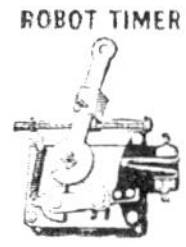

100% Accurate and Guaranteed. Price $2.50 Postpaid

Without Automatic Attach. $1.95 Postpaid

Clockwork timers had a brief life before Austin and other pneumatic ones appeared but never completely disappeared.

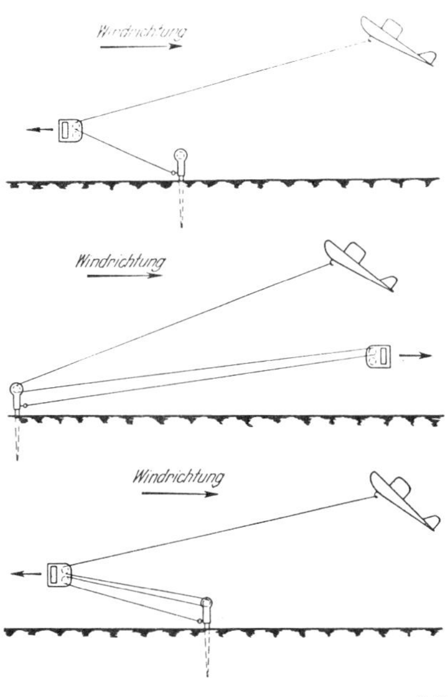

Launching with one, two or three pulleys illustrated in Modellflug

PENDULUM CONTROL FOR PETROL 'PLANES

By C. R. JEFFRIES

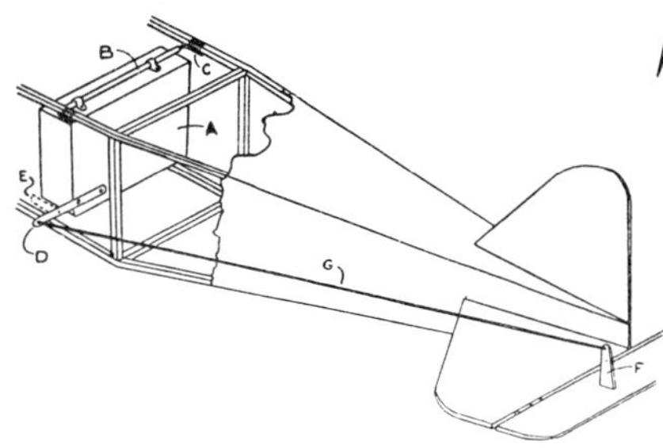

Pendulum controls have, over the years, exercised many minds, two of the best-known being Rupert Moore and P. E. Norman. One of the earliest articles was this in January 1939 *Aeromodeller*. A stop prevents rearward movement of the pendulum so that it only operates when the model is nose down. The sketch at left suggests using a power model's ignition battery as the pendulum.

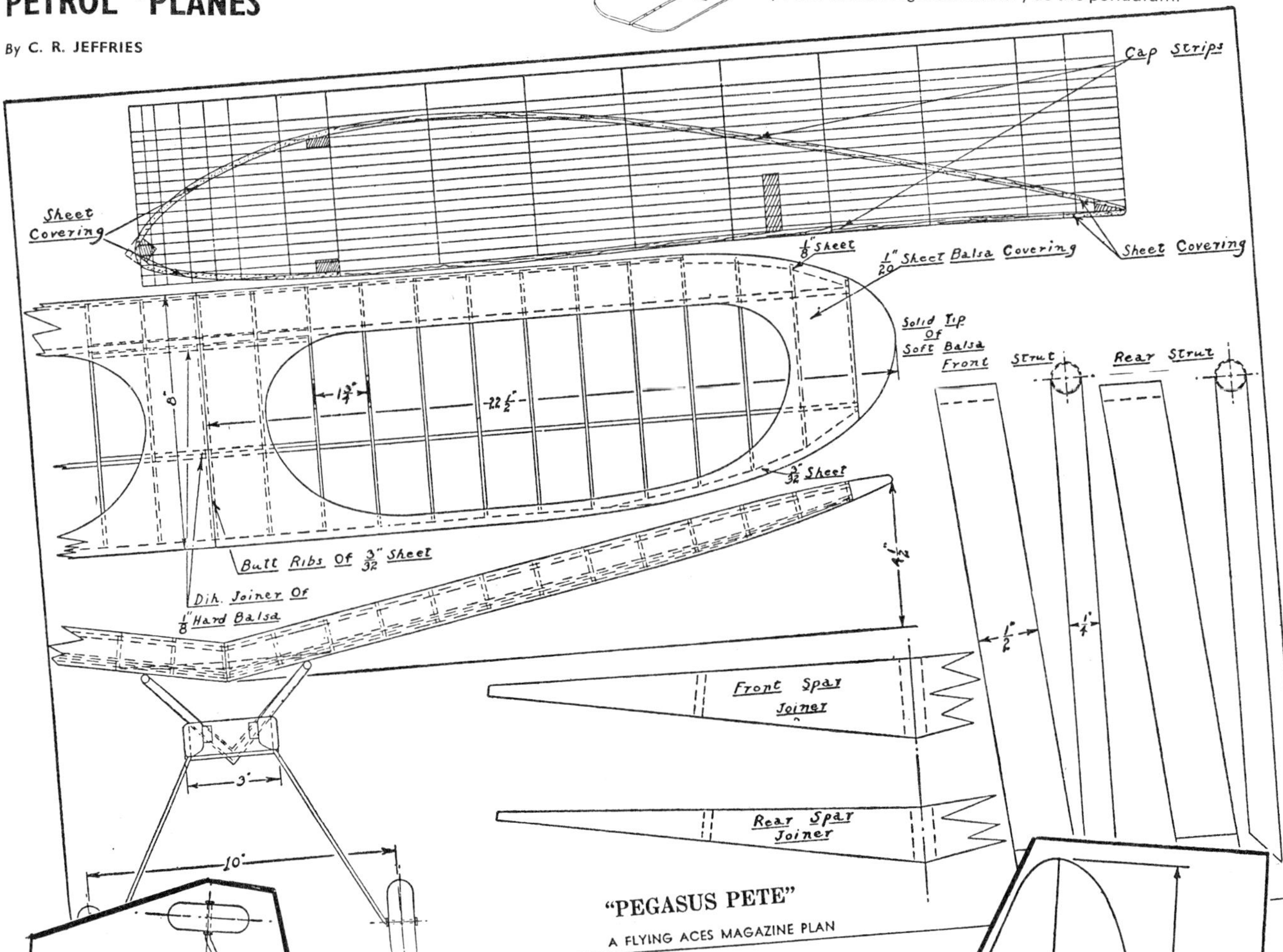

"PEGASUS PETE"
A FLYING ACES MAGAZINE PLAN

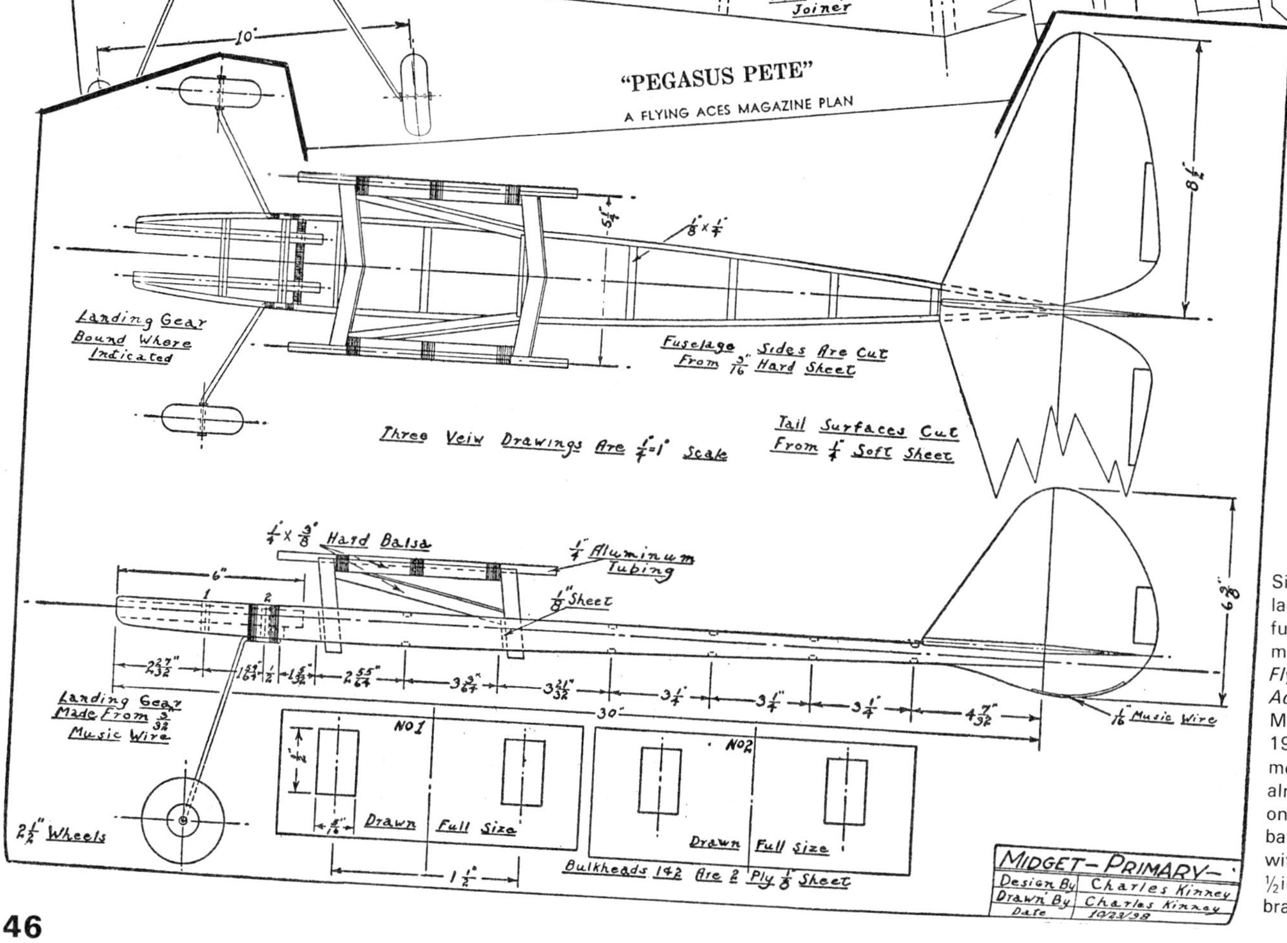

Simple ladder-type fuselage model from *Flying Aces*, March 1939; wing mount is almn. tube on ½ × ¼in. balsa struts with ⅜ × ½in. bracing.

1939

By 1939 radio control had become more practical, though still heavy, largely because of the batteries needed. Howard McEntee started a series in *M.A.N.* May, 1939

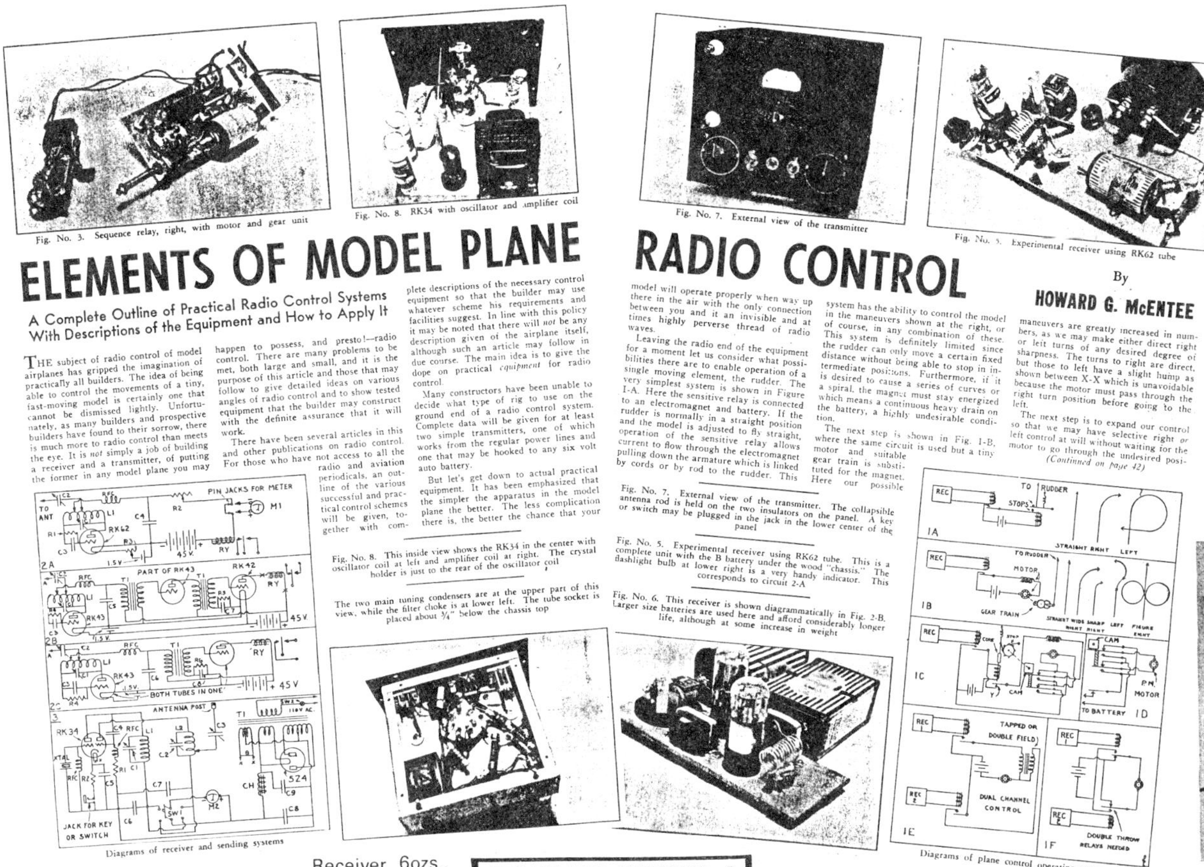
Fig. No. 3. Sequence relay, right, with motor and gear unit

Fig. No. 8. RK34 with oscillator and amplifier coil

Fig. No. 7. External view of the transmitter

Fig. No. 5. Experimental receiver using RK62 tube

ELEMENTS OF MODEL PLANE RADIO CONTROL

A Complete Outline of Practical Radio Control Systems With Descriptions of the Equipment and How to Apply It

By HOWARD G. McENTEE

THE subject of radio control of model airplanes has gripped the imagination of practically all builders. The idea of being able to control the movements of a tiny, fast-moving model is certainly one that cannot be dismissed lightly. Unfortunately, as many builders and prospective builders have found to their sorrow, there is much more to radio control than meets the eye. It is *not* simply a job of building a receiver and a transmitter, of putting the former in any model plane you may happen to possess, and presto!—radio control. There are many problems to be met, both large and small, and it is the purpose of this article and those that may follow to give detailed ideas on various angles of radio control and to show tested equipment that the builder may construct with the definite assurance that it will work.

There have been several articles in this and other publications on radio control. For those who have not access to all the radio and aviation periodicals, an outline of the various successful and practical control schemes will be given, together with complete descriptions of the necessary control equipment so that the builder may use whatever scheme his requirements and facilities suggest. In line with this policy it may be noted that there will *not* be any description given of the airplane itself, although such an article may follow in due course. The main idea is to give the dope on practical *equipment* for radio control.

Many constructors have been unable to decide what type of rig to use on the ground end of a radio control system. Complete data will be given for at least two simple transmitters, one of which works from the regular power lines and one that may be hooked to any six volt auto battery.

But let's get down to actual practical equipment. It has been emphasized that the simpler the apparatus in the model plane the better. The less complication there is, the better the chance that your model will operate properly when way up there in the air with the only connection between you and it an invisible and at times highly perverse thread of radio waves.

Leaving the radio end of the equipment for a moment let us consider what possibilities there are to enable operation of a single moving element, the rudder. The very simplest system is shown in Figure I-A. Here the sensitive relay is connected to an electromagnet and battery. If the rudder is normally in a straight position and the model is adjusted to fly straight, operation of the sensitive relay allows current to flow through the electromagnet pulling down the armature which is linked by cords or by rod to the rudder. This system has the ability to control the model in the maneuvers shown at the right, or of course, in any combination of these. This system is definitely limited since the rudder can only move a certain fixed distance without being able to stop in intermediate positions. Furthermore, if it is desired to cause a series of curves or a spiral, the magnet must stay energized which means a continuous heavy drain on the battery, a highly undesirable condition.

The next step is shown in Fig. 1-B, where the same circuit is used but a tiny motor and suitable gear train is substituted for the magnet. Here our possible maneuvers are greatly increased in numbers, as we may make either direct right or left turns of any desired degree of sharpness. The turns to right are direct, but those to left have a slight hump as shown between X-X which is unavoidable because the motor must pass through the right turn position before going to the left.

The next step is to expand our control so that we may have selective right *or* left control at will without waiting for the motor to go through the undesired posi-

(*Continued on page 42*)

Fig. No. 8. This inside view shows the RK34 in the center with oscillator coil at left and amplifier coil at right. The crystal holder is just to the rear of the oscillator coil

The two main tuning condensers are at the upper part of this view, while the filter choke is at lower left. The tube socket is placed about ¾" below the chassis top

Fig. No. 7. External view of the transmitter. The collapsible antenna rod is held on the two insulators on the panel. A key or switch may be plugged in the jack in the lower center of the panel

Fig. No. 5. Experimental receiver using RK62 tube. This is a complete unit with the B battery under the wood "chassis." The flashlight bulb at lower right is a very handy indicator. This corresponds to circuit 2-A

Fig. No. 6. This receiver is shown diagrammatically in Fig. 2-B. Larger size batteries are used here and afford considerably longer life, although at some increase in weight

Diagrams of receiver and sending systems

Diagrams of plane control operating systems

Receiver available in kit form from Nathan R. Smith Mfg. Co., L.A., weighing 5½ozs. (Rx) or 8ozs. total. Suggested use was as a motor shut-off or similar device.

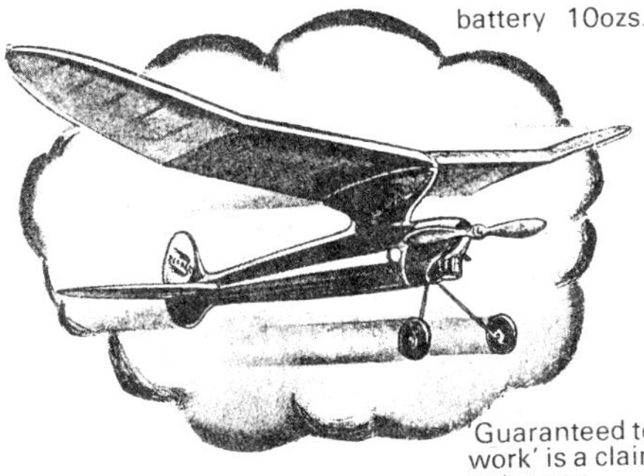

Receiver 6ozs., battery 10ozs.

'Guaranteed to work' is a claim not many modern radio kit suppliers would care to make!

M.A.N. July/August 1939

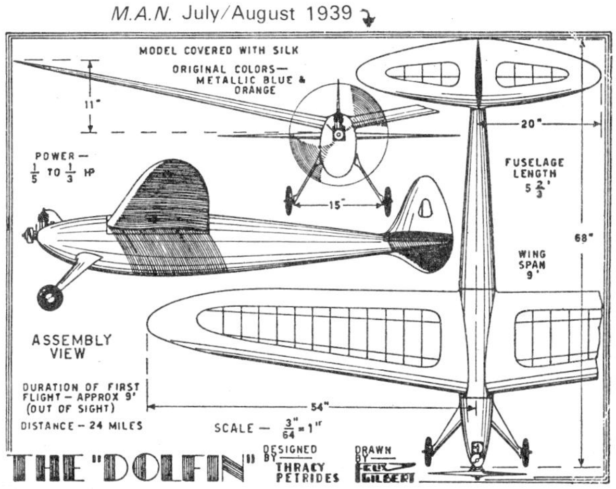

Thracy Petrides' 'Dolfin' or, as published, 'Dolphin'

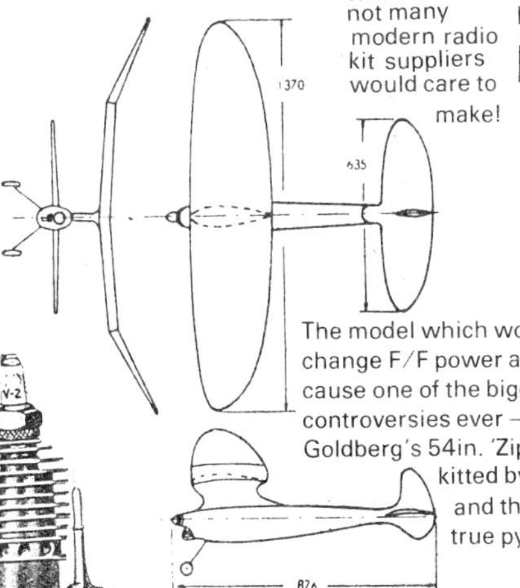

The model which would change F/F power and cause one of the biggest controversies ever — Carl Goldberg's 54in. 'Zipper', kitted by Comet and the first true pylon model.

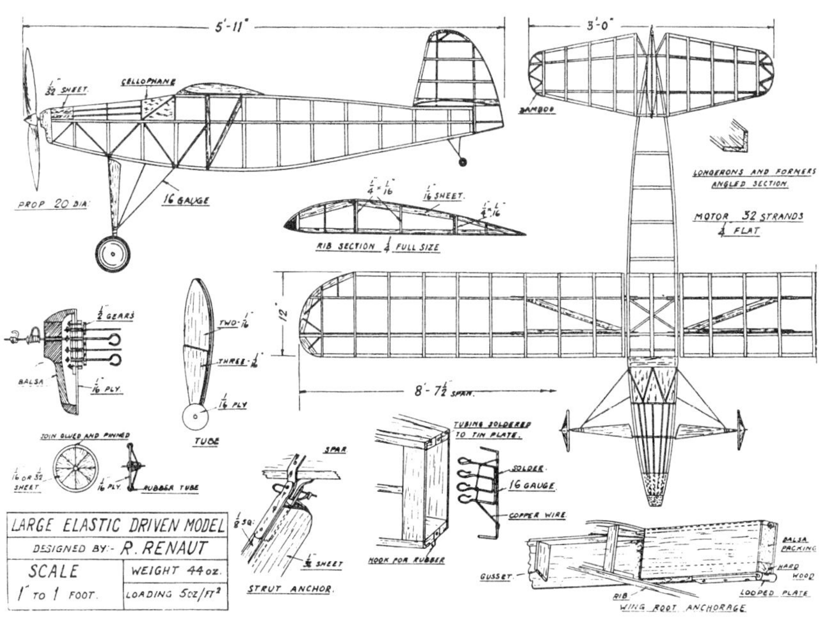

Big, but not the biggest. An 8ft. 7½in. rubber model (*Aeromodeller*, February 1939). Cheaper today to buy an engine rather than all that rubber!

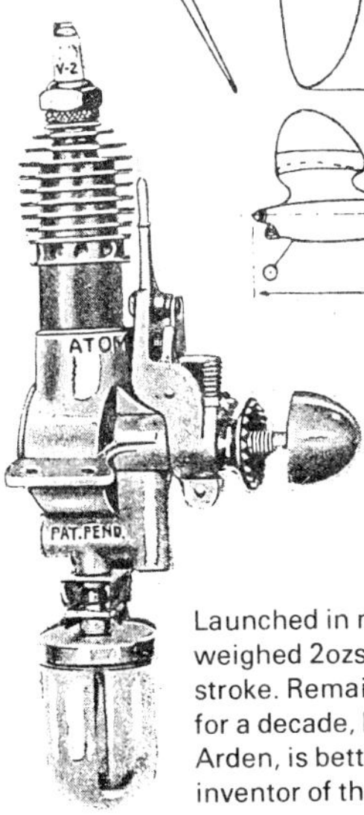

Launched in mid-1939, the 'Atom' engine weighed 2ozs. and was ½in. bore and stroke. Remained available and widely used for a decade, but the man behind it, Ray Arden, is better known as, effectively, the inventor of the glowplug model engine.

"DEBBY"

A smart 63in. design by Nick Limber, originally published in *Flying Aces* in July 1939. Very typical of the period.

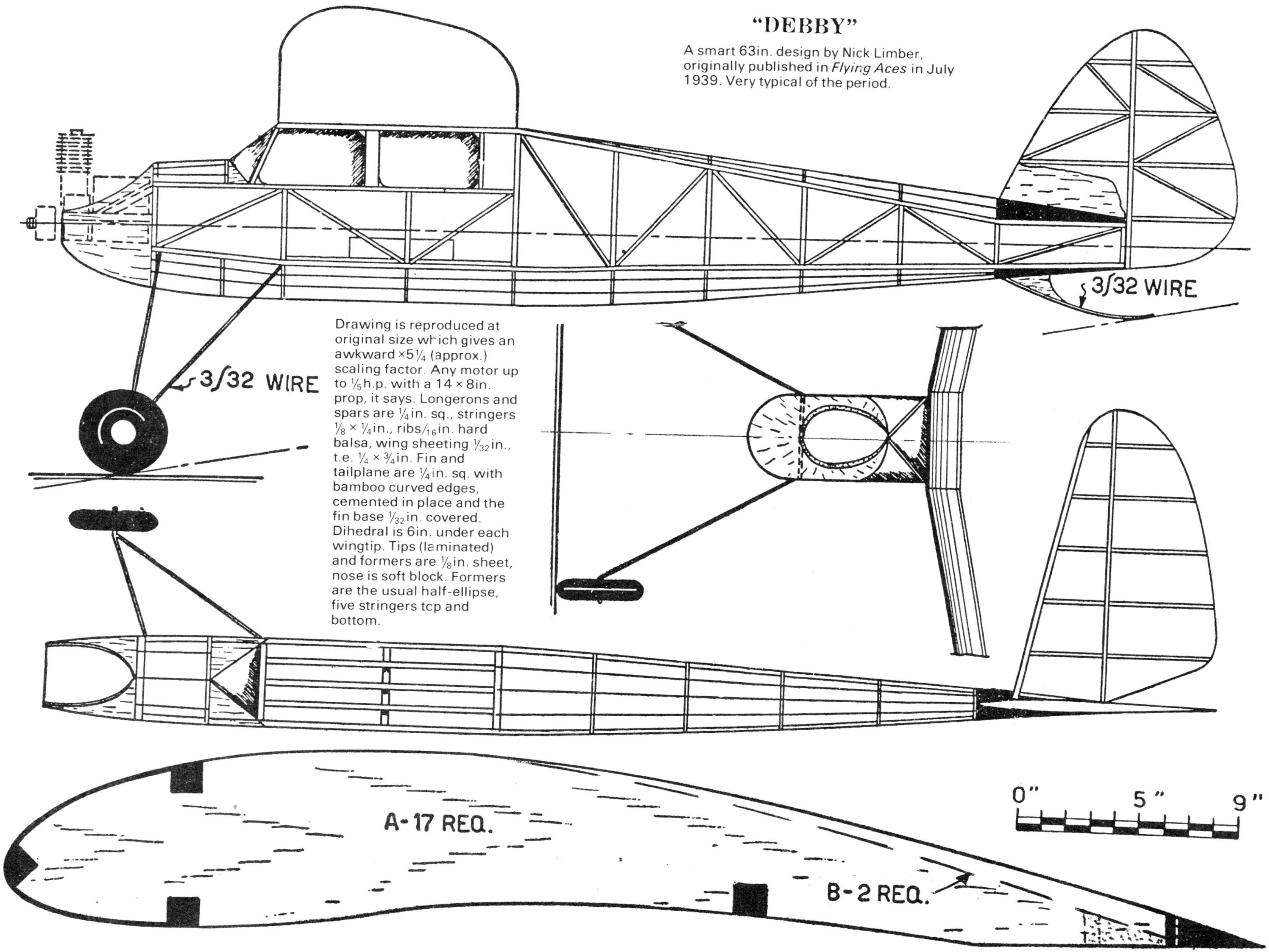

Drawing is reproduced at original size which gives an awkward ×5¼ (approx.) scaling factor. Any motor up to ⅕h.p. with a 14 × 8in. prop, it says. Longerons and spars are ¼in. sq., stringers ⅛ × ¼in., ribs/₁₆in. hard balsa, wing sheeting ¹⁄₃₂in., t.e. ¼ × ¾in. Fin and tailplane are ¼in. sq. with bamboo curved edges, cemented in place and the fin base ¹⁄₃₂in. covered. Dihedral is 6in. under each wingtip. Tips (laminated) and formers are ⅛in. sheet, nose is soft block. Formers are the usual half-ellipse, five stringers tcp and bottom.

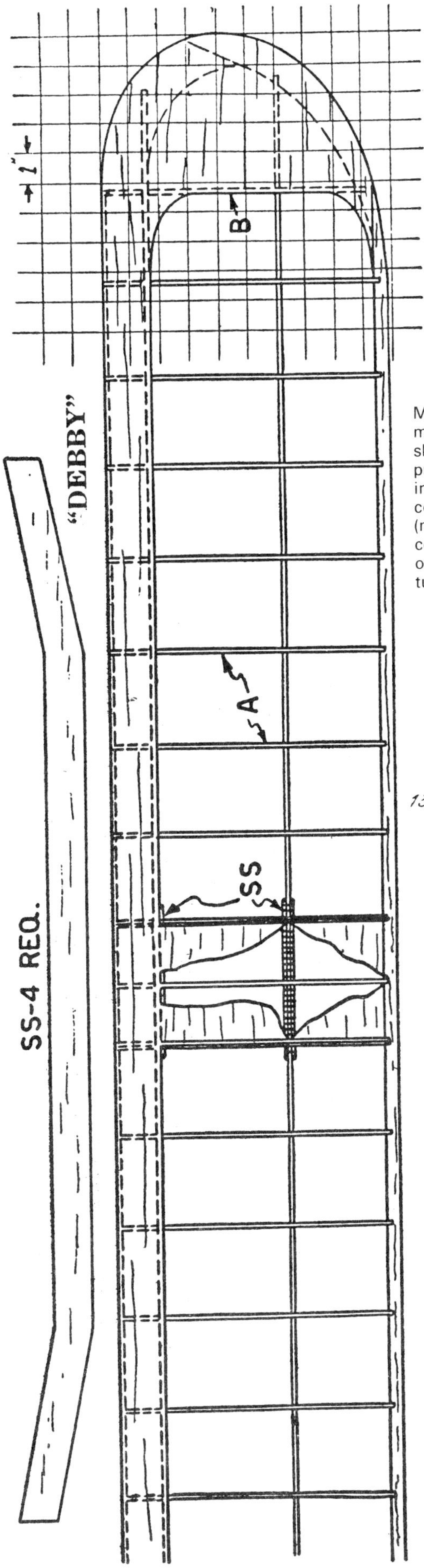

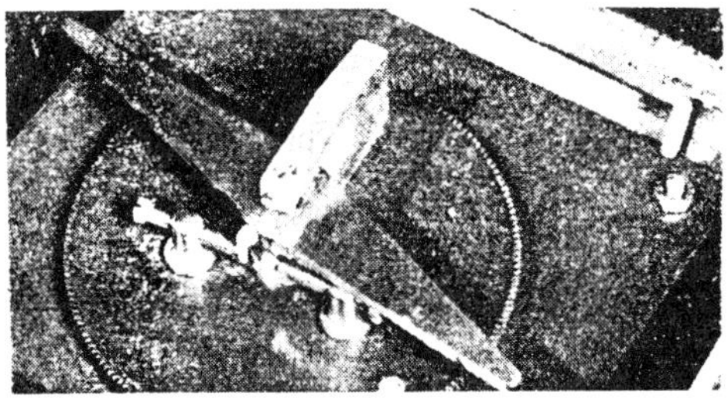

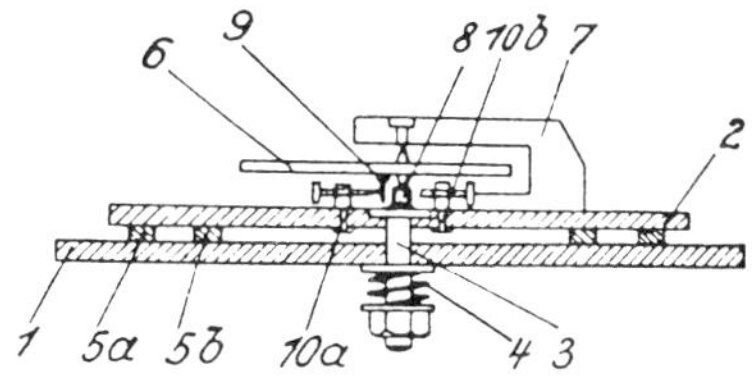

Magnet steering is a method of flying a slope-soaring glider practised quite widely in mountainous countries. A compass (magnet) is set between contacts with the model held on the desired heading. Any turn away from that heading

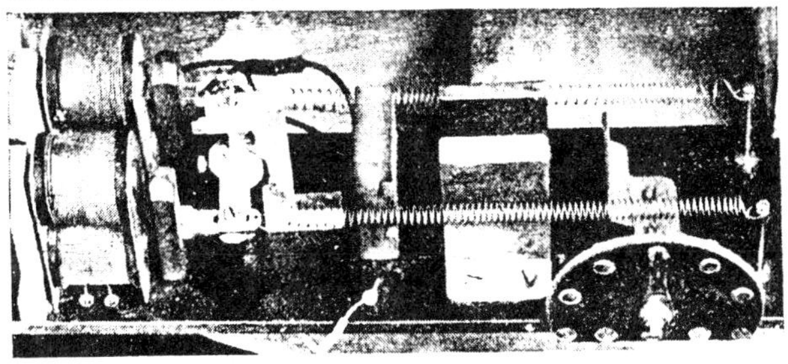

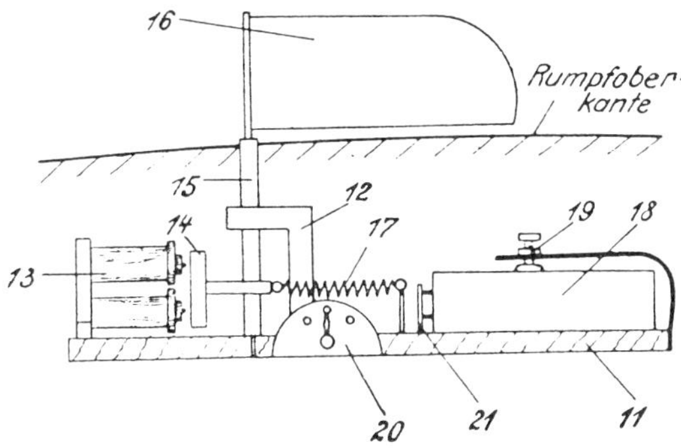

will mean that the model's contacts touch the compass contacts, completing a circuit and energising a solenoid motor, applying corrective rudder. When back on course the contacts break and straight flight is resumed. A nose rudder has been found effective. This (first?) description of the system appeared in *Modellflug* September 1939, written by Rolf Jauch.

Model humour is common world-wide. This is H. Pownug's idea of magnet steering illustrated by Hermann Kegel.

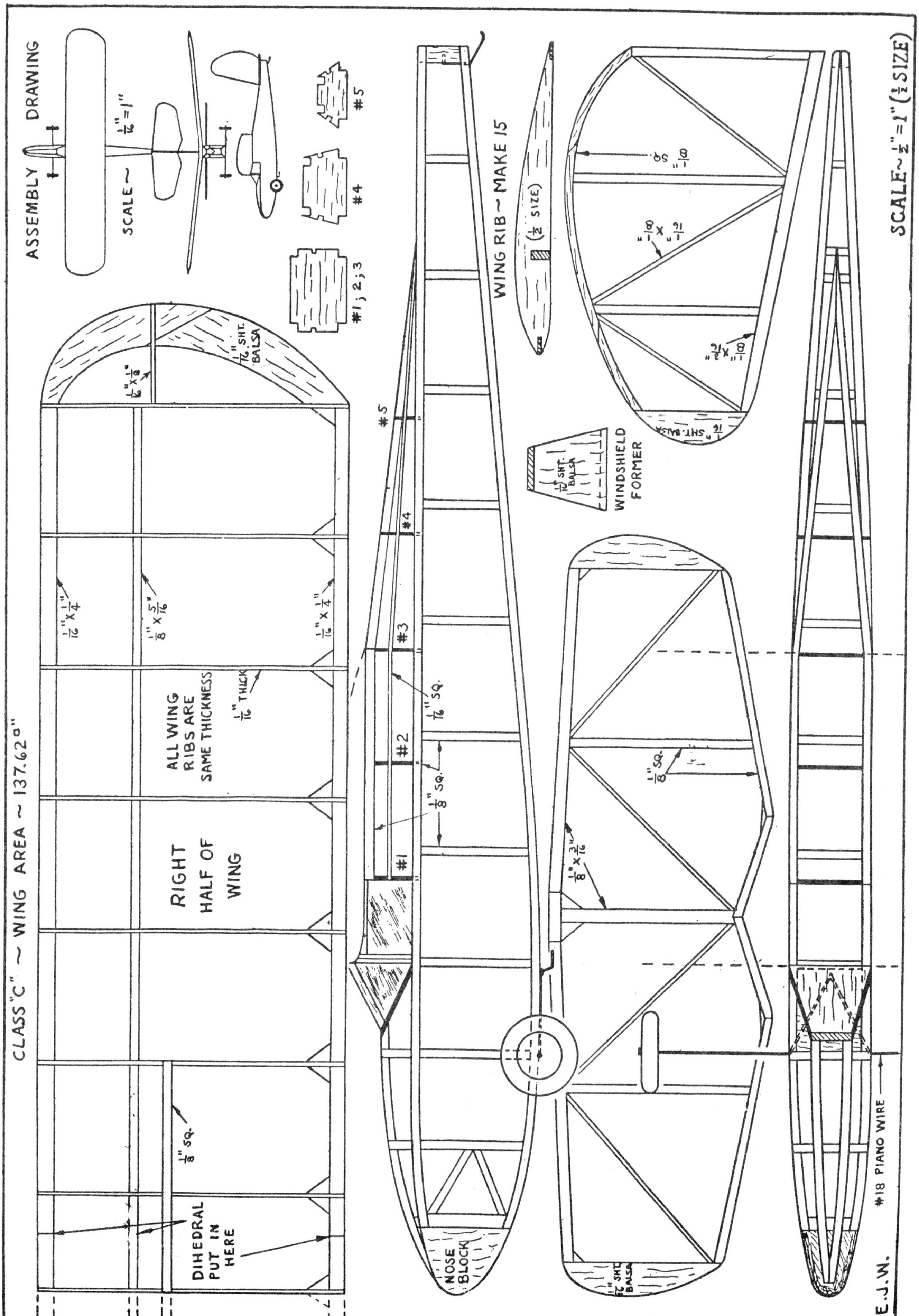
CLASS "C" ~ WING AREA ~ 137.62□"
RIGHT HALF OF WING
ALL WING RIBS ARE SAME THICKNESS
DIHEDRAL PUT IN HERE
ASSEMBLY DRAWING
WING RIB ~ MAKE 15
WINDSHIELD FORMER
NOSE BLOCK
#18 PIANO WIRE
E.J.W.
SCALE ~ 1/2" = 1" (1/2 SIZE)

Gliders were not widely used outside Germany, but this little American model, first built by Elbert J. Weathers in 1933, published in Frank Zaic's 1938 Yearbook and subsequently in *Flying Aces*, August 1939, appeared to have exceptional performance. E. J. W. built five, due to flyaways on 100ft. towline. Flights of 21, 28 and 33 minutes were timed. Dihedral is 1½in. under each tip, incidentally.

Radio control attracted increasing attention in 1939. This set of plans was advertised by Johnson Smith & Co. of Detroit.

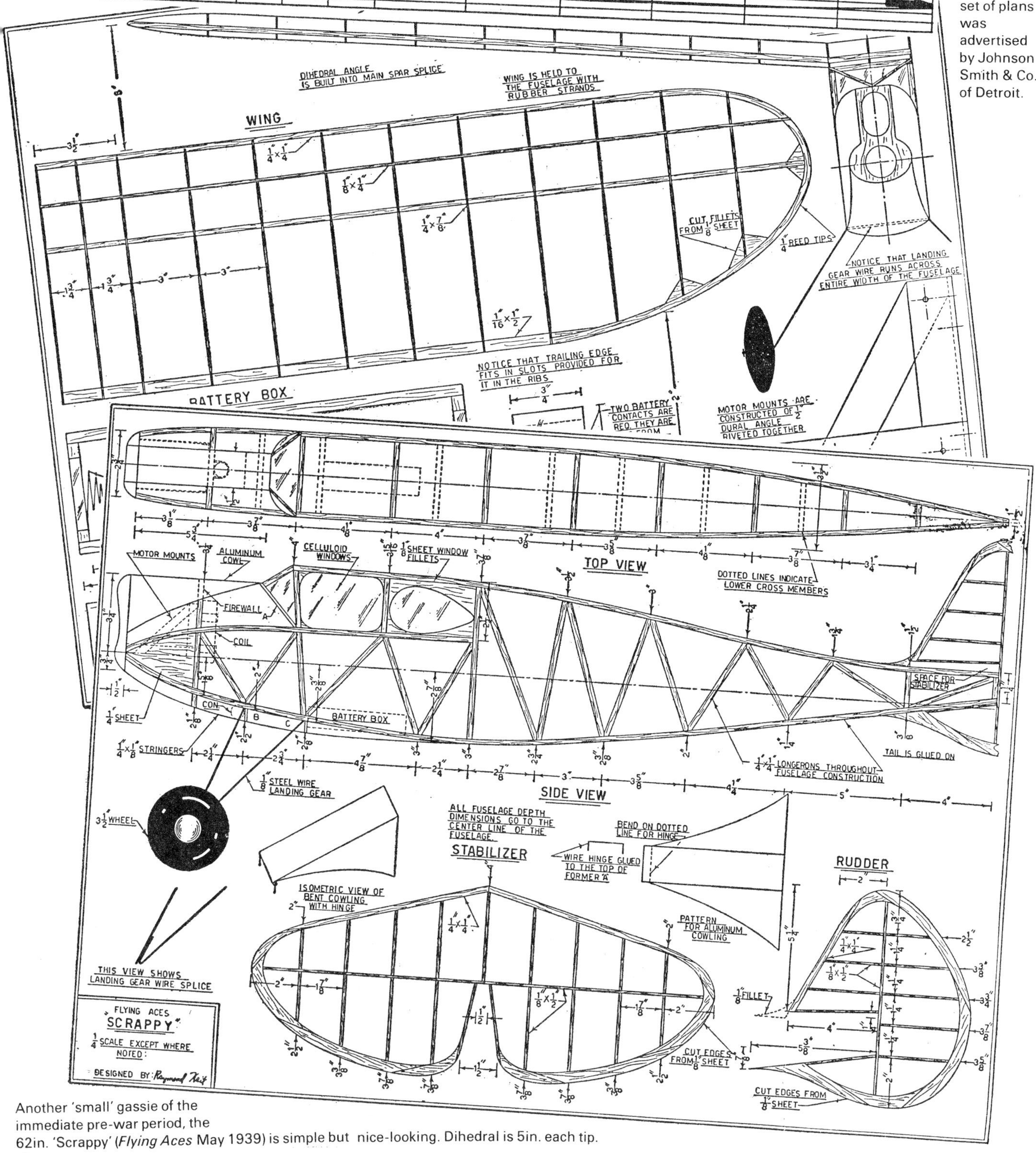

Another 'small' gassie of the immediate pre-war period, the 62in. 'Scrappy' (*Flying Aces* May 1939) is simple but nice-looking. Dihedral is 5in. each tip.

1939-40

Dick Korda's simple-seeming cabin slabsider had tremendous performance and a very slightly different model was flown by him for the staggering average of 15:50.2 to win the 1939 Wakefield competition. His first flight was 43:29, breaking the world-record set up the previous year by Bob Copland (GB)

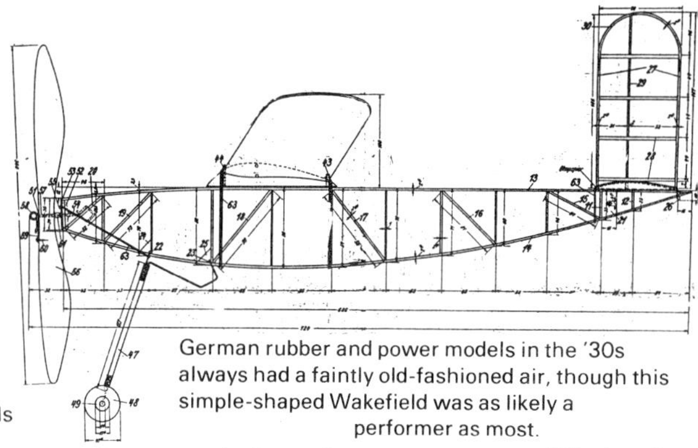

German rubber and power models in the '30s always had a faintly old-fashioned air, though this simple-shaped Wakefield was as likely a performer as most.

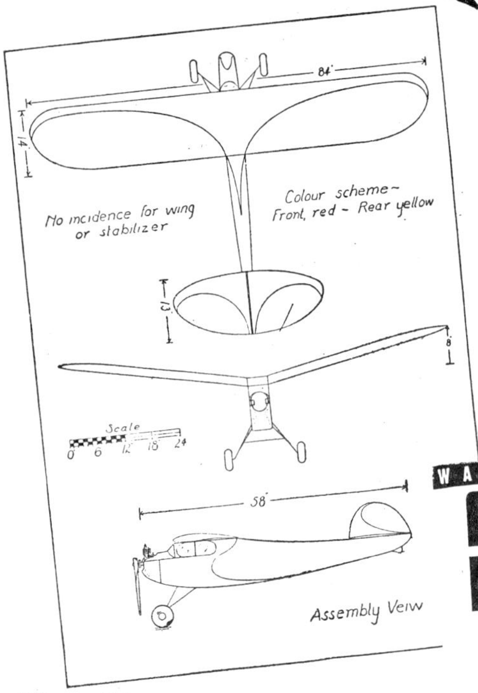

Sal Taibi's 'Powerhouse' (*M.A.N.* November 1939) was a successful approach to bigger models

Standard cover of *Modellflug*, which continued to be published through most of the war years.

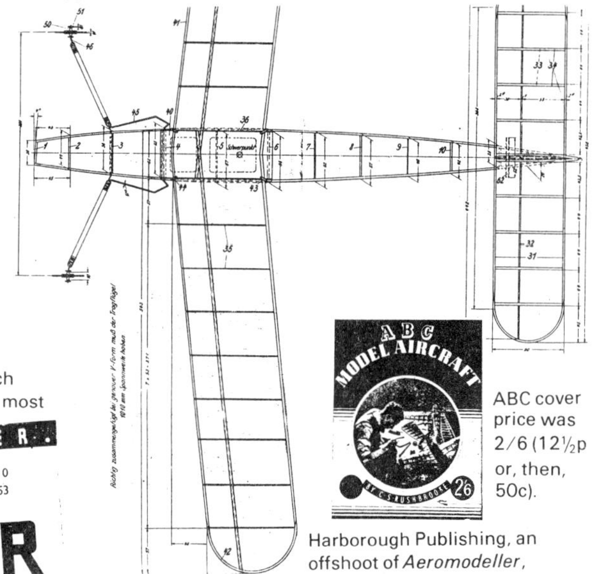

ABC cover price was 2/6 (12½p or, then, 50c).

Harborough Publishing, an offshoot of *Aeromodeller*, started to produce books in 1940.

All-metal construction was developed in Germany, notably by the Heller brothers of Schmalkalden. It was revived briefly in the 1950s but failed, probably because of the difficulty of repair.

Radio Control

Model Pilotless 'Planes

By John E. Lodge

THIS year, some 3,000,000 Americans will invest nearly £200,000 for strips of silk, bundles of balsa wood and tiny motors capable of lifting their model airplanes into the air for sustained flight. But model flying no longer is a plaything for children only, and now experiment in radio control is sweeping across the country. On the airport at Los Angeles, where 300 members of the Gas Model Airplane Association of Southern California fly their 'planes from a 600' square field, you will find a new fleet of the radio ships, rolling, diving and circling under absolute control of their owners, who merely touch buttons operating two radio sets to enforce their will.

These experimenters claim no extraordinary qualities for their handiwork. Each new 'plane is an "experimental job." It barely tips the scale at 4½ lb., when loaded with 2 oz. of fuel and radio equipment. The single wing, weighing 11 oz. and held in place by four rubber bands, stretches 8' from tip to tip.

This model 'plane is known as a "single," the radio control being hooked up through a tiny electric motor and worm reduction gear to the rudder tab. Designed to fly in a straight line, the little 'plane takes off after a brief run and may be manœuvred right and left. At other times, springs hold the controls in neutral. Now the two builders are working on a new model to combine rudder, elevator and motor-speed controls, all operated through a single transmitter.

No one has measured the rate at which these models climb after leaving the ground. Seldom does one remain aloft longer than half-an-hour. Yet one club member lashed two tomato cans into the narrow fuselage of a high-wing monoplane, filled them with gasoline, sent it up one mornin and watched the silver wing flash around th 8½ hours before, fuel exhausted, the sh earth. During this fligh altitude of 5,000'.

Newest of the " a high-wing operate tic

Model Engineer had more or less lost touch with aircraft models but included this dreadful piece of uninformed journalese in October 1939.

Wakefields were still No. 1 interest and if you couldn't build one in 1940 at least you could read about them!

Simulated blue-print for a 13½in. Avro Commodore was a double-page spread in 1940.

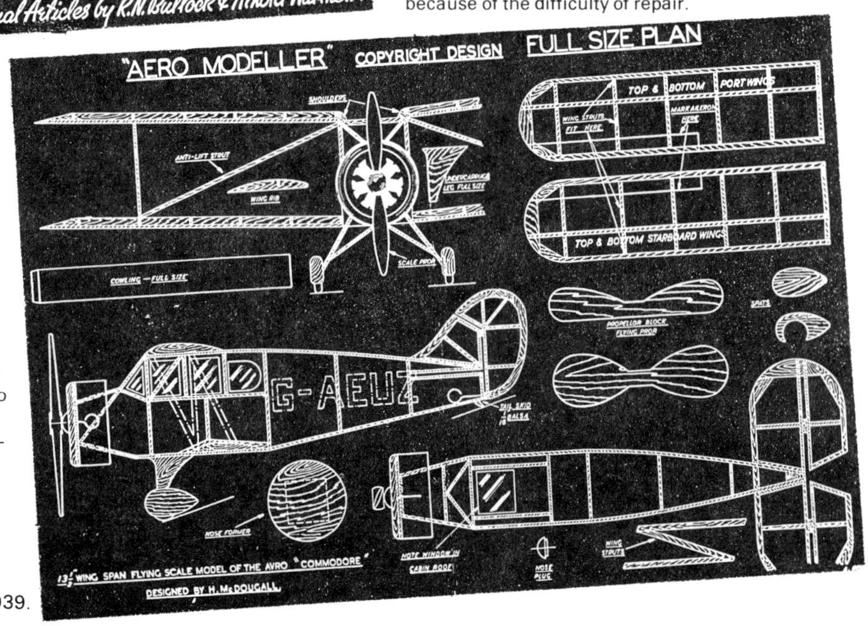

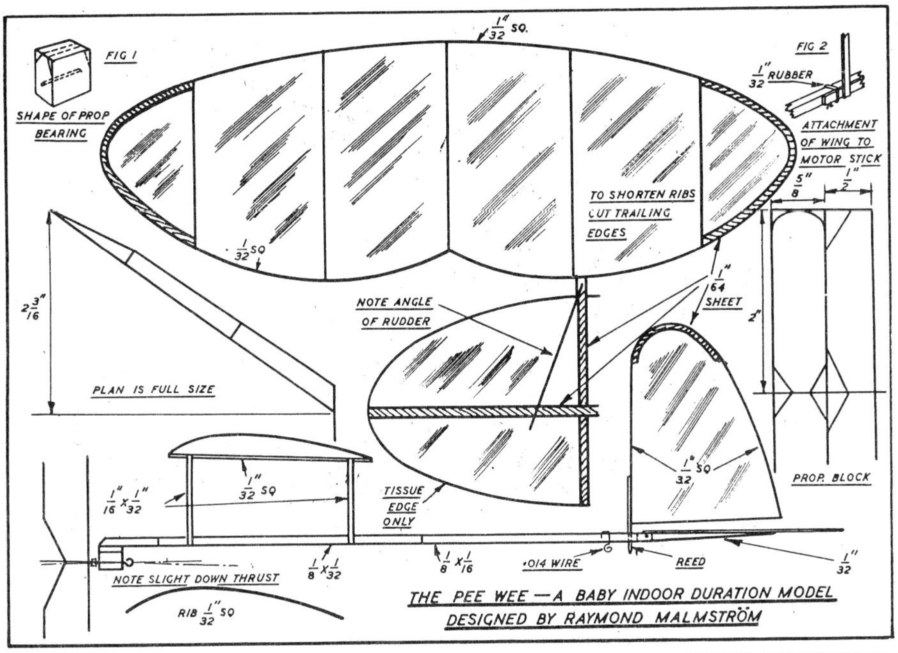

The first of Ray Malmström's ingenious and 'flyful' models appeared in *Aeromodeller* in 1940 and they are still appearing in 1987

1940

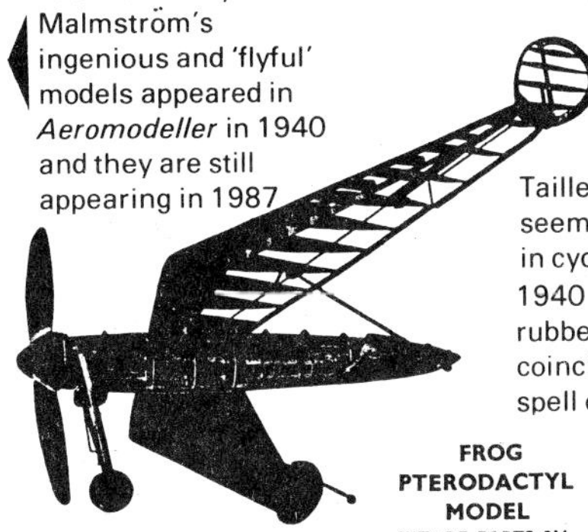

Tailless models seem to appear in cycles. This 1940 Frog kit for rubber power coincided with a spell of activity.

FROG PTERODACTYL MODEL
KIT OF PARTS 21/-

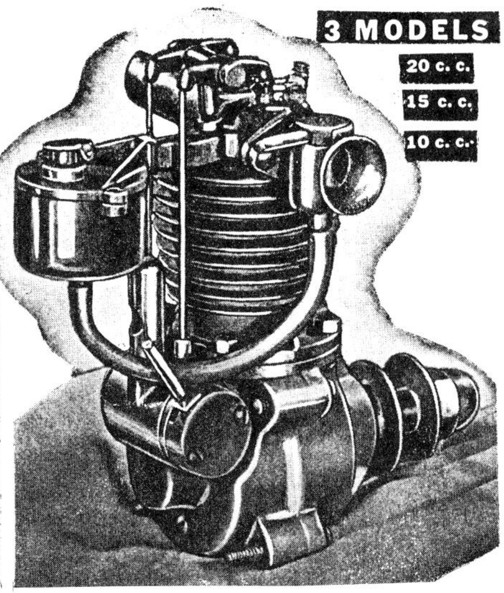

April 1940 saw the launch of the Feeney o.h.v. 4-stroke engines of 10, 15 and 20cc, any of which cost $29.50, or in kit form $13.75.

Use **GUILLOW'S** Exclusive

PANEL-CUT BALSA

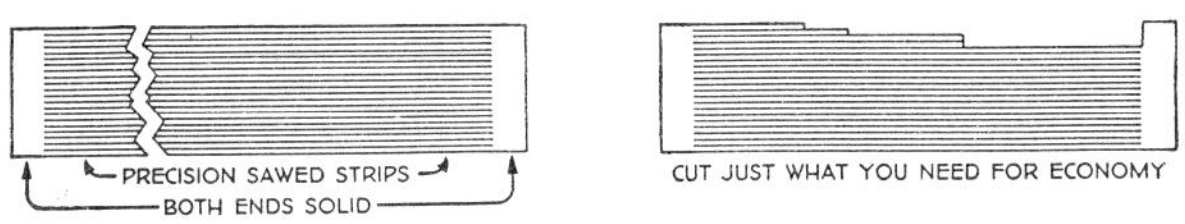

Guillow Panel-Cut Balsa is accurately sawed to give true dimensions of strips, and is supplied in flat sheets with *both* ends solid. Strips are thus held safely in place until ready for use. There are no loose ends! Breakage is eliminated! Cut whatever you need to work with . . . the remainder is protected and kept in perfect condition without waste . . . that's economy!

Something new (and still sometimes seen) was panel-cut or comb balsa, introduced by Guillows in January 1940.

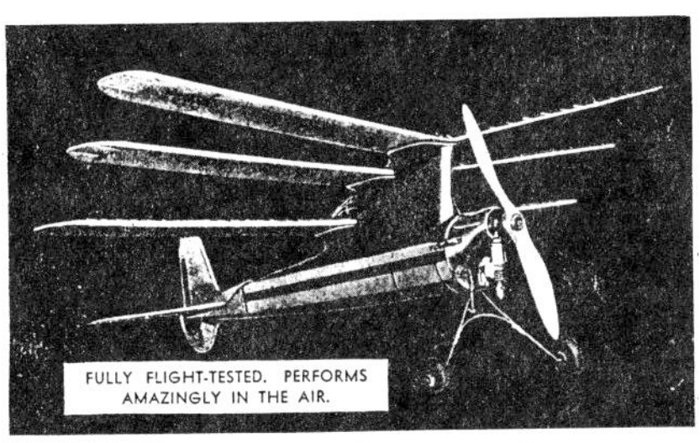

'The most amazing performance of any design ever flown' was the modest claim for the 42in. Nemeth Hi-Triwing.

Originally 1937-built, this model flew 40min. o.o.s. in 1938. Published *M.A.N.* June 1940.

MODEL AIRCRAFT FLYING

EXTRACT FROM
AIR MINISTRY ORDER
DATED 3rd AUGUST, 1940

"THE FLYING OF ALL TYPES OF PETROL ENGINED MODEL AIRCRAFT IS PROHIBITED."

"THE FLYING OF ELASTIC-DRIVEN MODEL AIRCRAFT AND GLIDERS EXCEEDING SEVEN FEET IN WING SPAN IS ALSO PROHIBITED."

THE SOCIETY OF MODEL AERONAUTICAL ENGINEERS DESIRE TO POINT OUT THAT THERE ARE NO RESTRICTIONS IN FORCE RELATING TO THE FLYING OF MODEL AIRCRAFT BEYOND THOSE DETAILED IN THE ABOVE AIR MINISTRY ORDER, CONTRAVENTION OF WHICH IS A PUNISHABLE OFFENCE.

P.T.O.

No engine-powered models or any model over 7ft. could be flown in Britain, probably to avoid radar confusion. This ban lasted four years, but wartime shortages of building materials limited activity anyway.

"G" LINE FLYING
Sensational - New - Thrilling

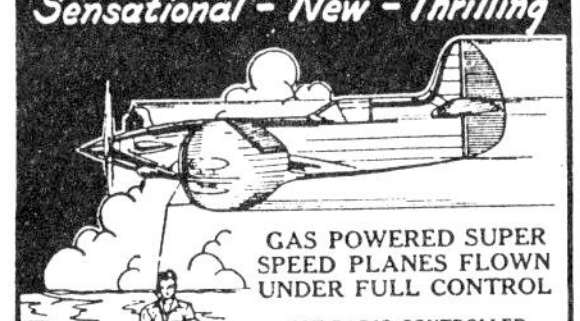

Protected By Patent Pend.

TIGER SHARK
SPEED DEMON

$4.95 KIT COMPLETE

The Super Streamlined TIGER SHARK, "G" Line Speed Plane, roars through the air at the unbelievable speed of from 50 to 75 M.P.H. Due to the inherent "G" Line Flying Stability, engineered into this design, the Shark is unusually stable and easy to control. May be powered with any 1/7 or 1/5 H.P. Motor.

$4.95 KIT COMPLETE

Texas Ranger Combination Model

The TEXAS RANGER, a beautiful 45" High-Wing Cabin Sport Job, is a specially designed Combination FREE-FLIGHT and "G" LINE Flying Model. Has unusual climbing and flat gliding qualities in free flight and is very stable and easy to fly with the new "G" Line Control. May be powered with any 1/7 H.P. Motor.

DELUXE CONSTRUCTION KITS are most complete, including plenty of Balsa and Printed Parts. M & M Air Wheels, Prop Blank, Bamboo Paper, Spring Steel Wire, Ignition Wire, 50 Ft. "G" Line, generous supply of Cement and Dope, etc. Together with large fully detailed plans and instructions for building the ships and new "G" Line Control.

Rush in your order today. Enjoy New Thrills in Gas Model Flying.

DEALERS WRITE FOR SPECIAL DISCOUNTS
VICTOR STANZEL & CO.,
Schulenburg, TEXAS

G-line flying was an extension of r.t.p. with the flyer using a long pole to which the model was tethered by perhaps 30ft. of line. Introduced January 1940.

Note fuselage similar to one on p.20.

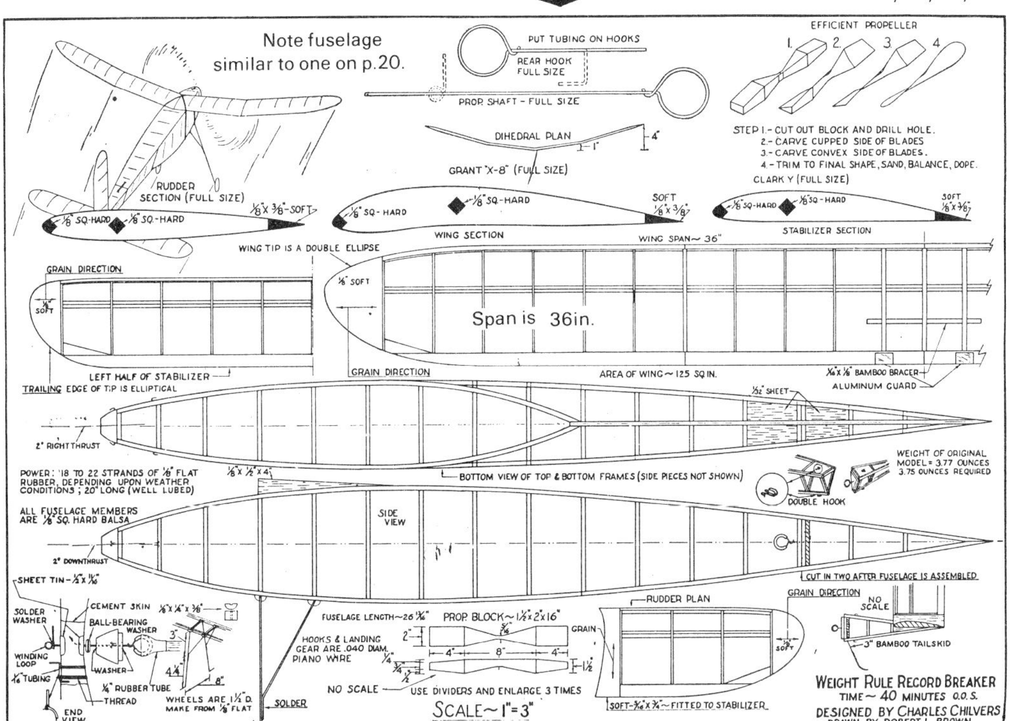

Span is 36in.

1940

"THERE ARE FAIRIES AT THE BOTTOM OF MY GARDEN . . ."

Possibly the best-remembered of all Freddie's cartoons. Most people had heard of the Fairey Battle aircraft.

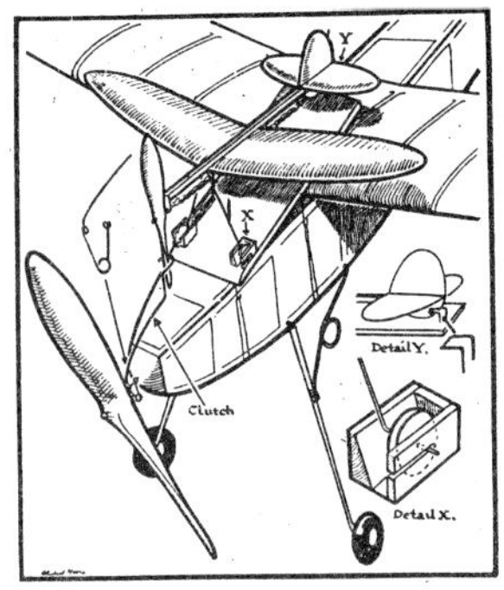

The Short Mercury/Maia composite had generated interest in pick-a-back flight

The 'Zipper', advertised as the most sensational, popular and imitated gas model in history, continued to win contests under the rules brought in for 1940, where it fitted Class C (.29 to .65 cu. in. motors). It was joined by the 'Zipper A' (top picture below) for Class A and the 'Mercury', a parallel chord version for Class B all kitted by Comet and all with remarkable contest records.

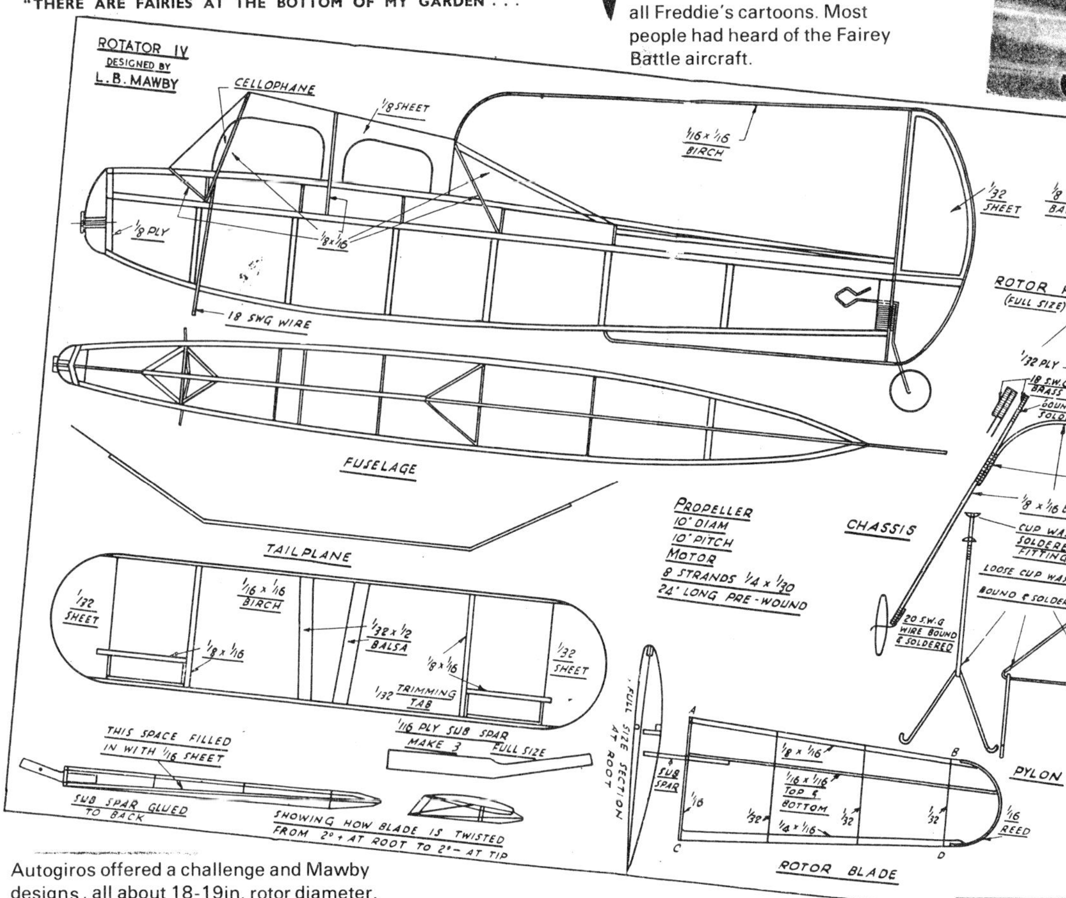

Autogiros offered a challenge and Mawby designs, all about 18-19in. rotor diameter, gave consistent flights. (*Aeromodeller* 1940)

The RC-1, designed for R/C but safe for free-flight, was a 10ft. model by Ben Shereshaw for a New York firm, Radio Control Headquarters.

Twelve-foot target aircraft launched by catapult, twin props from single motor, parachute retrieval. Tx used a telephone dial.

This model by Sal Taibi looks quite small when drawn or photographed. In fact it was 7ft. 4in. span and gave consistent 1:40 flights from a 10sec. motor run.

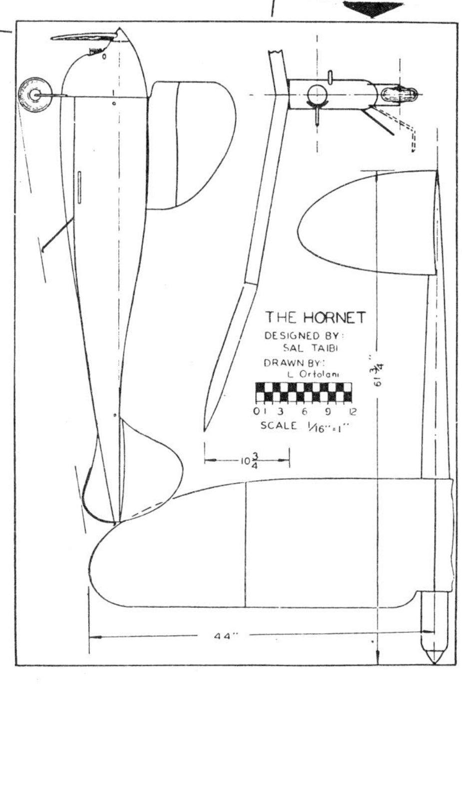

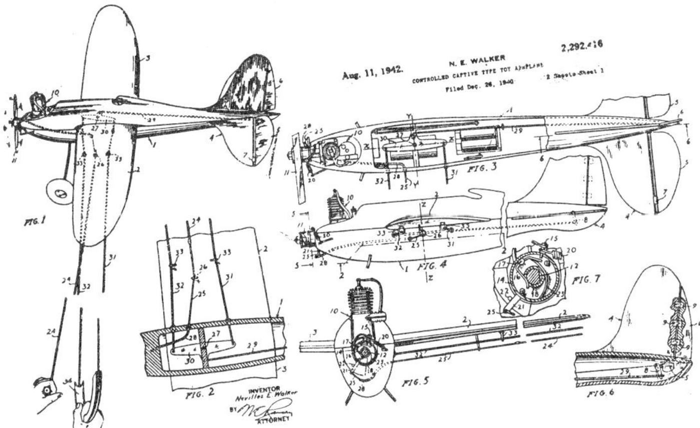

Ready to Fly in 6 Hours

Designed for "B" motors, the A-J Fireball is sturdily constructed to withstand terrific punishment. *All parts come ready-cut to shape. All-balsa fuselage is completely carved* and specially designed in two halves for added strength and easy access to motor and batteries (which nest in a special compartment). Hardwood mounts for metal parts are built into fuselage. Wing span 36 inches. Wing is built of sheet balsa over high speed ribs. No silk or paper. No doping required.

Due to semi-finished form, plane can be made ready for flight in less than 6 hours. All-balsa surfaces can be beautifully finished in glossy enamel. All materials complete except for liquids.

Complete Patent Protection.

The methods of flying power models, especially in competitions, which had developed in the U.S. led to a high proportion of lost models and it was a search for a method of operation which reduced flyaway chances which brought about G-line, U-Control and Remote Control. Of these U-Control was the survivor on which all control-line flying has been based for some 45 years. Although first advertised in September 1940, the advert states 'After two years of ready-built production and research . . . ' but Jim Walker did not apply for patent cover until Dec. 26 1940. The application covered not only the bell-crank but third-line motor control, lead-out guides, offset rudder and other details since shown to be required for top-class performance. Walker appears to have been highly protective of his patent rights and few kits or plans used or showed a bell-crank control in the early days of C/L flying.

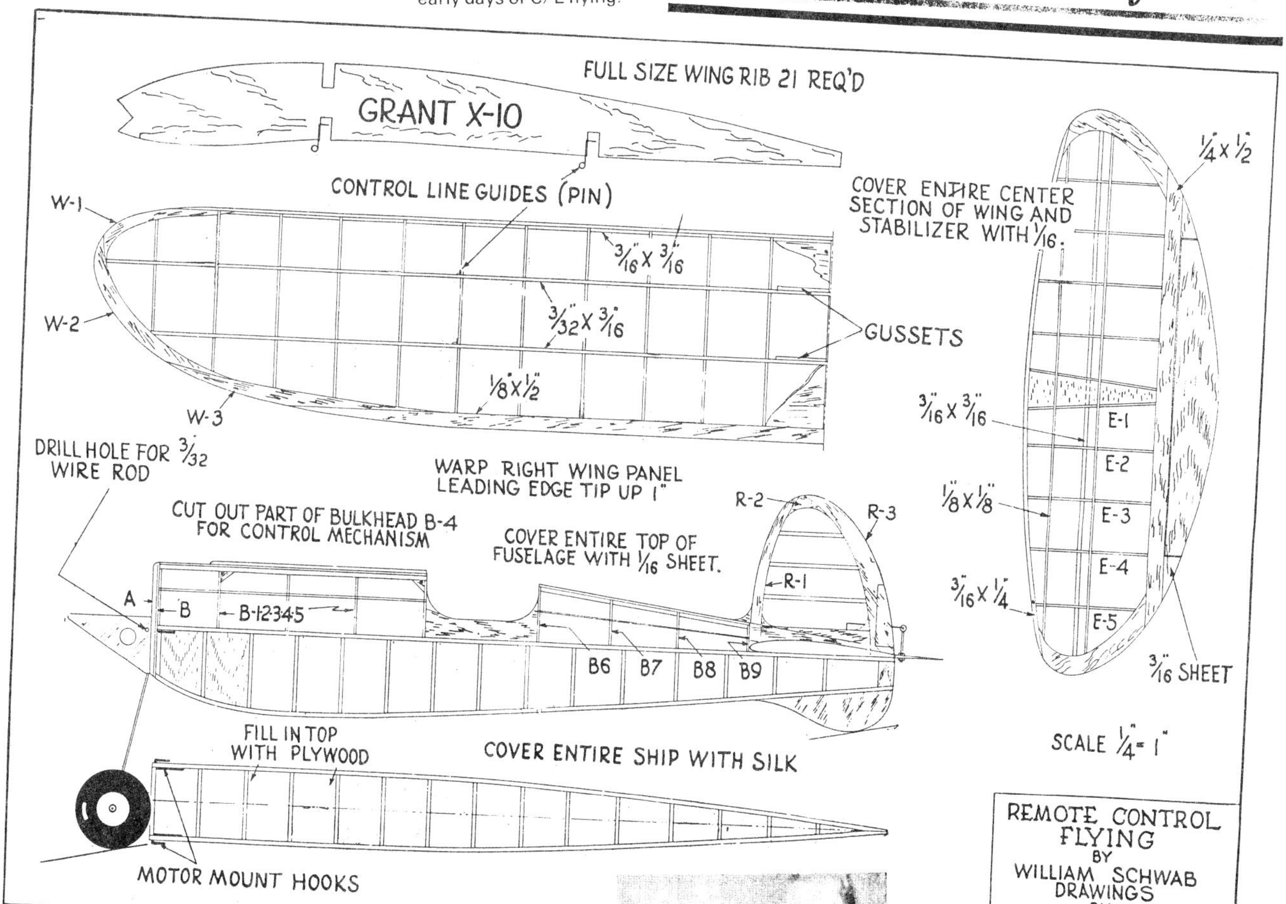

Virtually everyone to whom tethered powered flying appealed and who was able to follow American developments (and, of course, owned or could obtain an engine!) built Schwab's 'Controlled Lightning' in April 1941 *M.A.N.* This 48in. Ohlsson 23 model flew on 30ft. fishing lines. Schwab spoke of 'a few years ago' and 'some time later' but gives no approximate date.

1941

Elf Four
May, 1941

$49.50

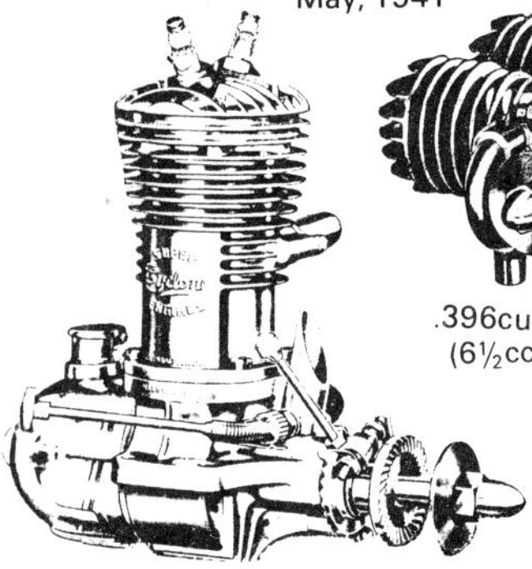

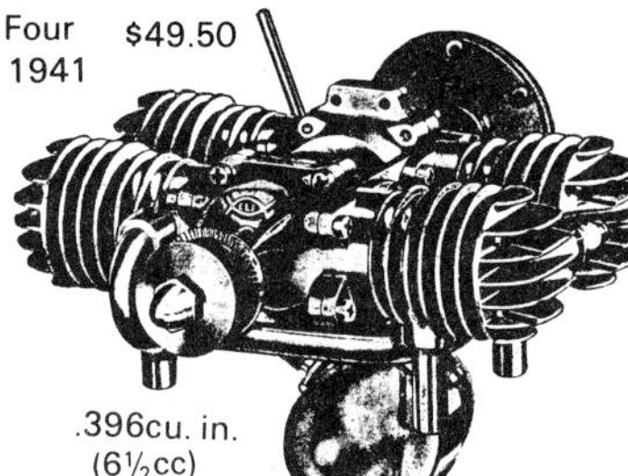

.396cu. in.
(6½cc)

Motor development continued in the war period, notably in the U.S. The dual ignition Super Cyclone was a favourite for years.

Twin
$31.50

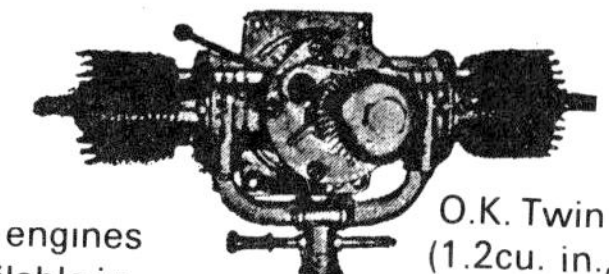

O.K. Twin
(1.2cu. in., 20cc) $40

Surprisingly, model engines continued to be available in Germany (1941 advert.).

New! Plus-Power SUPERCHARGER

● Now you can equip your model airplane, race car or boat with a Supercharger like the ones used on full-size motors. Plus-Power Supercharger is easy to mount on Class B, C, and larger Class A motors. High speed replaceable bearings assure long, efficient operation. Centrifugal rotor creates pressure on air intake. Special baffles produce whirling motion of air resulting in better mixture. Increases power, pep and speed amazingly. Weight complete, only 2¼ oz. Only $3, postpaid, or C.O.D. Send for yours today.

Pat. applied for

Efficiency Engineering Co.
701 Grand Ave., Des Moines, Iowa
DEALERS: Inquire on letterhead for discounts.

Supercharging has been tried many times.

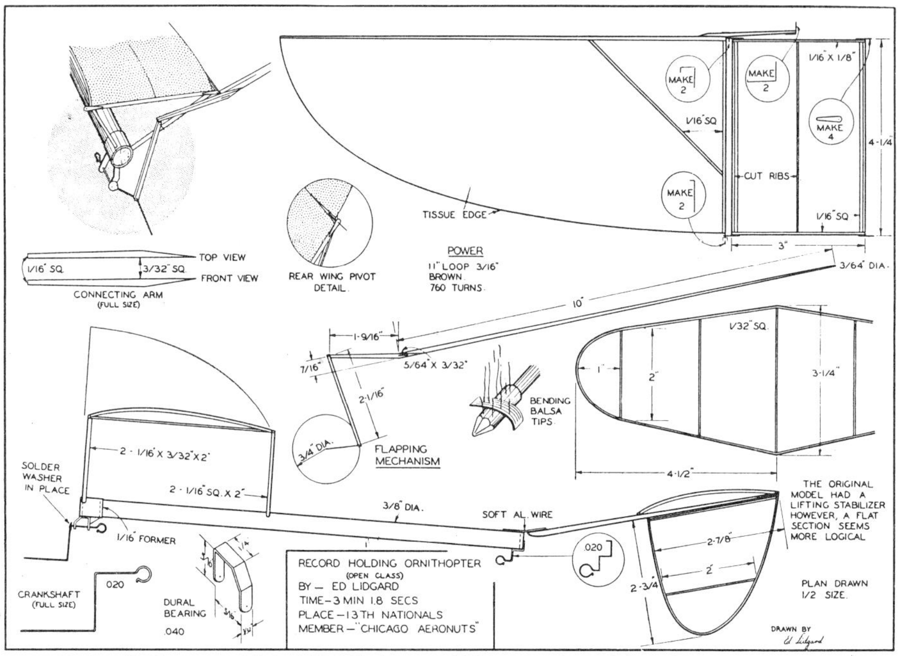

Ed Lidgard's flapper, published *M.A.N.* January 1941, arose from club ornithopter craze in 1937. Dennis Turner flew a similar model for 2:21.6 and later a remarkable 4:19.2 for a senior record.

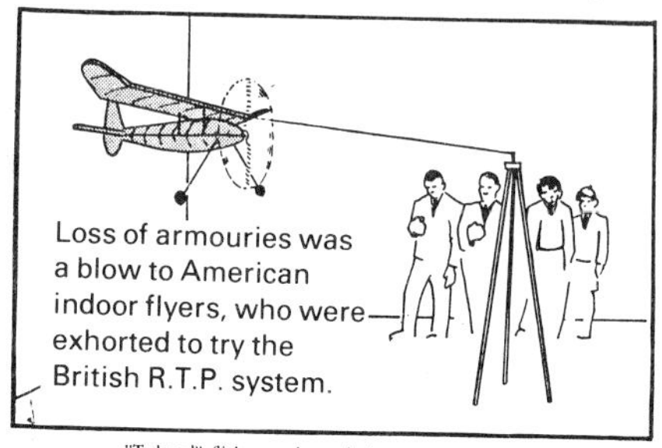

Loss of armouries was a blow to American indoor flyers, who were exhorted to try the British R.T.P. system.

"Tethered" flights can be made in small space indoors

FLY THEM ON A STRING

THERE'S been a little matter troubling us during the past few months—in fact, ever since the National Defense program slipped into high gear.

With the armories in use by our National Guard and army units, many indoor enthusiasts have found they can no longer fly in accustomed spots. So what do some of these modelers do? Do they go out and look for a smaller hall? No, sir, they sit around, wring their hands, bemoan their luck and decide to give up modeling.

Aren't they a bunch of sissies, though?

Take a look at England. Do you think the model flyers over there have a bed of roses? Not a bit of it! Even before the outbreak of hostilities there were few large buildings available to indoor modelers or outdoor flyers when inclement weather prohibited flying of outdoor craft.

So the English lads developed "round-the-post" flying. By setting up "pylons" in small halls they were able to fly outdoor rubber-powered craft and heavier types of indoor models inside small halls. Everybody had a lot of fun and many interesting problems presented themselves, such as proper type of pivot to use and how to attach the tether to the model. And should one or two point attachment be used? The accompanying sketch adapted from a cover illustration of "The Aeromodeler" best explains pylon flying.

Now let's demonstrate that we are as resourceful as our fellow flyers across the water. Pylon flying can be the basis for some very fine competition with plenty of fun assured participants. In addition to such flying, it is suggsted that "stunt" and "stunted" events be held. Stunt events can be run off for original design models which perform unorthodox gyration "stunted" events can be held for "mod of models—twin-pushers of 10 or 12

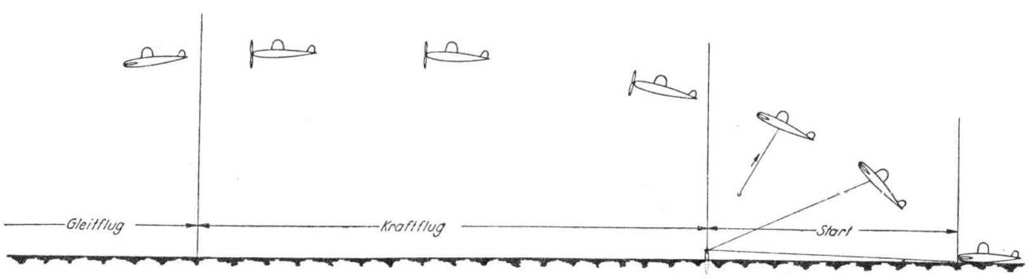

Very attractive 'Stratosphere' contest model by Henry Cole (*M.A.N.* May 1941) with good performance (we built one!) Original was lost after 35mins. Fuselage is teardrop cross-section,

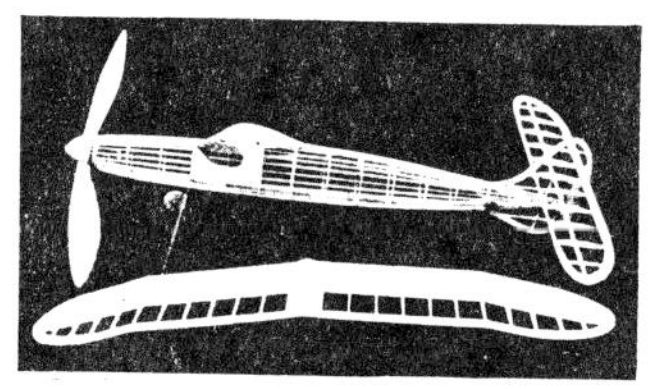

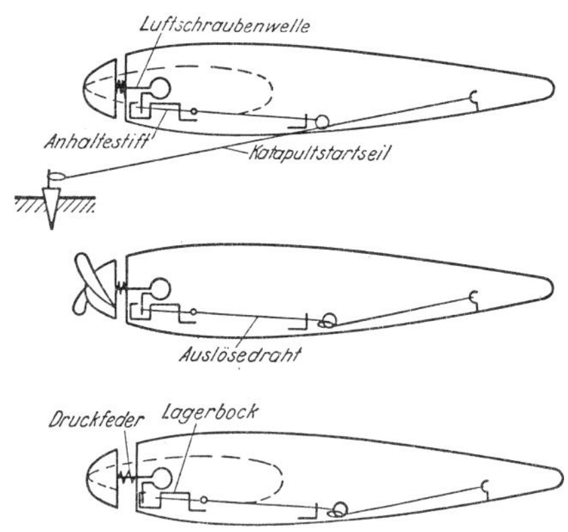

Another bright idea from *Modellflug* (January 1941) — catapult launching for rubber models. Prop unfolds and starts running on release of catapult line, refolds in normal way for glide.

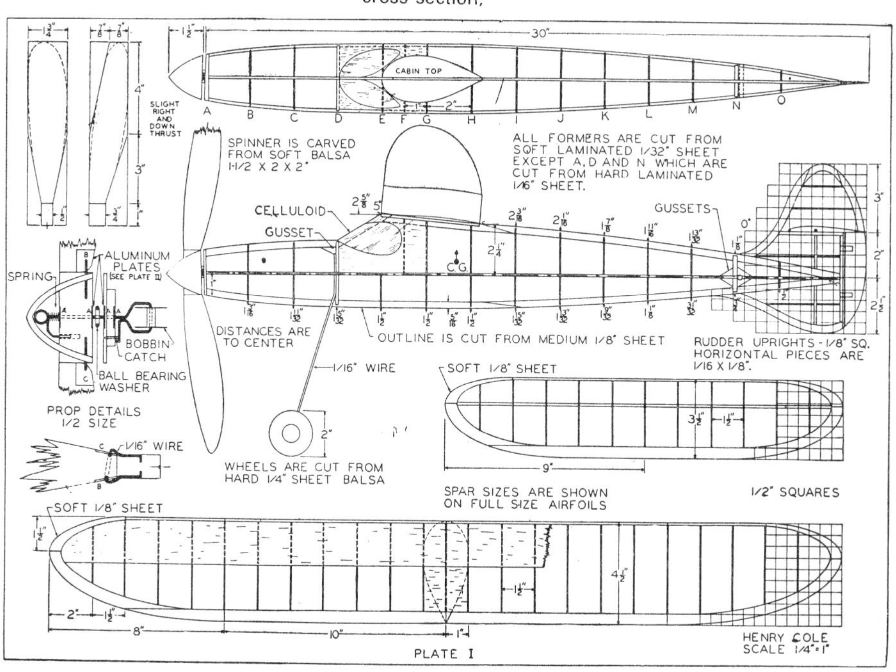

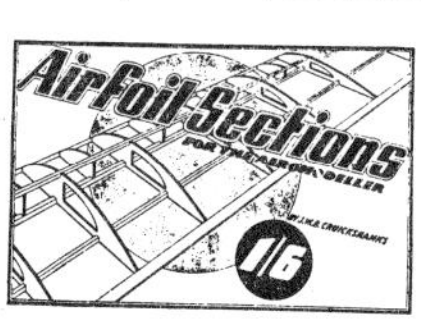

'Airfoil Sections for the Aeromodeller' cost 1/6 (7½p) and included 36 useful airfoils, with ordinates and characteristics.

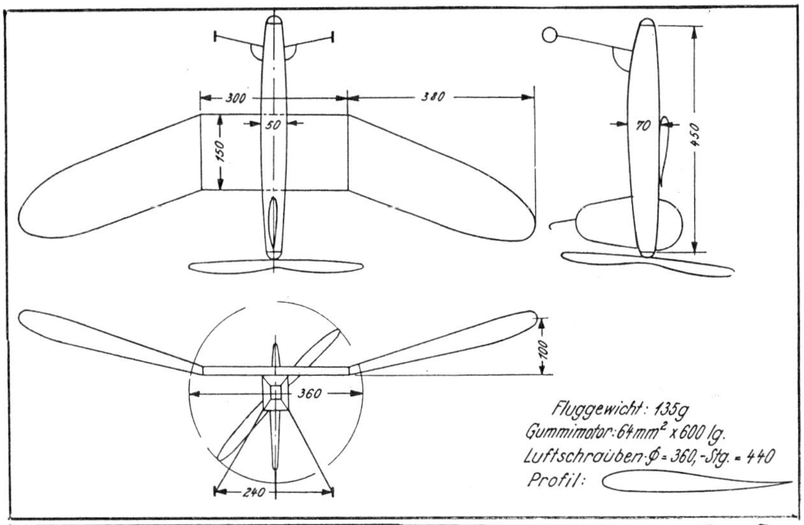

Pusher by H. Kiernow actually appeared in May/June 1944 *Modellflug* though built earlier.

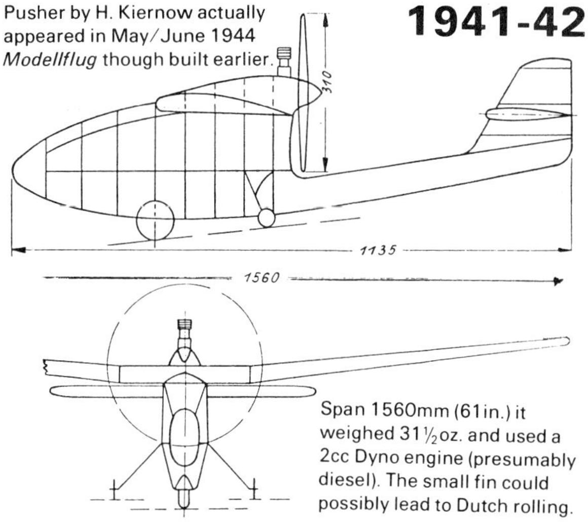

Span 1560mm (61in.) it weighed 31½oz. and used a 2cc Dyno engine (presumably diesel). The small fin could possibly lead to Dutch rolling.

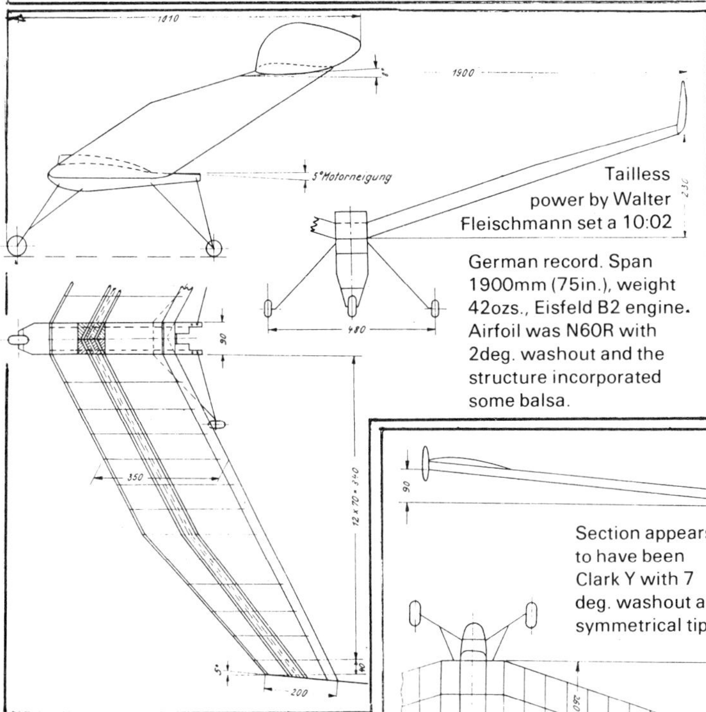

Tailless power by Walter Fleischmann set a 10:02 German record. Span 1900mm (75in.), weight 42ozs., Eisfeld B2 engine. Airfoil was N60R with 2deg. washout and the structure incorporated some balsa.

MN2 tailless rubber model by G. Sult (*Modellflug* November 1941).

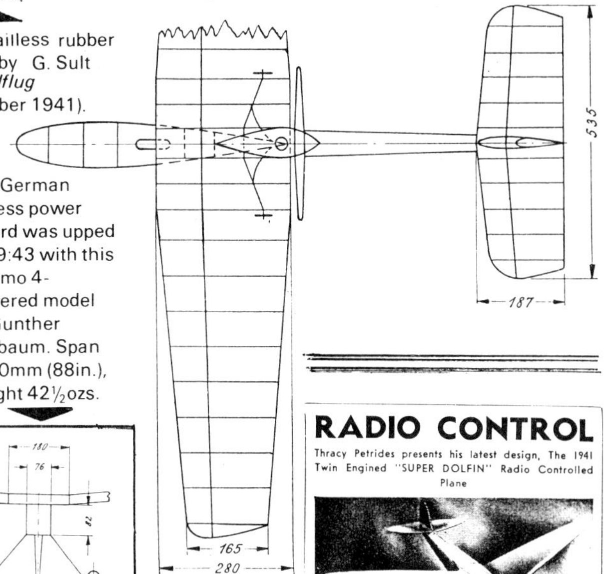

The German tailless power record was upped to 19:43 with this Kratmo 4-powered model by Gunther Maibaum. Span 2240mm (88in.), weight 42½ozs.

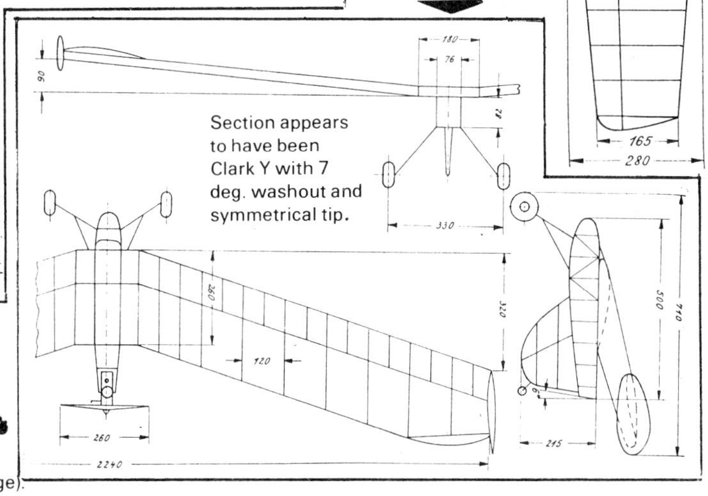

Section appears to have been Clark Y with 7 deg. washout and symmetrical tip.

The 'Stratosphere' model (see opposite page).

Thracy Petrides moved into radio with his 'Dolfin' and a development, the twin 'Super Dolfin'.

A forerunner of the present-day open class fly-off model was Chet Lanzo's 55in. design which produced some remarkable flight times.

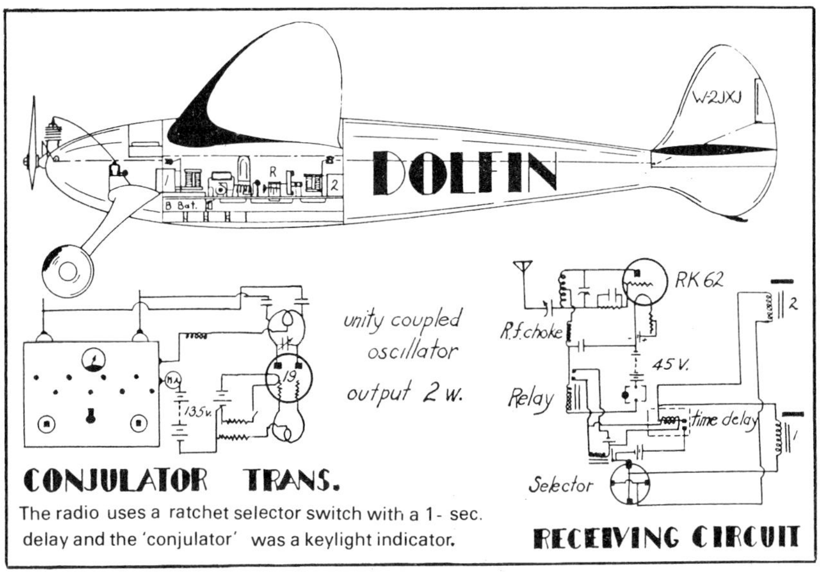

The radio uses a ratchet selector switch with a 1- sec. delay and the 'conjulator' was a keylight indicator.

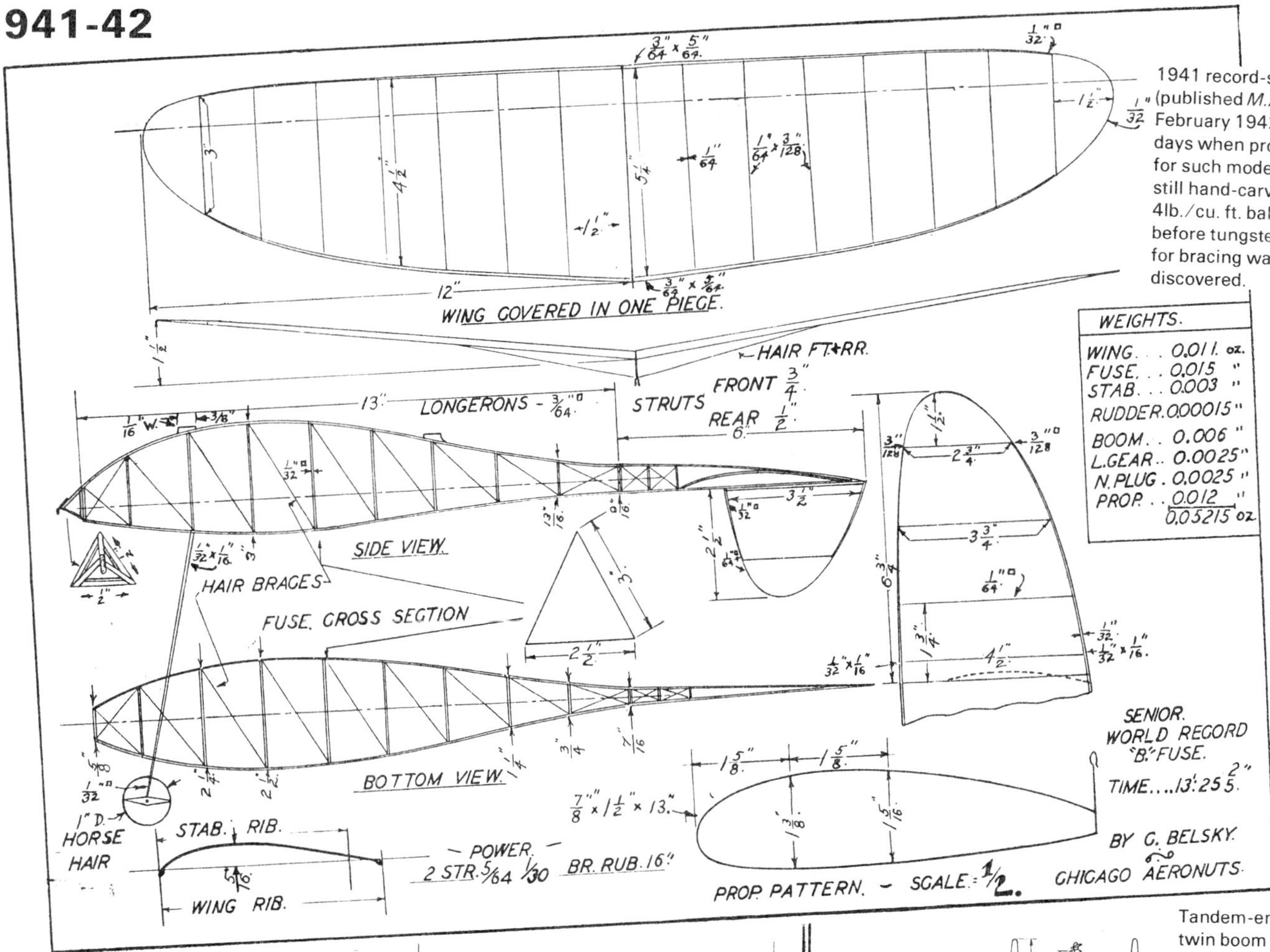

1941 record-setter (published *M.A.N.* February 1942) in the days when propellers for such models were still hand-carved from 4lb./cu. ft. balsa and before tungsten wire for bracing was discovered.

Not all Americans went in for pylon power models. Sal Taibi was very successful with machines like the 53in. 'Pacer B' and the .35-powered 80in. 'Pacer C'

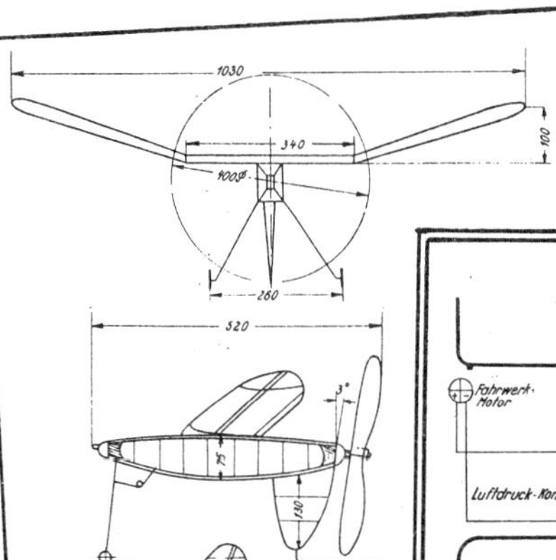

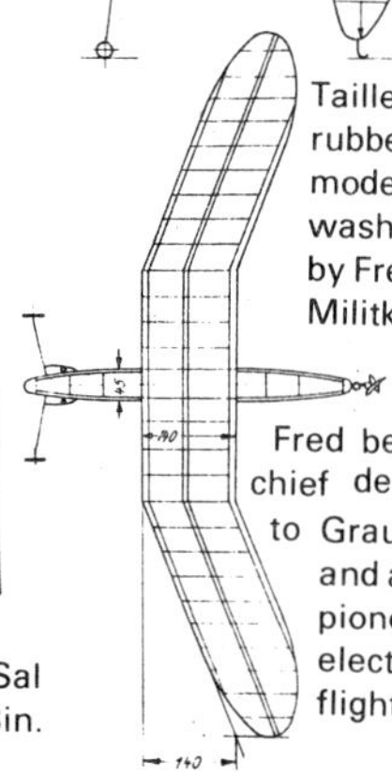

Tailless rubber model (9° washout) by Fred Militky in 1942. Fred became chief designer to Graupner and a pioneer in electric flight.

Tandem-engined twin boom power model with novel

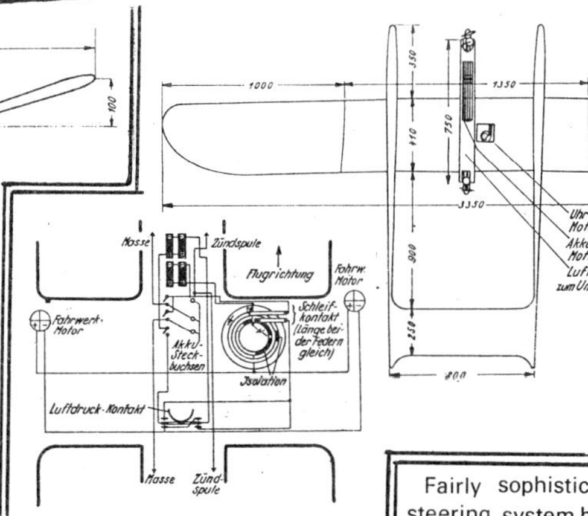

flight path control by Sigfried Keuchel (*Modellflug* September 1942).

Fairly sophisticated compass steering system by Peter Baverman *Modellflug* December 1942.

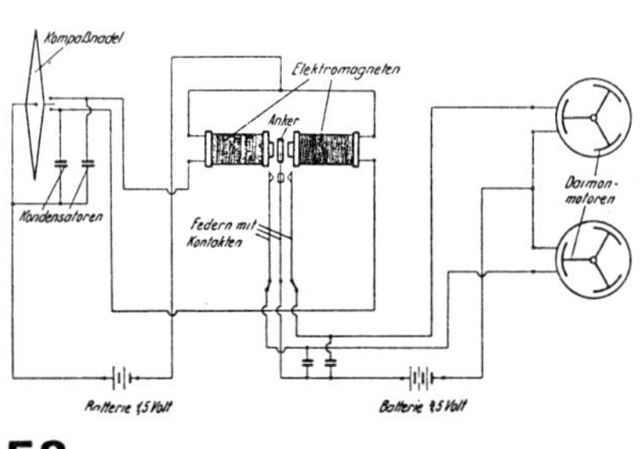

Stück-Zahl	Benennung	Teil-Nr	Werkstoff	Abmessungen in mm
1	Batterieanschluß	12		
2	Niete	11	Aluminium	2⌀x8
2	Lötstelle als Fuß	10	Zinn	
2	Klemmschraube	9	Messing	1,5⌀x12
2	Kontaktfeder	8	Silberdraht	10(nur a.d.Spitzen)
1	Kontaktstift	7	Platin	2⌀x3
1	Manschette	6	Messing	1x10x25
2	Achsenhälfte	5	Stahl	2⌀x6
1	Magnetnadel	4	"	3x5x80
2	Lagerschraube	3	"	3⌀x6
1	Lagerbügel	2	Aluminium	2x9x77
1	Grundplatte	1	Isolierstoff	3x100⌀

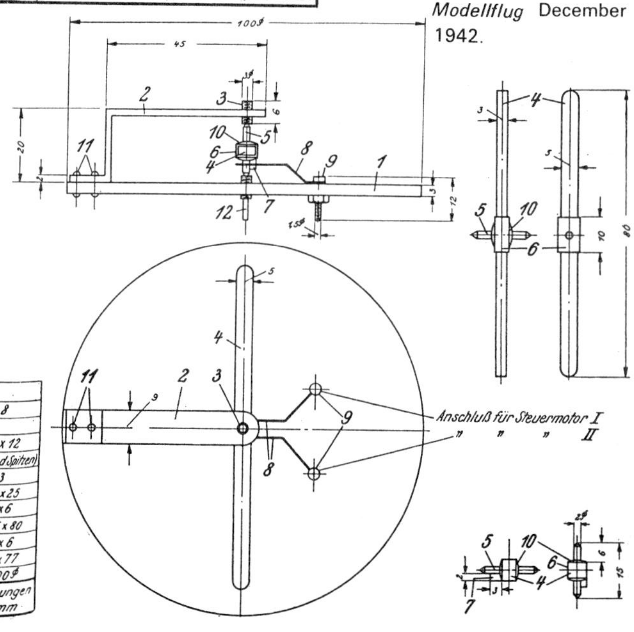

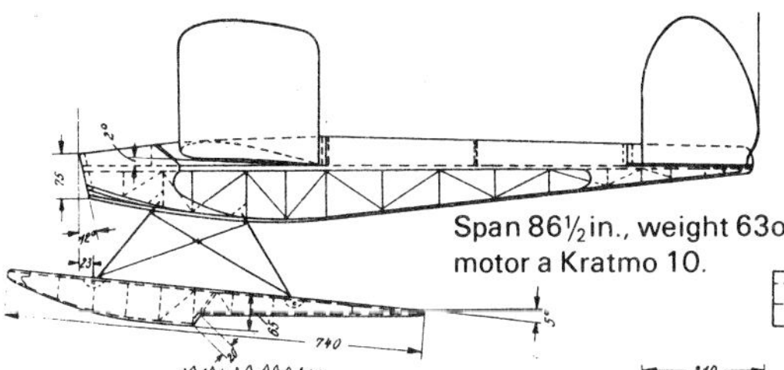

Span 86½in., weight 63ozs., motor a Kratmo 10.

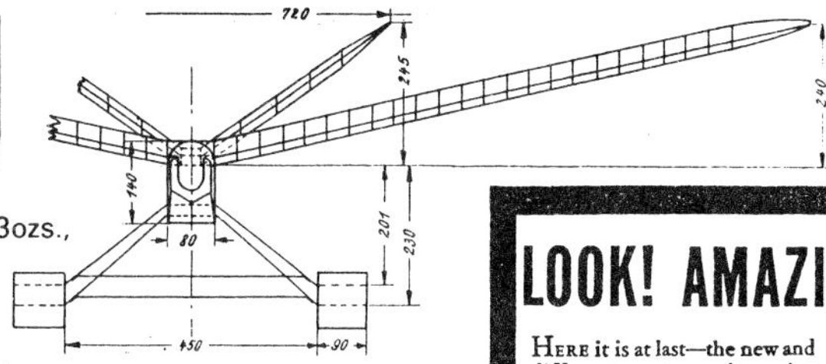

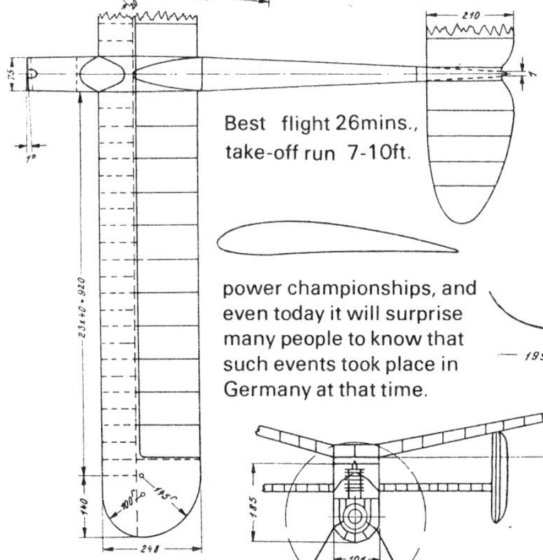

Best flight 26mins., take-off run 7-10ft.

Walter Fleischmann (*Modellflug*, February 1942) used butterfly tail and floats which were very much state-of-the-(model)-art.

In the same issue as the model above, this 78½in. 62oz. model by Max Tewes also used a Kratmo 10. Interesting use of heavily toed-in fins. Model was winner at the 1941 national power championships, and even today it will surprise many people to know that such events took place in Germany at that time.

LOOK! AMAZING NEW MOTOR UNIT!

HERE it is at last—the new and different motor unit you've been waiting for at a price any model builder can afford. It's SKY-KOIL 32! A light, sturdy spring-type motor that gives you more power, longer life and a brand new idea in model building—suspended power!

With a SKY-KOIL 32 Motor you no longer need excessive bracing or complicated design. Longitudinal stress is not exerted by SKY-KOIL 32 Motor.

With a SKY-KOIL 32 there is little danger in crash landings. The motor distributes the shock, and protects the fusilage.

Easy to mount in any model. Quickly transferred from one model to another. Any type propeller easily attached to motor. Comes complete ready to install, with complete tested plans for a light-weight, *easy-to-build model*, all for only $1.29!

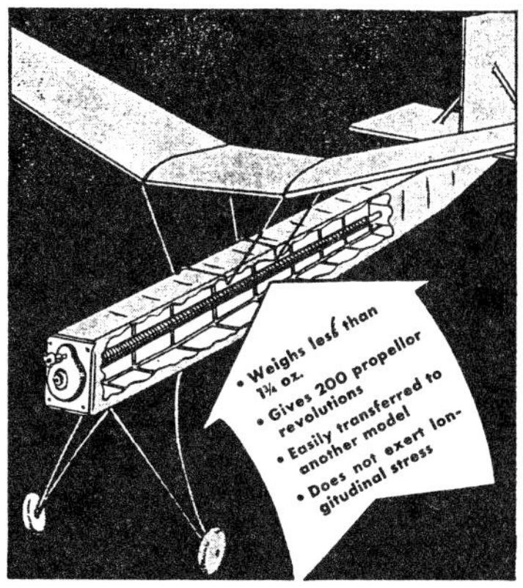

MAIL COUPON WITH ONLY $1.29

ROKWELL INDUSTRIES, Inc.
3924 Juniata Street, St. Louis, Missouri
Enclosed find $1.29. Please send me one complete SKY-KOIL 32 motor unit. If I am not thoroughly satisfied I will return motor within ten days for a full refund.
Name
Address
(Missouri purchasers please include Sales Tax)

SKY-KOIL 32 MOTORS are precision built from quality materials. Your money back if not satisfied!

A variation on the clockwork motor which resurfaced for a time some years later. 200 revolutions of the prop sounds a very short power run.

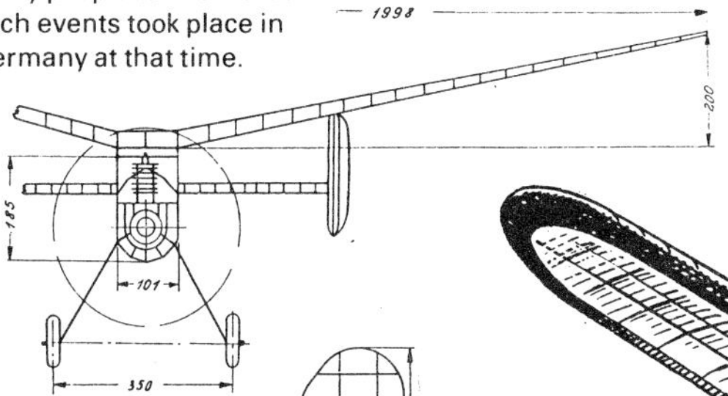

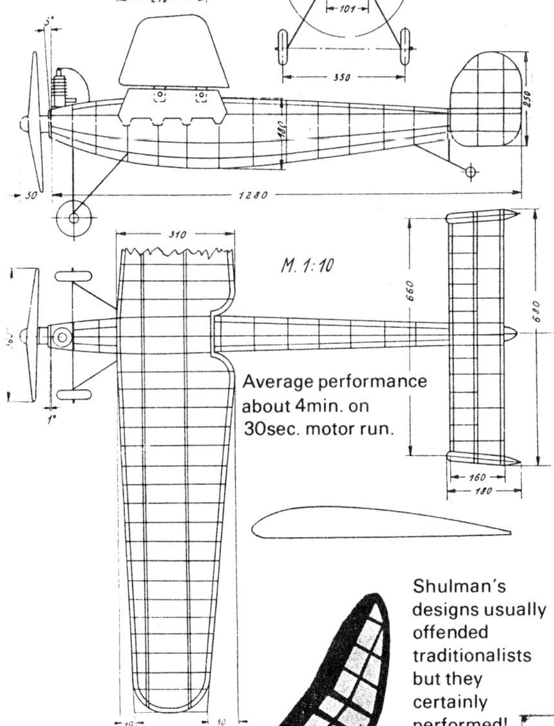

M. 1:10

Average performance about 4min. on 30sec. motor run.

The Cleveland 'Playboys' (A at 33in., B at 54½ and C at 80) were up with the top performers and have recently been reissued for electric power or vintage flying. The P6E and P26-A were two of the most colourful U.S. military machines and among the most popular subjects for 1930s/40s kits.

HAWK P6E

BOEING P26-A FIGHTER

The Sky's The Limit . . .

Build big—build small . . . tail-less, wing-less, pusher or tractor — with versatile Super ATOM as your collaborator, your operating range is limitless! SPECIFICATIONS: Bore and stroke ½"; Displacement .097 cu. in.; Weight, complete 2 ounces. Develops the greatest-power-to-weight ever achieved. $15.50.

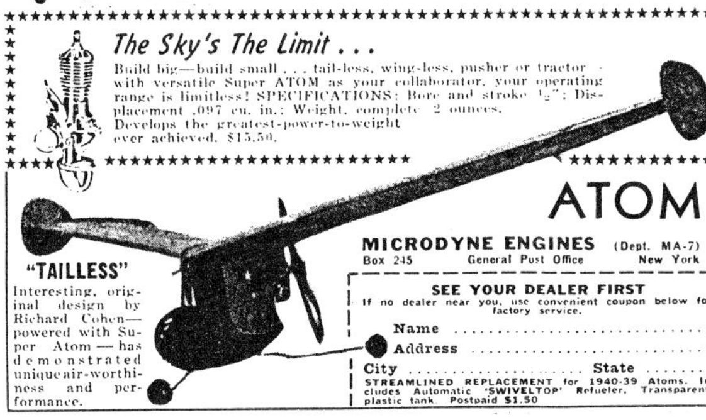

ATOM

MICRODYNE ENGINES (Dept. MA-7)
Box 245 General Post Office New York

"TAILLESS"
Interesting, original design by Richard Cohen—powered with Super Atom — has demonstrated unique air-worthiness and performance.

SEE YOUR DEALER FIRST
If no dealer near you, use convenient coupon below for factory service.
Name
Address
City State
STREAMLINED REPLACEMENT for 1940-39 Atoms. Includes Automatic 'SWIVELTOP' Refueler, Transparent-plastic tank Postpaid $1.50

This advertisement is not for the tailless model but for the Atom engine!

Shulman's designs usually offended traditionalists but they certainly performed!

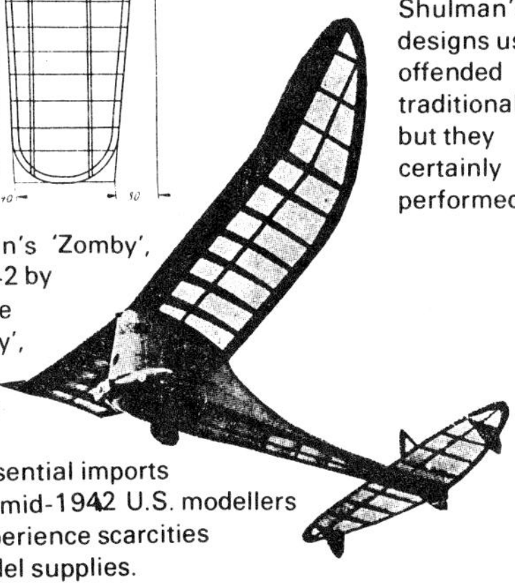

Leon Shulman's 'Zomby', kitted in 1942 by Megow as the 'Super Zomby', was 44in., class A or B

With less essential imports reduced by mid-1942 U.S. modellers began to experience scarcities of some model supplies.

ATTENTION!!!
JASCO CUSTOMERS

• No more balsa after May 15th! Special Brown Rubber out of stock since April 10th. —Cause: Not enough balsa or rubber to meet the needs of the arming industry. If we are fortunate in obtaining more balsa or rubber, we will let you know. In the meantime please do not order Balsa, Thermic Kits or Rubber.

• Working on domestic products to meet the needs of model builders. Will announce when ready.

• P.S. What did you do today to beat the Japs?

JUNIOR AERONAUTICAL SUPPLY CO.
100 East 10th Street New York City

Title of a book put out by Austin-Craft!

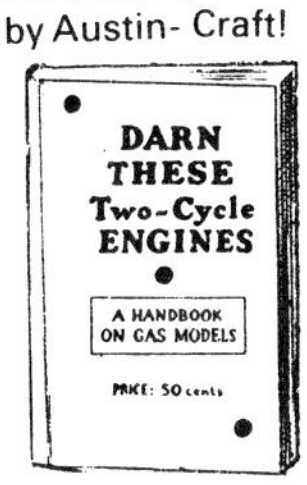

NATIONALS FLYING SCALE "CHAMP"

Not Only National Winner but a Realistic Plane That Flies Two Minutes Consistently

By HENRY STRUCK

Most full-size plans in model magazines ran to four or five pages, restricting the number of examples we can show in this book. We have picked this 36in. 'Interstate Cadet' as being capable of adaptation to engine power as well as an attractive rubber model of good performance and distinctively American in its approach.

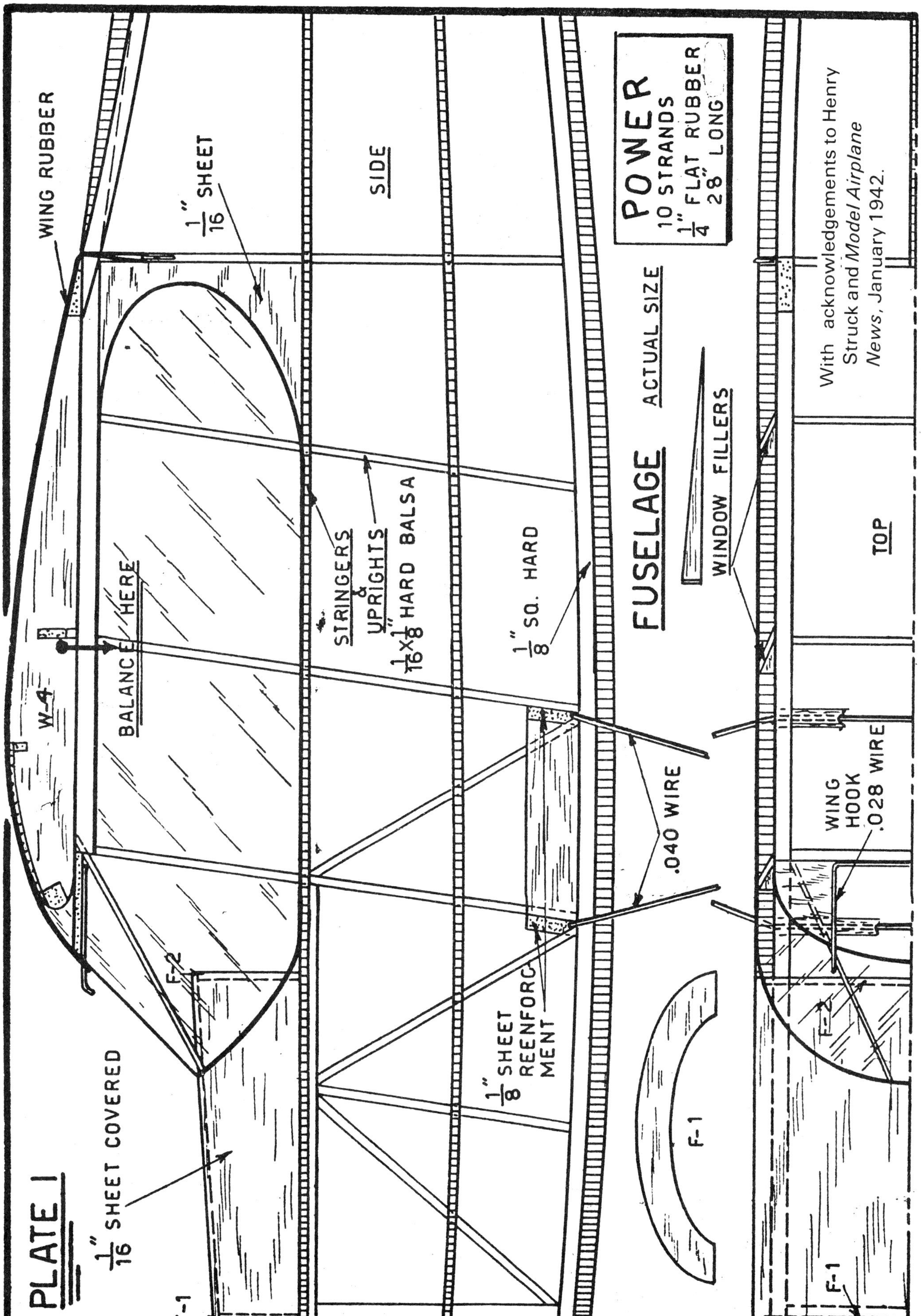
PLATE I
1/16" SHEET COVERED
F-1
F-2
WING RUBBER
1/16" SHEET
SIDE
W-4
BALANCE HERE
STRINGERS & UPRIGHTS 1/16x1/8" HARD BALSA
1/8" SQ. HARD
1/8" SHEET REENFORCEMENT
.040 WIRE
F-1
FUSELAGE
ACTUAL SIZE
POWER
10 STRANDS
1/4" FLAT RUBBER
28" LONG
WINDOW FILLERS
TOP
WING HOOK .028 WIRE
F-2
F-1
With acknowledgements to Henry Struck and Model Airplane News, January 1942.

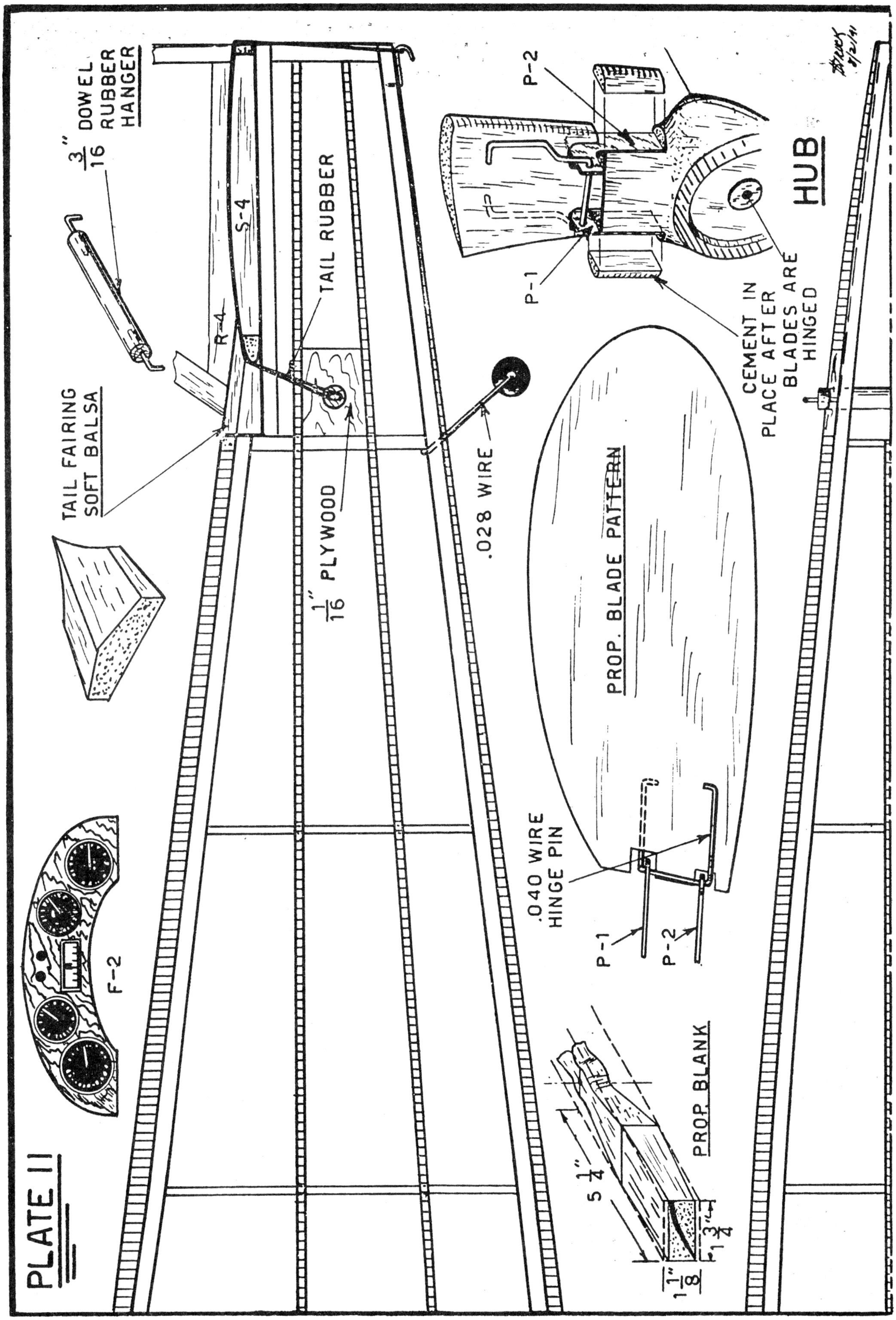
PLATE II
F-2
TAIL FAIRING SOFT BALSA
$\frac{3}{16}$" DOWEL RUBBER HANGER
R-4
S-4
TAIL RUBBER
$\frac{1}{16}$" PLYWOOD
.028 WIRE
PROP. BLADE PATTERN
.040 WIRE HINGE PIN
P-1
P-2
PROP. BLANK
5 $\frac{1}{4}$"
1 $\frac{1}{8}$"
1 $\frac{3}{4}$"
HUB
P-2
P-1
CEMENT IN PLACE AFTER BLADES ARE HINGED

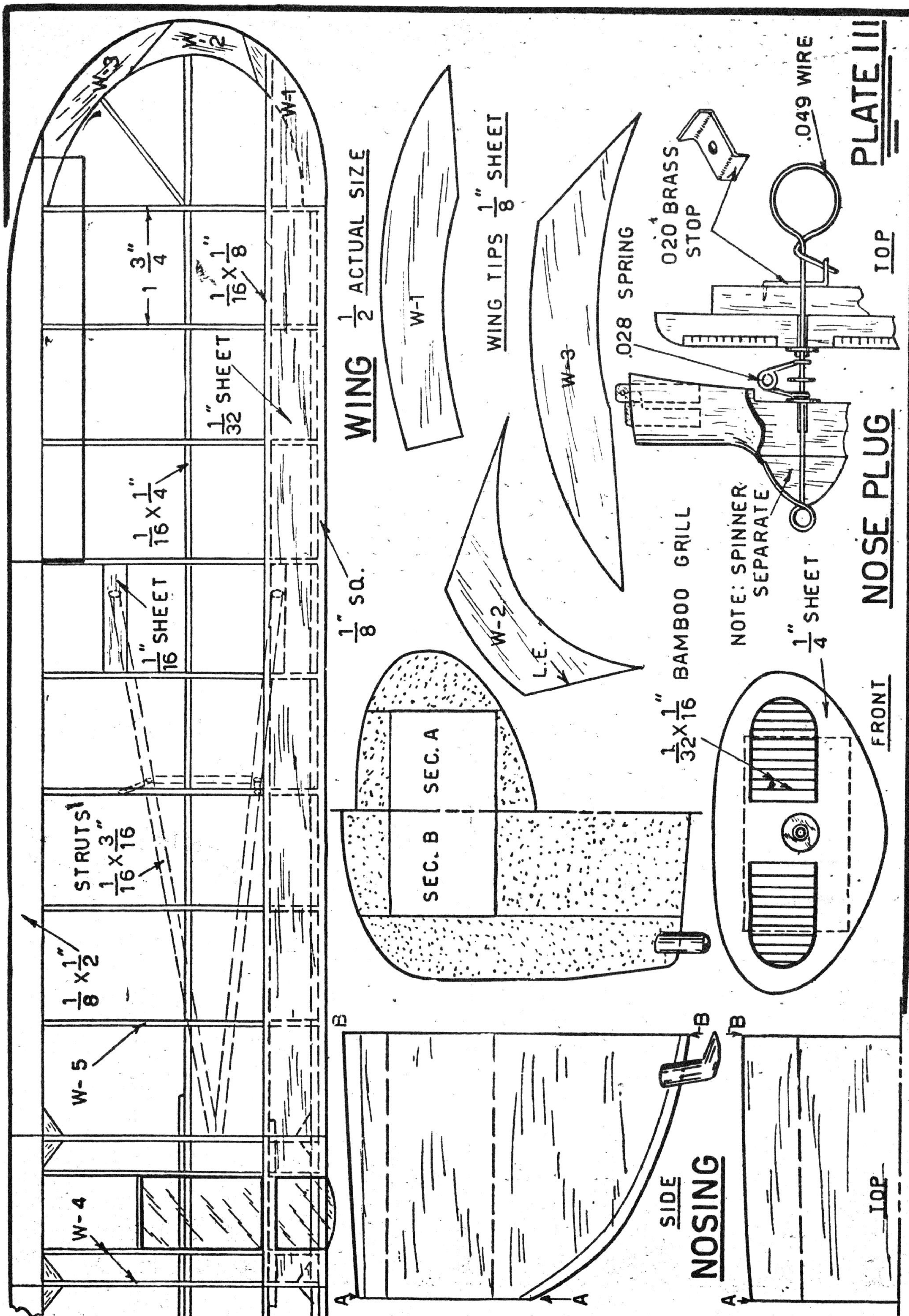
PLATE III
WING 1/2 ACTUAL SIZE
WING TIPS 1/8" SHEET
W-1
W-2
W-3
W-4
W-5
L.E.
1 3/4"
1/16 X 1/8"
1/32" SHEET
1/16 X 1/4"
1/16" SHEET
1/8" SQ.
STRUTS 1/16 X 3/16"
1/8 X 1/2"
SEC. A
SEC. B
.049 WIRE
.020 BRASS STOP
.028 SPRING
TOP
NOTE: SPINNER SEPARATE
BAMBOO GRILL
1/4" SHEET
1/32 X 1/16"
FRONT
NOSE PLUG
SIDE
NOSING
TOP
A
B

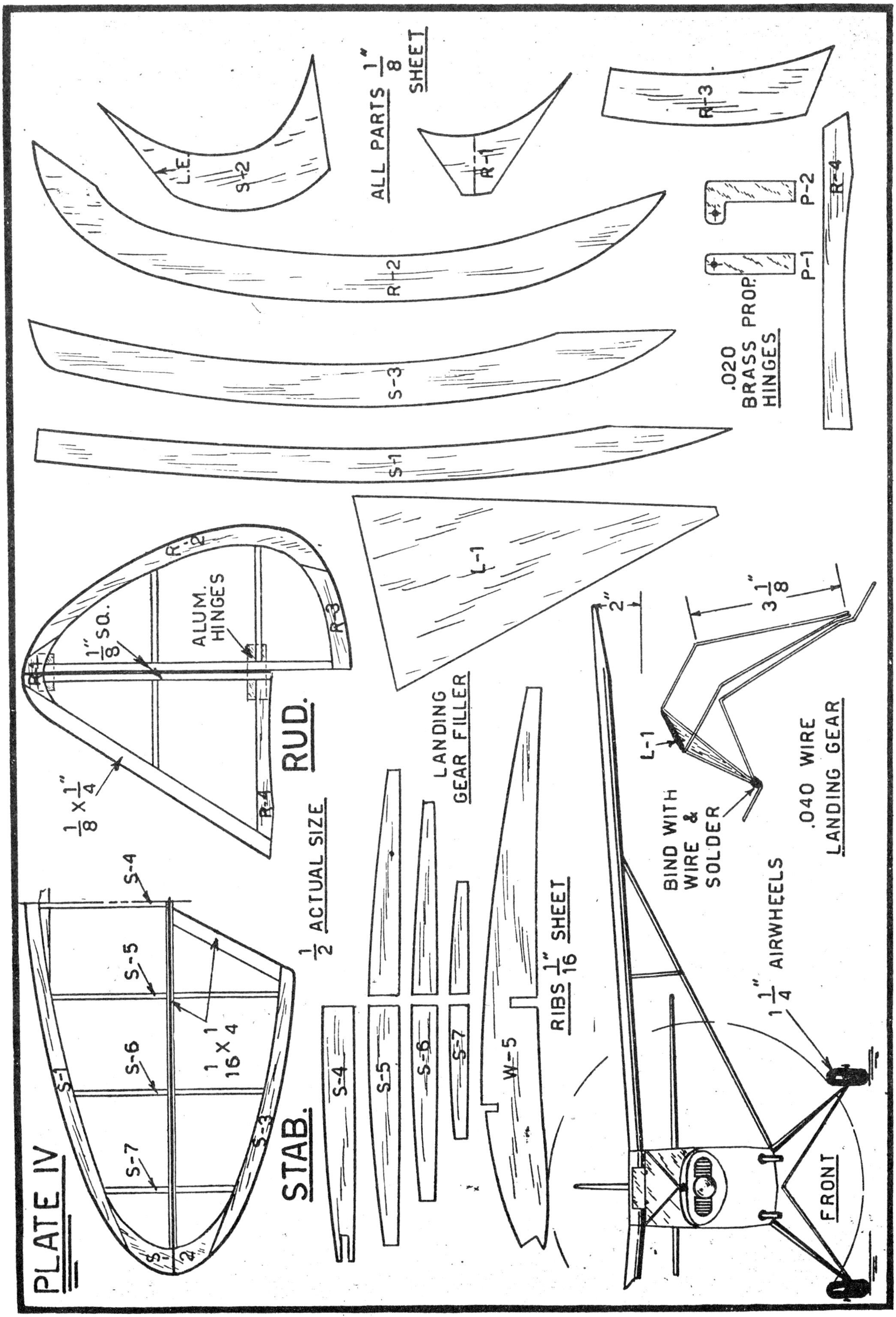
PLATE IV
STAB.
S-1
S-3
S-4
S-5
S-6
S-7
1/16 X 1/4
RUD.
R-2
R-3
R-4
1/8" SQ.
ALUM. HINGES
1/8 X 1/4
1/2 ACTUAL SIZE
RIBS 1/16" SHEET
W-5
LANDING GEAR FILLER
L-1
ALL PARTS 1/8" SHEET
R-1
S-2
L.E.
.020 BRASS PROP. HINGES
P-1
P-2
2"
3 1/8"
BIND WITH WIRE & SOLDER
.040 WIRE LANDING GEAR
1 1/4" AIRWHEELS
FRONT

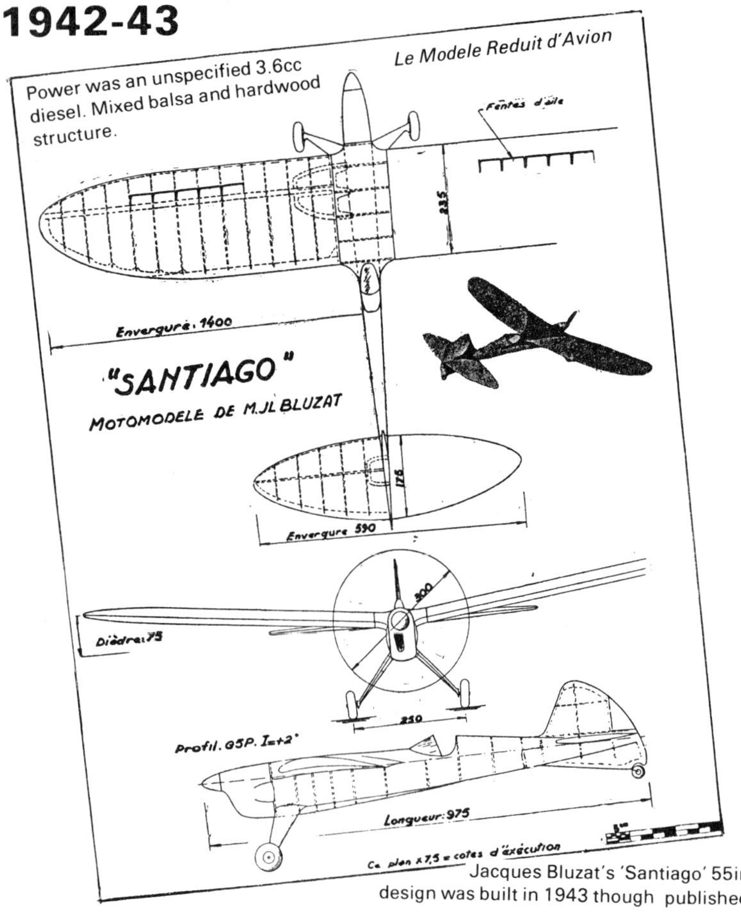

Jacques Bluzat's 'Santiago' 55in. F/F design was built in 1943 though published in 1951.

Joe Ott kits were apparently produced in phenomenal quantities but very few seem to have been exported.

Carl Goldberg's 'Interceptor', Class A or B, was kitted in 1942 and was the first kit to incorporate tip-up tail dethermaliser details. A constant-chord 50in. development of the 'Mercury' type of model, the Interceptor was possibly the apogee of the straight pylon layout.

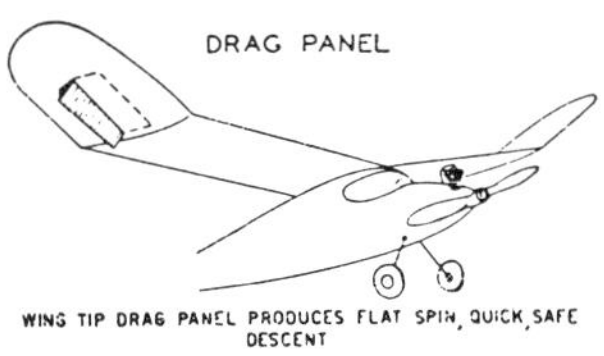

Hillcrest, who with Austin were the source of 'airdraulic' timers, made a notable contribution to dethermaliser development.

INSURE YOUR MODEL AGAINST LOSS

Now, with motors and materials becoming more difficult to obtain, a gas model floating away out of sight on an updraft is not a pleasant sight to any model builder.

DETHERMALIZERS

(Consisting of flaps, spoilers, parachutes, etc., which open at a predetermined time to increase the sinking speed of the model) offer the only present solution to this problem. With this idea in mind HILLCREST engineers have developed a new timing device especially suited to this purpose. Light, compact, and reliable it may be easily installed on any model. This timer unit, similar in appearance to the HILLCREST Flight Timer, works on a vacuum principle instead of air pressure. After a great deal of experimenting it was found that with the air pressure type of timer as the ship gained altitude the decreased air pressure made the time run many minutes longer than it had been set for originally. So by reversing the action this was not only eliminated but gave the very desirable feature of decreasing the timer duration the higher the plane flew.

SPECIFICATIONS

Weight 3/8 oz. Length overall 2". Maximum duration guaranteed over 10 minutes. A simplified air valve gives an easy adjustment to any duration setting you care to use.
Comes complete with detailed instructions and plans for several types of dethermalizing flaps, etc.

HILLCREST DETHERMALIZER .... only 75c
Be the first with the latest, order yours today.
HILLCREST FLIGHT TIMER $1.25
HILLCREST PLASTIC BATTERY BOXES. Two sizes Penlite and Med. .... .45
HILLCREST 2-in-1 PLUG WRENCHES. Fits in any cylinder head. Two sizes; V & V-2; Blue Crown & V-3 .... 15c ea.
Write for FREE catalog. Please include stamp for postage.

HILLCREST
TOPS IN MODEL AVIATION
1279 THIRD AVE. LOS ANGELES, CALIF.

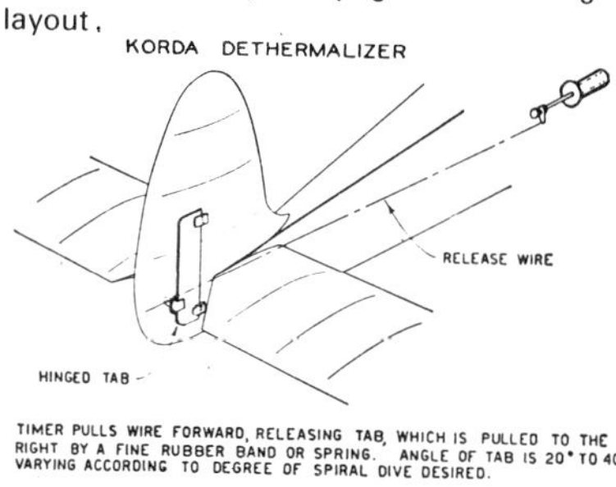

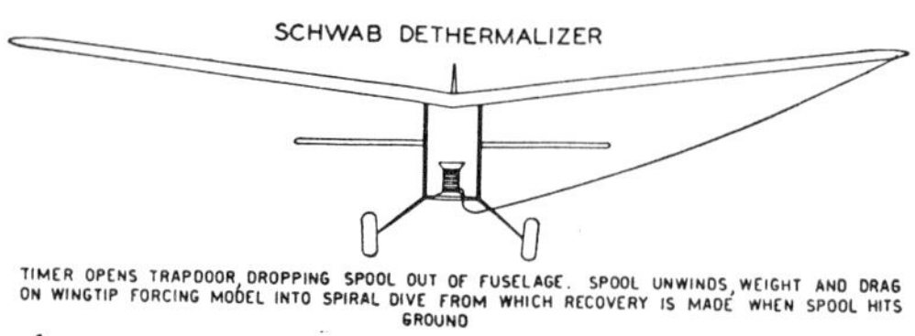

Many famous names were experimenting with methods of bringing models down in 1941/2 as the numbers lost increased.

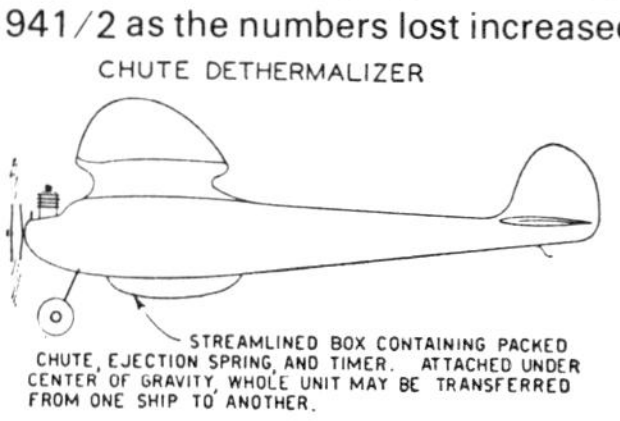

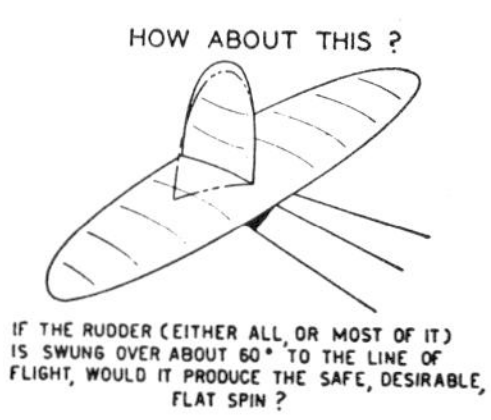

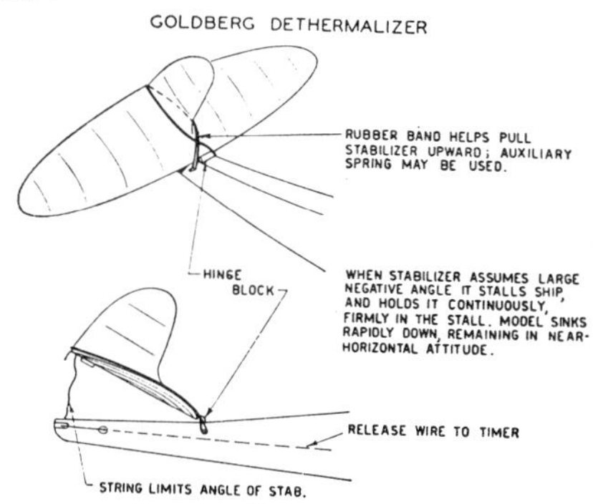

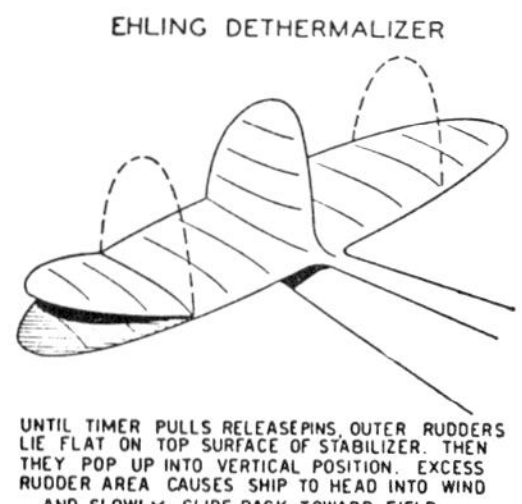

'Bring Them Down Safely', *M.A.N.* September 1943, was the first exhaustive survey of dethermalising techniques in the magazine. Goldberg's tip-up tail dates from observations of a damaged chuck glider in 1931 applied to an Interceptor flown by volunteer Hershey Goldenberg (Chicago) in February 1942, but Korda and other Cleveland flyers had been using a spiral dive tab for some months. New rules for 1942 proposed a maximum time of 4min., anything over deducted and 6mins. or over scoring zero. Dethermalisers thus became essential to contest flying.

Jim Walker came up with whip-power early in 1943, where profile models are towed by a line up to 40ft. long on an 8ft. pole. Speeds of nearly 100m.p.h. and all sorts of stunts are possible, but the idea never really caught on.

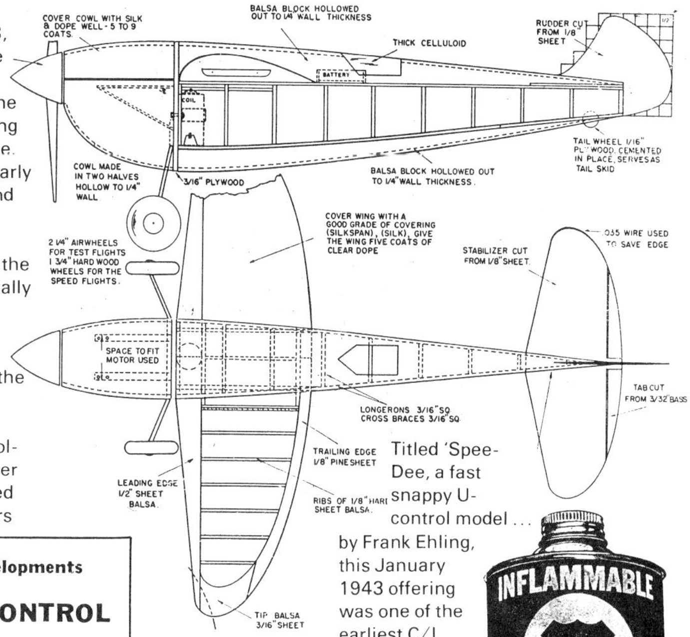

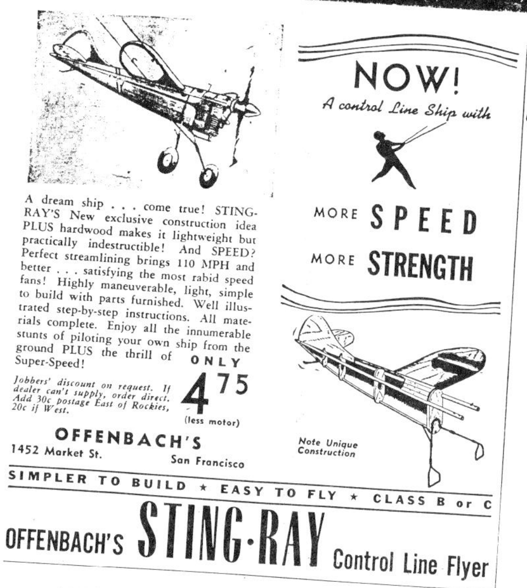

August, 1943, and the 'Sting-Ray' kit advertised with the expression 'a control-line ship'. The Walker patents had inhibited many manufacturers

Titled 'Spee-Dee, a fast snappy U-control model . . . by Frank Ehling, this January 1943 offering was one of the earliest C/L models published.

Latest Developments in RADIO CONTROL

A booklet written for the radio control beginner as well as the expert. Features—"The theory of the RK-62 receiver"—"Control devices"—Table of A and B batteries—A comparison of the electrical and mechanical qualities of bakelite, steatite and polystyrene—Self-neutralizing escapement—sequence solenoid—rubber powered selector—rubber powered indicator—"Radio Control Circuits," an article written in answer to the questions asked in the thousands of letters received by us. Eighteen different radio control hook-ups with hook-up diagrams.

Ask your dealer or send twenty-five cents for Illustrated Instruction Manual. Stamps will not be accepted.

Radio Control Headquarters
P. O. Box 214 Deal, New Jersey

Power Mist—Spitfire
Proven Superior Racing Fuels

NEW—
"BLUE BLAZER" SOLVENT FUEL
Easy starting, castor lubrication. Will not gum or draw moisture; excellent for protective clean-up after using other prepared fuels.
Pints 45c Quarts 80c
For plastic or metal tanks
5 methanol base fuels, 2 blended fuels.
Full information, fuel facts and formulas
Please enclose stamped addressed envelope.

FRANCISCO LABORATORIES
3787 Griffith View Dr. Los Angeles, 26, Calif.

Radio transmission was barred but you could still read about it in 1943.

SPECIAL NOTE: *Government regulations now require a non-highway gasoline ration coupon to purchase Phillips Model Motor Blend. Coupon may be obtained from your local ration board.*

Even in the States coupons were needed to buy gasoline!

The bell-crank illustrated at last on P. W. Westburg's 'C/L Fighter', *M.A.N.* December 1943.

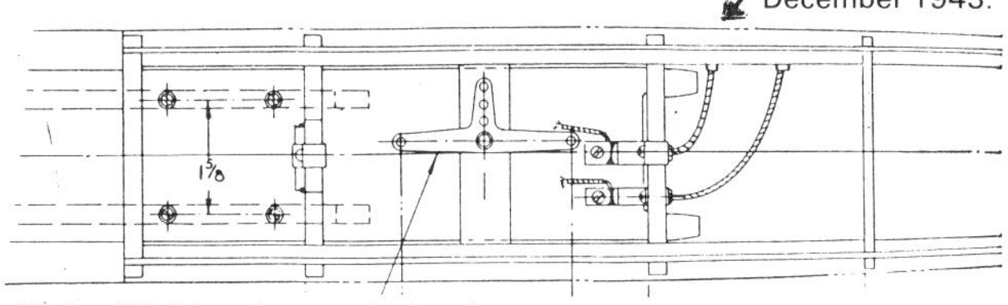

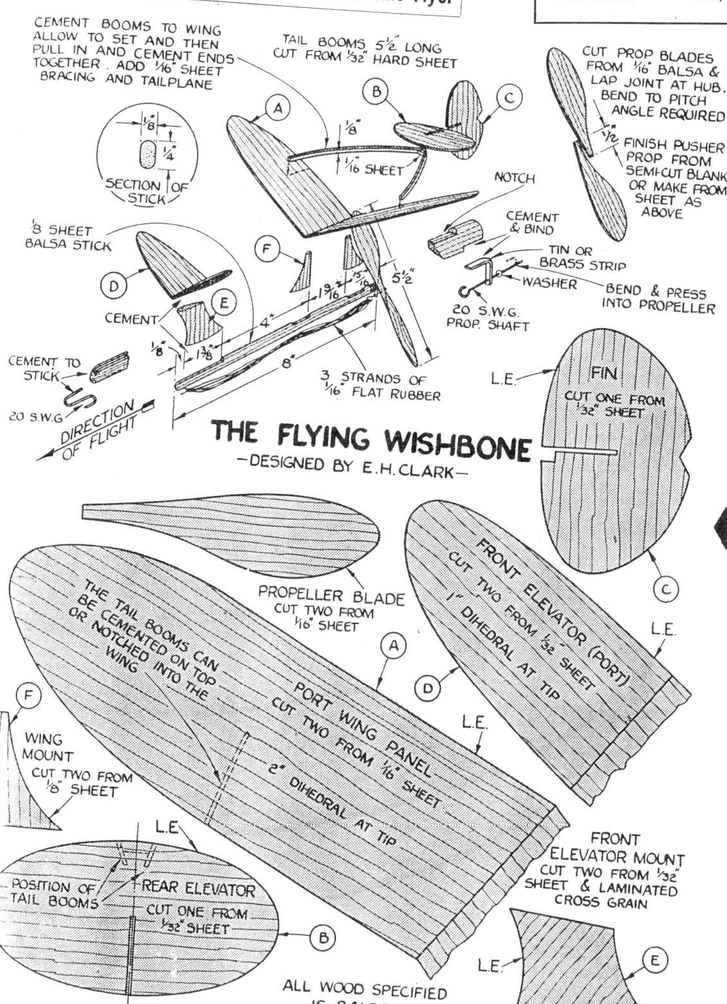

Span is 11in.

'Flying Wishbone' an early drawing by Bill Dean when he worked on the *Aeromodeller* (December 1942)

Another of those intriguing adverts. An aileron pendulum, perhaps?

A "PHANTOM PILOT" FOR YOUR PLANE

★ **NO SLIPS**
★ **NO SKIDS** ★ **NO SPINS**

A Device for FREE FLIGHT Models (No Control Lines)

Easy to build . . . Simple to install . . . Amazing results

Insure against unnecessary crackups with this sensational NEW device.

No need to be a "fair weather flyer" any longer!

This simple aerodynamic control system provides your ship with completely automatic lateral stability. You can fly in practically any type of weather . . . take off in heavy crosswinds safely . . . make perfectly banked turns.

No dihedral needed. You gain the extra lift of a straight wing. Eliminate the sinking effect of dihedral and polyhedral.

Entire mechanism is installed within plane. No external control necessary.

To fly well, your model need no longer be an "ugly duckling!" The "Phantom Pilot" is especially adaptable for true scale models, radio controlled planes, and low winged high speed unstable models . . . as well as all other types of gas and rubber ships.

Scientifically tested and proven on hundreds of test flights.

Send $1 postpaid for complete instructions, drawings, and photographs to:

High speed midwing model, with no dihedral, equipped with automatic control system.

AUTOMATIC REFLEX CONTROL

ECKSTEIN BROS., c/o Wilfon Research Laboratory, 1112 East California Street, GLENDALE, CALIFORNIA

One British power model pioneer was Dr. J. F. P. Forster His simple timer is shown, right.

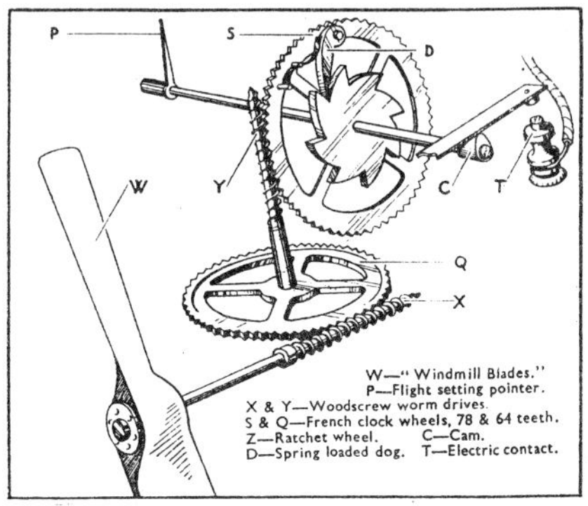

Bob Gosling's 'Ivory Gull' appeared in January 1943

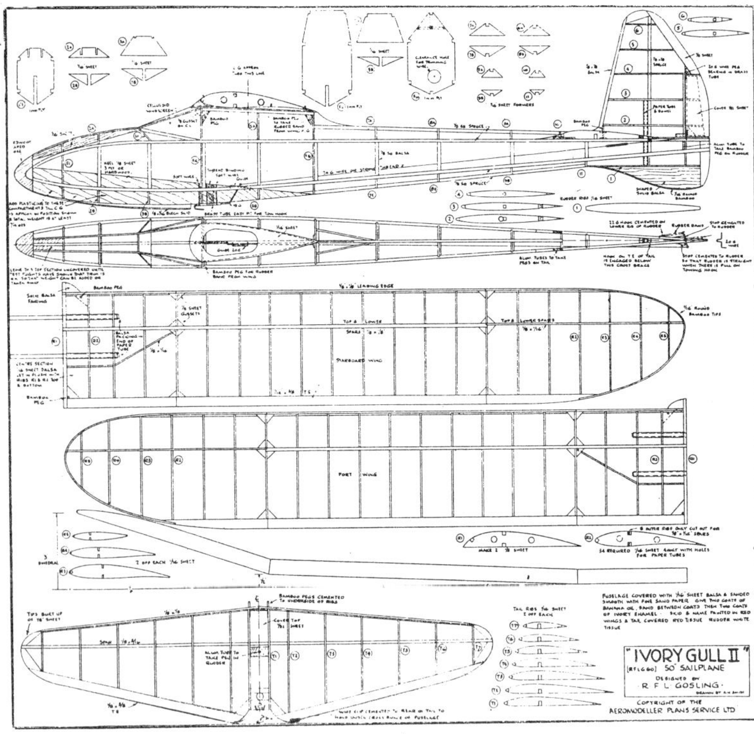

Slightly unusual in having a fuselage and undercarriage, 'Flapperoo' by Wilbur F. Tyler will R.O.G

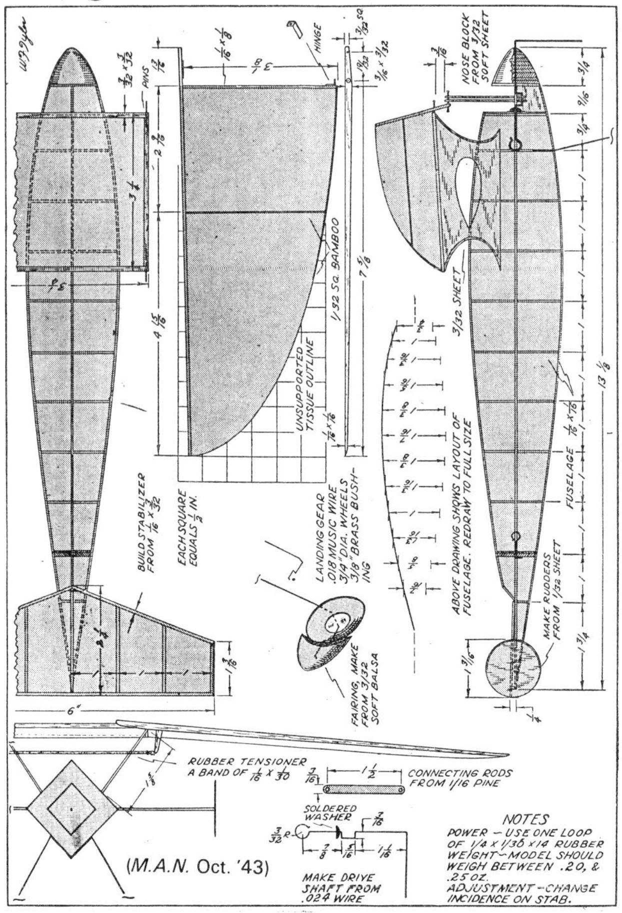

(M.A.N. Oct. '43)

During the '39-'45 war raw materials were in short supply. Balsa was used for Carley floats, Mosquito aircraft and sundry less obvious military uses

The 'Hardwood Flyer' by Caldwell Johnson (left) was paralleled by similar designs using little or no balsa in Britain.

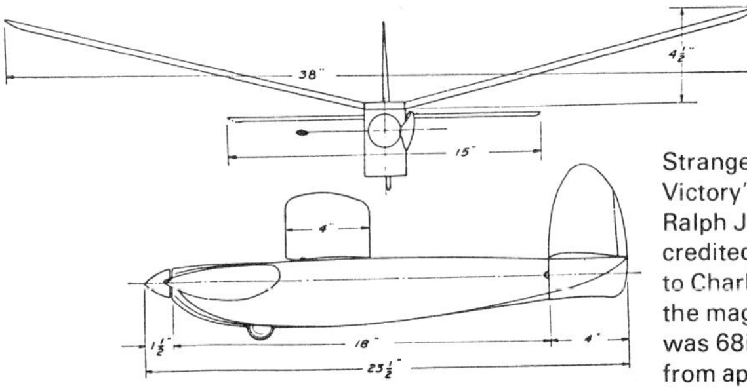

Strange machine is 'Vee for Victory' (*M.A.N.* June '43) by Ralph J. Ullenberg, who credited its successful flying to Charles Grant, still editing the magazine then. Wing was 68in. span, tapered from approx. 8½ to 5ins.

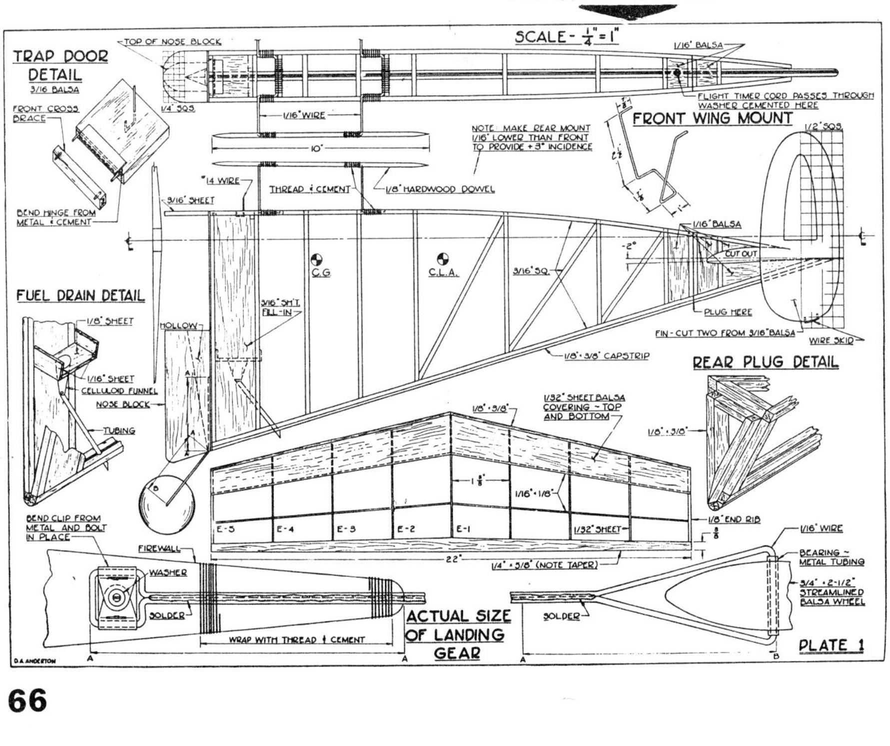

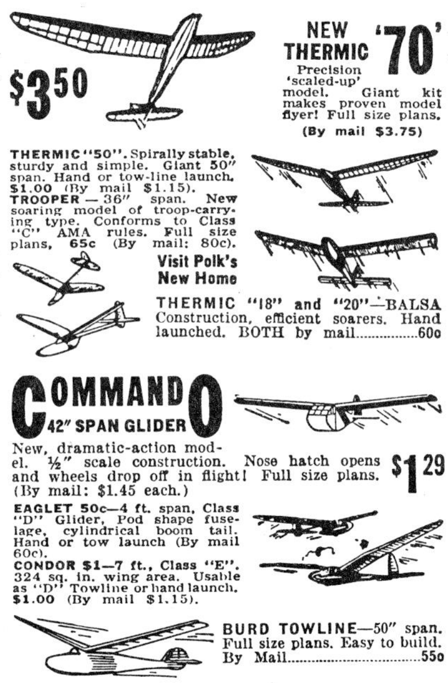

Gliders were catching on to an increasing extent. Well-known New York dealers Polk's offered this selection in 1943.

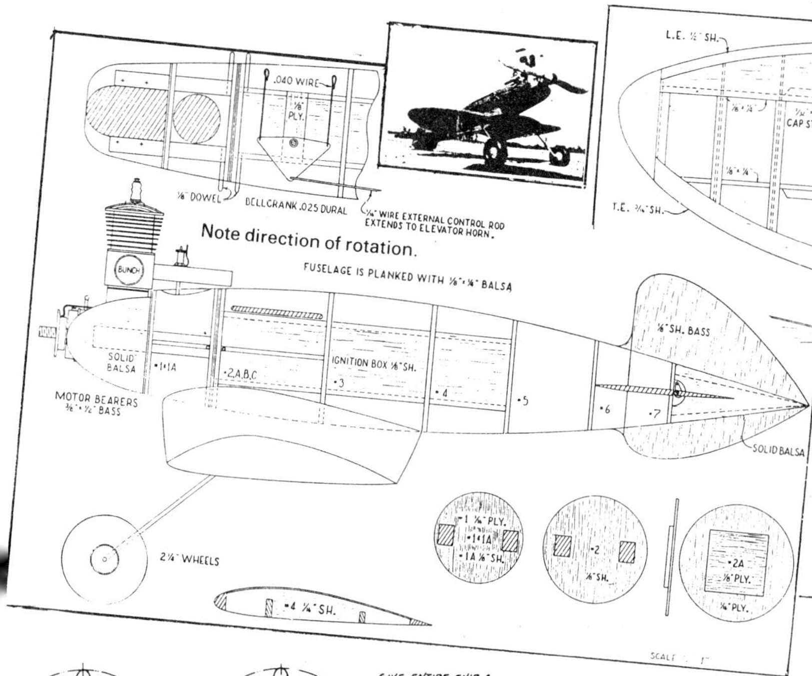

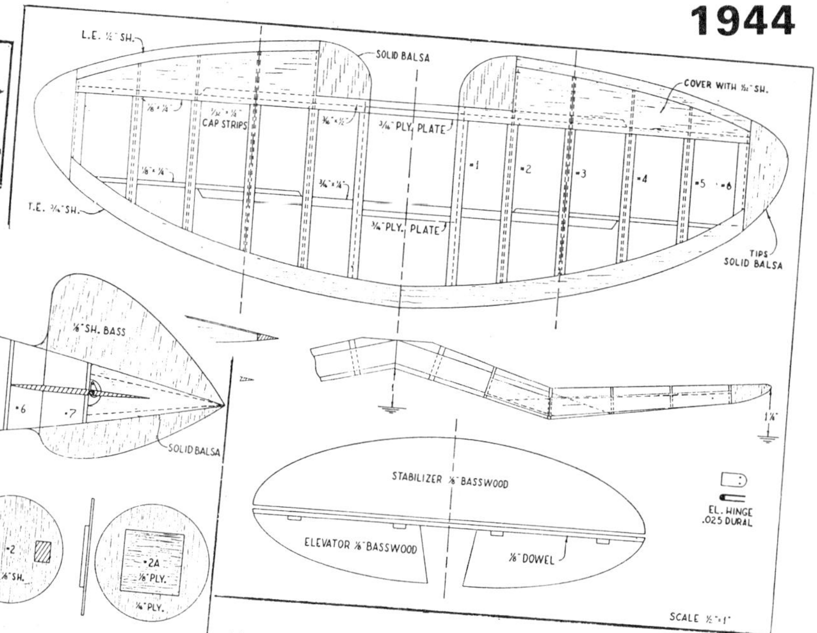

Harold de Bolt appeared on the C/L scene with 'Flying Tiger' (*M.A.N.* Feb. '44) and was to become a major figure.

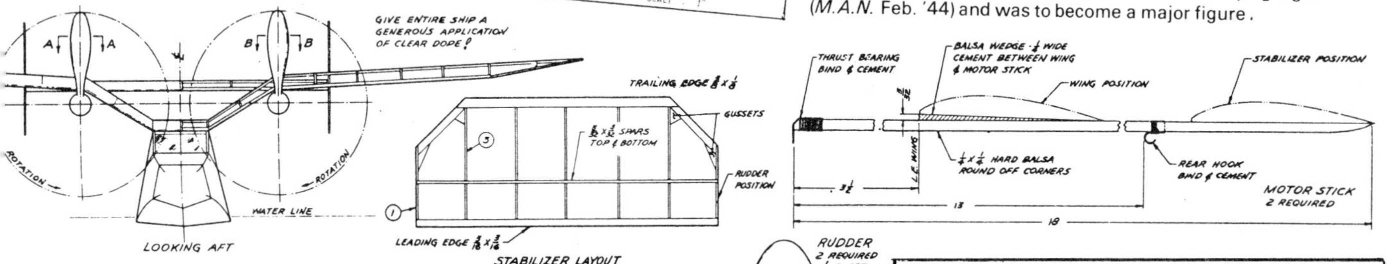

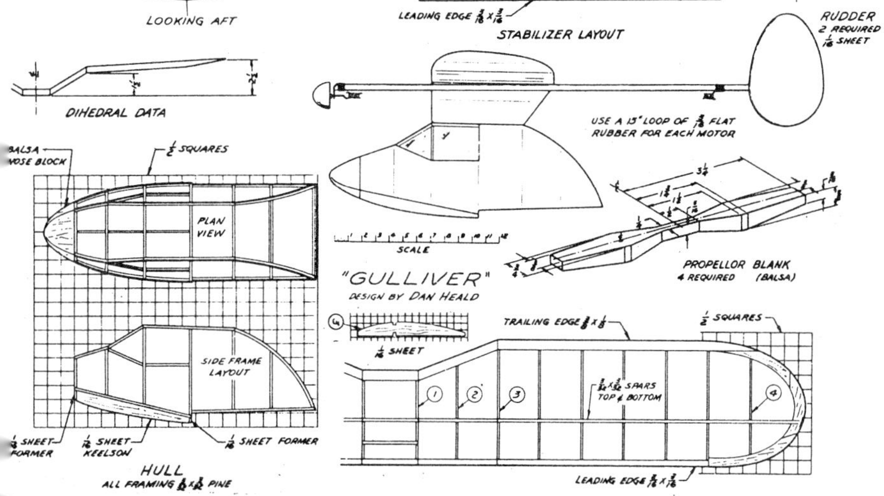

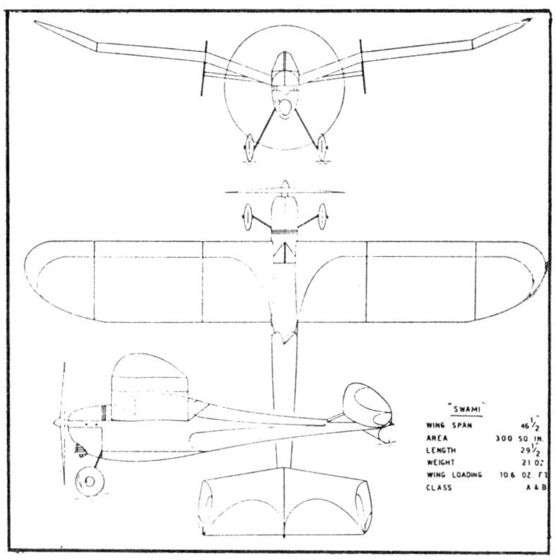

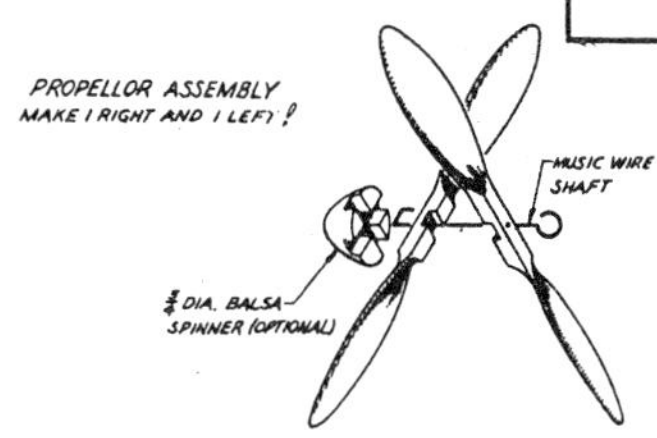

Jerry Stoloff's 'Swami' design was 46½in. span, Class A or B, i.e. engines just below and above .199cu. in. (3.3cc).

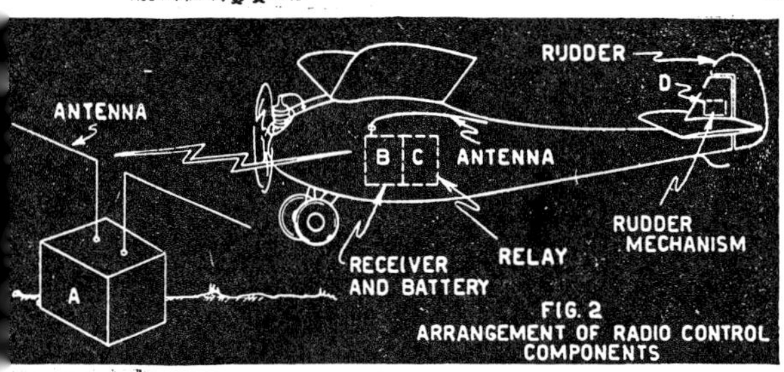

Something different — a 31in. twin rubber flying boat with opposed rotation four-blade 6½in. props. At least it only needed 5ft. of 3/16in. rubber!

Bill and Walter Good respectively with their 8ft. 'Big Guff' (published *Air Trails* 1941) which won the very low key R/C National category in 1938 and 39. (June 1944.

Flight of fancy by Robert Jamieson in a 1944 *Aeromodeller* — R/C combat with plastic models in a 'Things to Come' doodle. Wonder if he ever saw R/C pylon racing . . .?

1944

Speed seemed to be what most appealed to C/L flyers, at least at first. Modelcraft's 'Vee Gee' with the new Hornet or Torpedo motor was fast — 135m.p.h. at a later date.

More evidence of growing interest in gliders, although activity in English-speaking countries was slight compared with European countries.

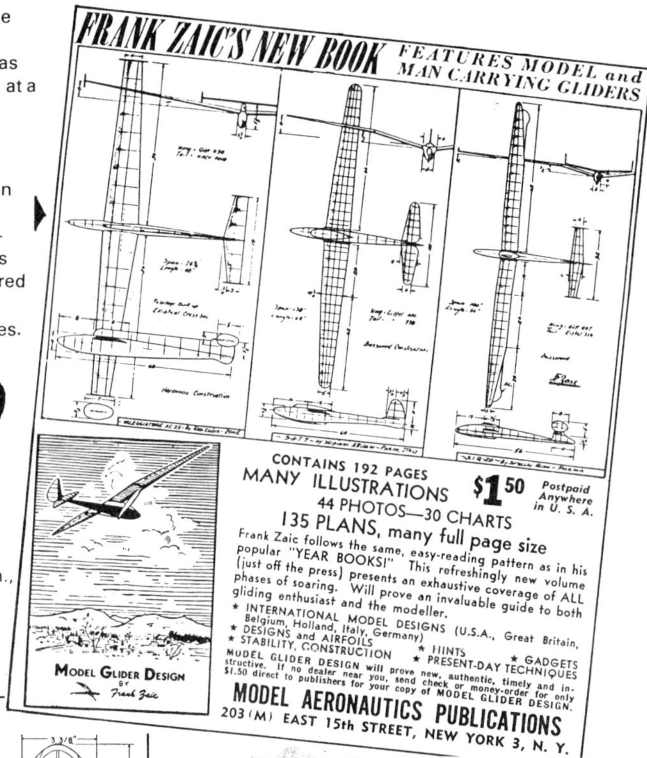

One of the few pusher models with contest performance was 'Yogi', an Eagle kit

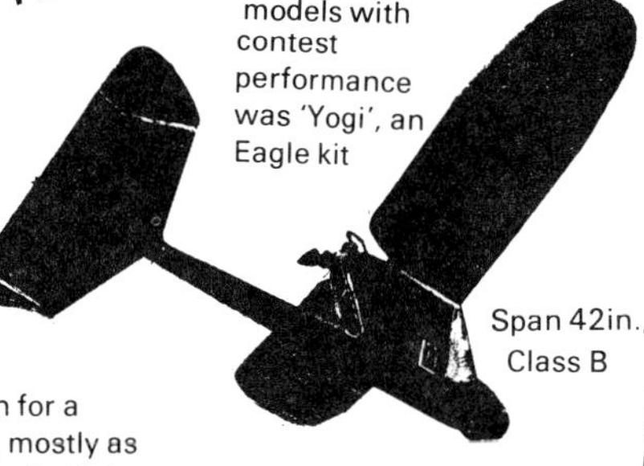

Span 42in., Class B

Earl Stahl, perhaps best-known for a stream of scale rubber models, mostly as full-size plans in *M.A.N.* but also in *Flying Aces* and *Air Trails*, also flew duration models. This is one example, from *M.A.N.* February 1944.

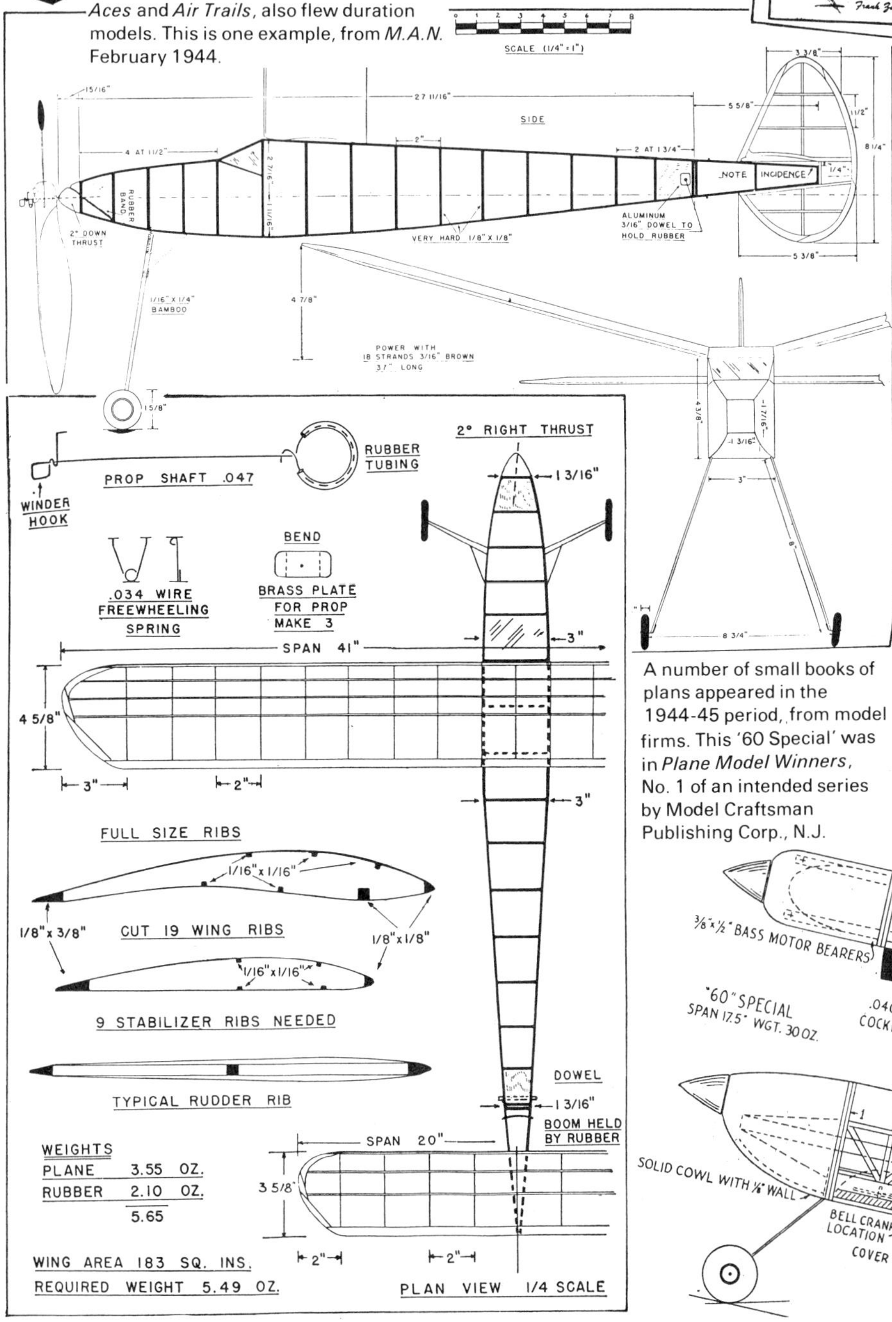

Victor Stanzel introduced a compromise C/L system by placing the handle on a stick, which was pointed up or down to climb or dive.

A number of small books of plans appeared in the 1944-45 period, from model firms. This '60 Special' was in *Plane Model Winners*, No. 1 of an intended series by Model Craftsman Publishing Corp., N.J.

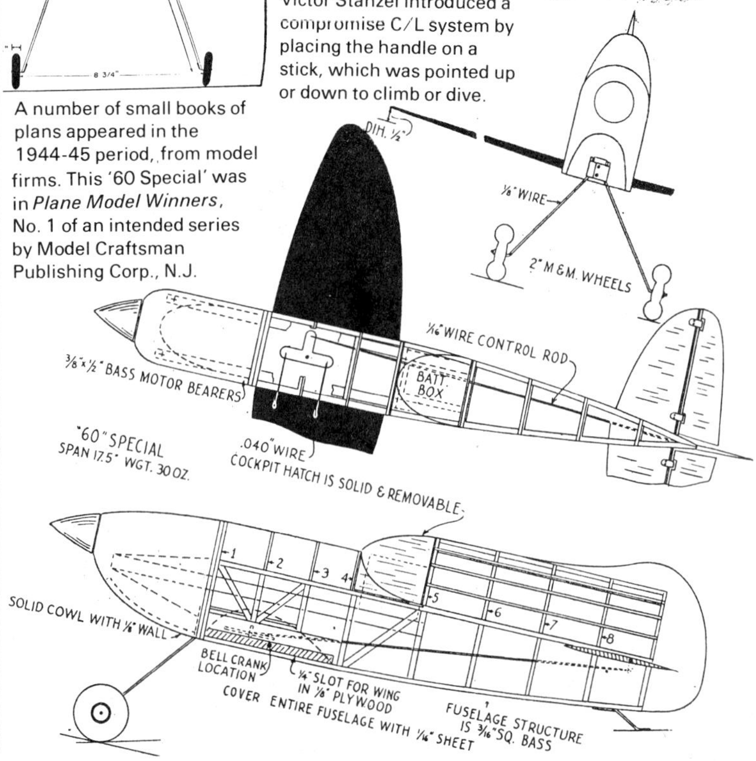

Drawings on this page are from the 1945 Swedish *Hobby- Boken.*

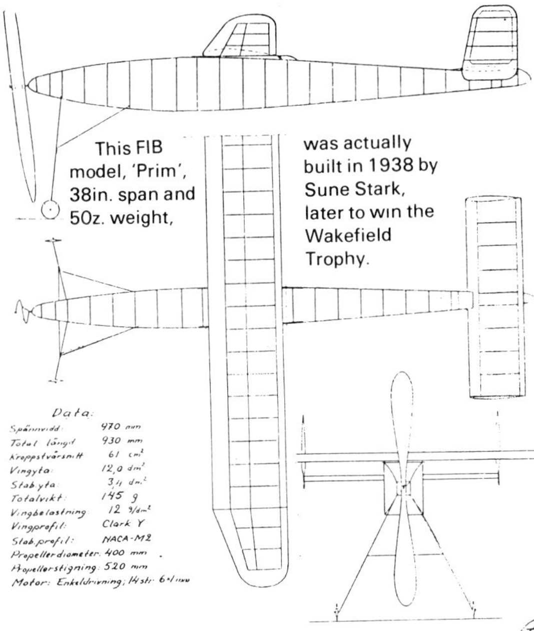

This FIB model, 'Prim', 38in. span and 5oz. weight, was actually built in 1938 by Sune Stark, later to win the Wakefield Trophy.

The ending of the war in Europe enabled modellers to catch up with what had been happening. In particular, the diesel had arrived on the scene (see later) and there had been much activity in unexpected places, especially in Scandinavia.

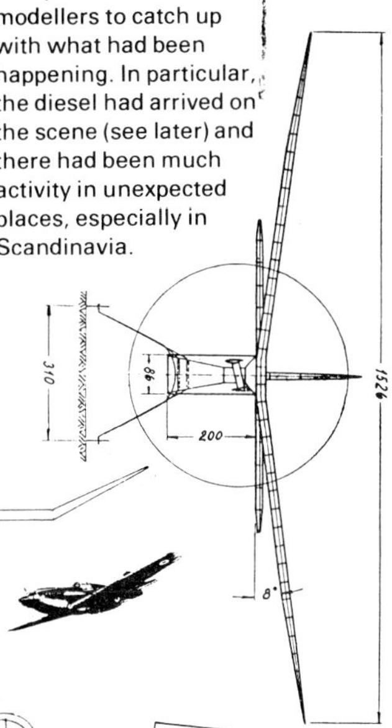

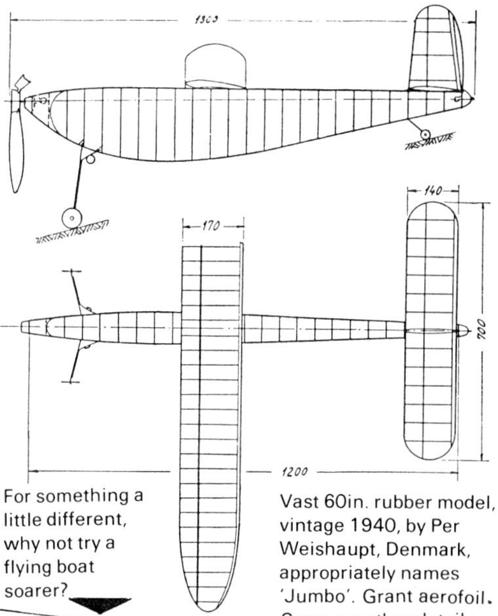

Vast 60in. rubber model, vintage 1940, by Per Weishaupt, Denmark, appropriately names 'Jumbo'. Grant aerofoil. Sorry, no other details.

For something a little different, why not try a flying boat soarer?

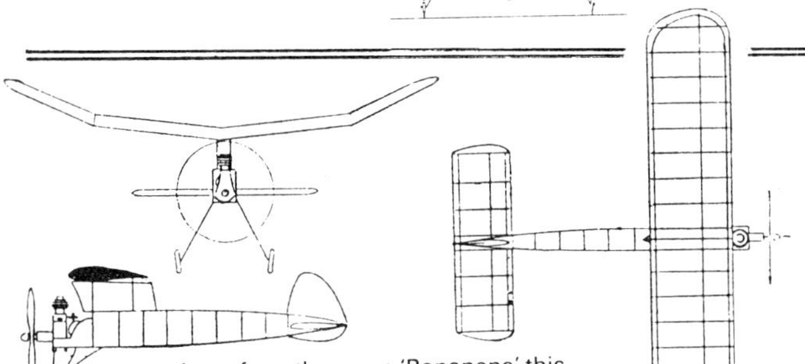

Apart from the name 'Bananens' this model has no information, but it looks like a Dyno diesel and span is therefore likely to be 1m.

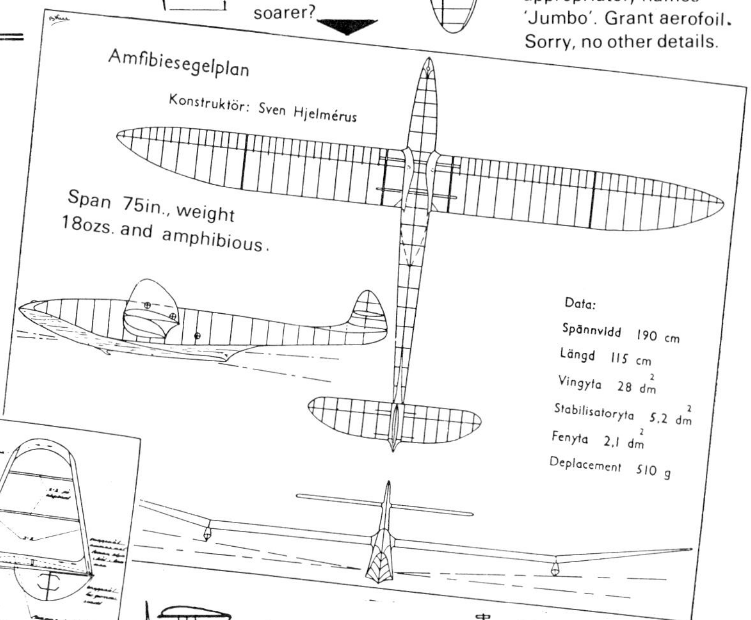

Span 75in., weight 18ozs. and amphibious.

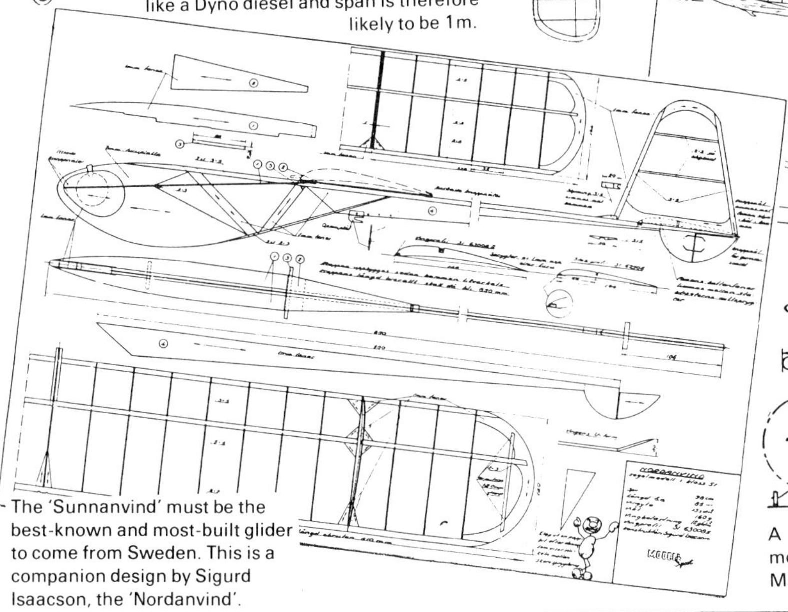

The 'Sunnanvind' must be the best-known and most-built glider to come from Sweden. This is a companion design by Sigurd Isaacson, the 'Nordanvind'.

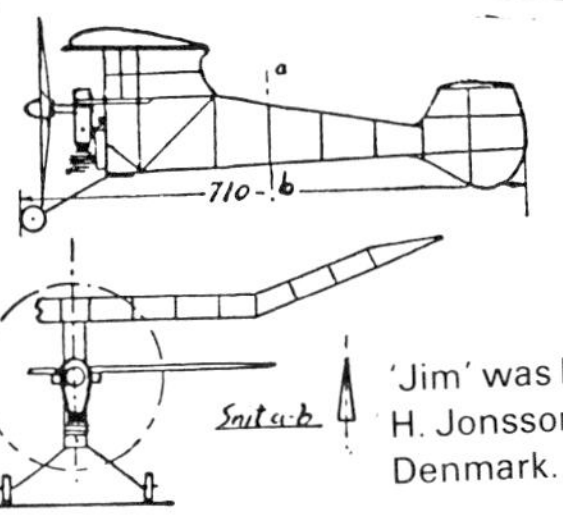

'Jim' was by H. Jonsson, Denmark.

A 1m pylon model for the Mikrodiesel,

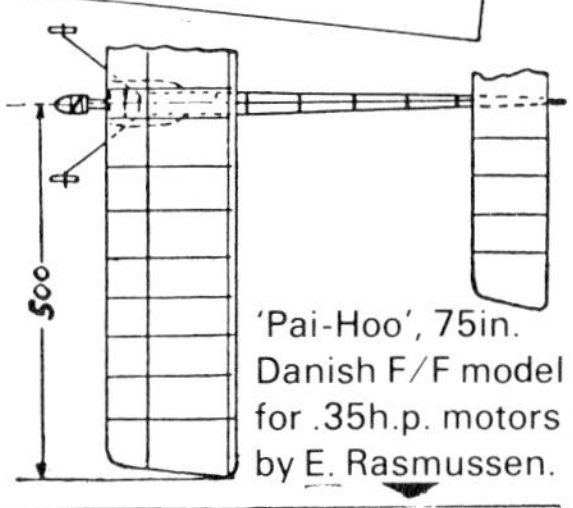

'Pai-Hoo', 75in. Danish F/F model for .35h.p. motors by E. Rasmussen.

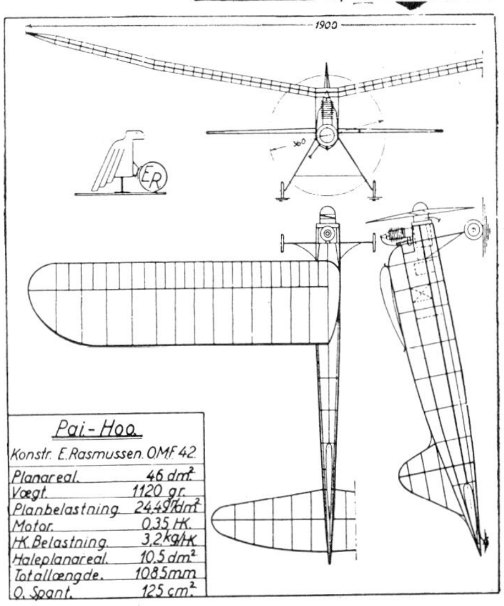

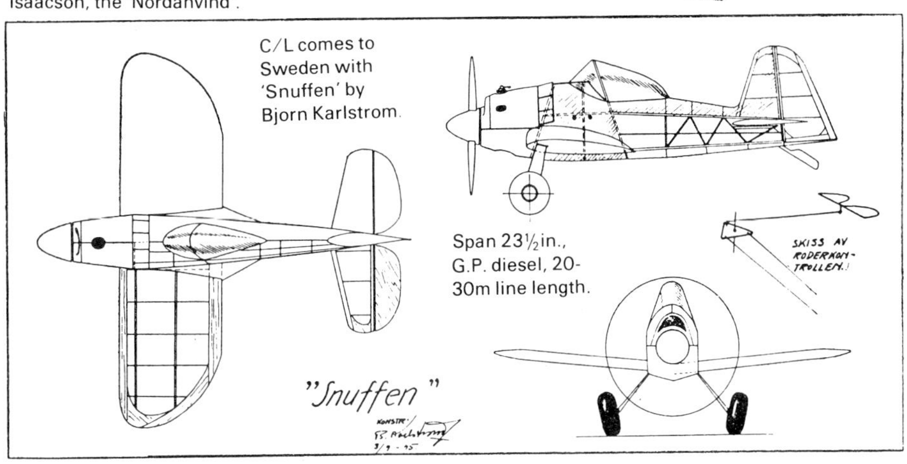

C/L comes to Sweden with 'Snuffen' by Bjorn Karlstrom.

Span 23½in., G.P. diesel, 20-30m line length.

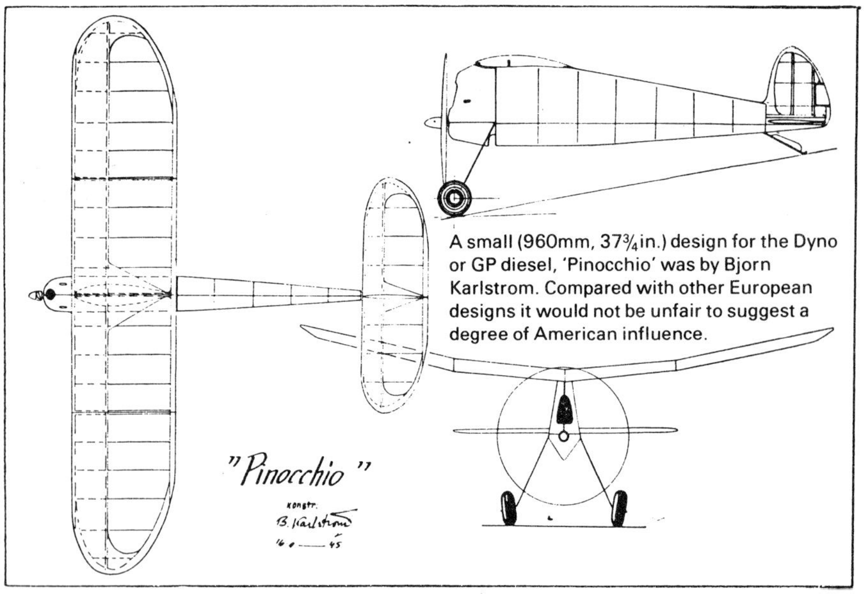

A small (960mm, 37¾in.) design for the Dyno or GP diesel, 'Pinocchio' was by Bjorn Karlstrom. Compared with other European designs it would not be unfair to suggest a degree of American influence.

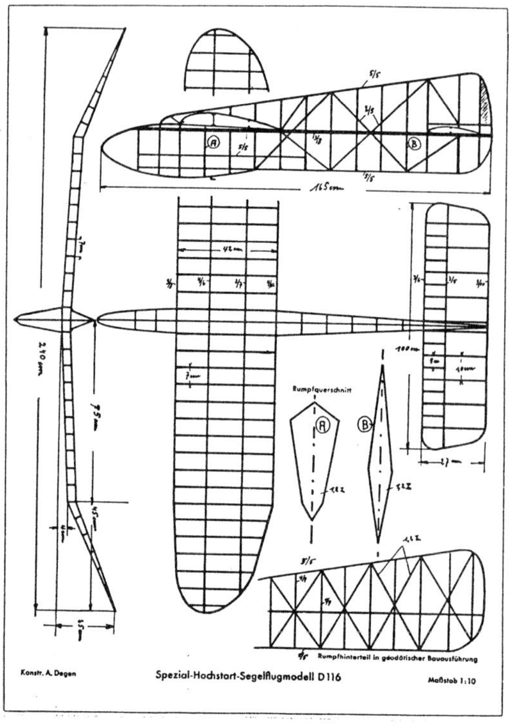

Swiss slope-soarer, almost 8ft. span, by Arnold Degen, for many years the leader of Swiss aeromodelling. Low aspect ratio and large tailplane are two factors not now used on gliders. Span 240mm, chord 420.

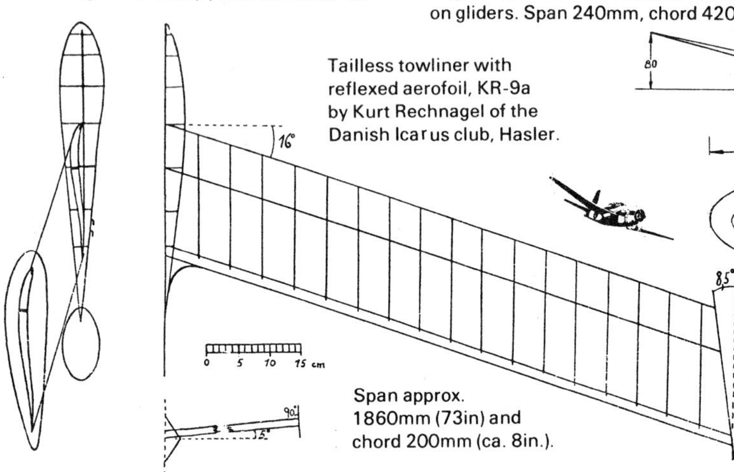

Tailless towliner with reflexed aerofoil, KR-9a by Kurt Rechnagel of the Danish Icarus club, Hasler.

Span approx. 1860mm (73in) and chord 200mm (ca. 8in.).

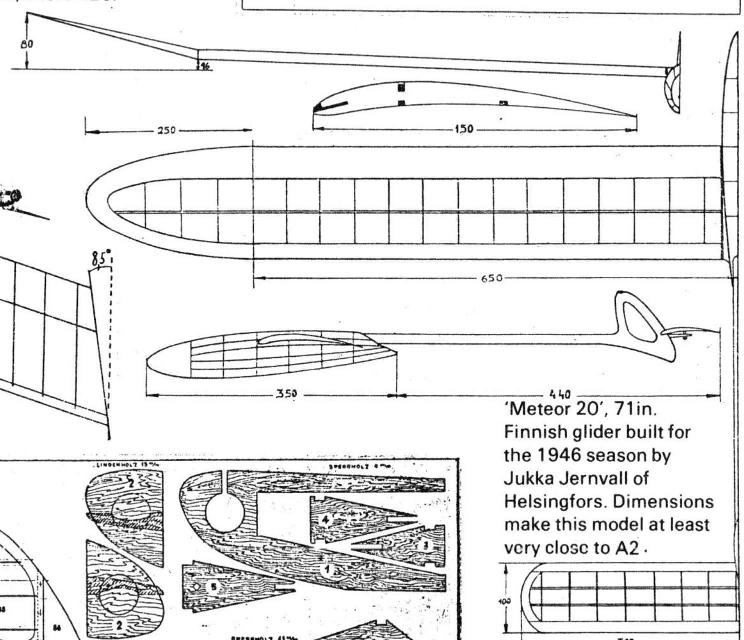

'Meteor 20', 71in. Finnish glider built for the 1946 season by Jukka Jernvall of Helsingfors. Dimensions make this model at least very close to A2.

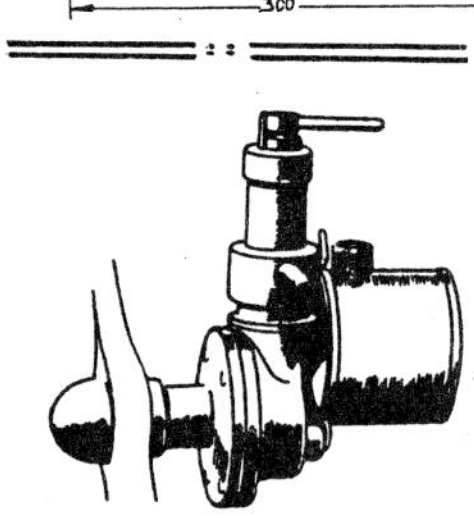

1945 was the year that the news about diesels broke. For many years it was thought that the Swiss Dyno was the first, in 1939, but it has since been shown that the Etha, above, preceded it by over two years.

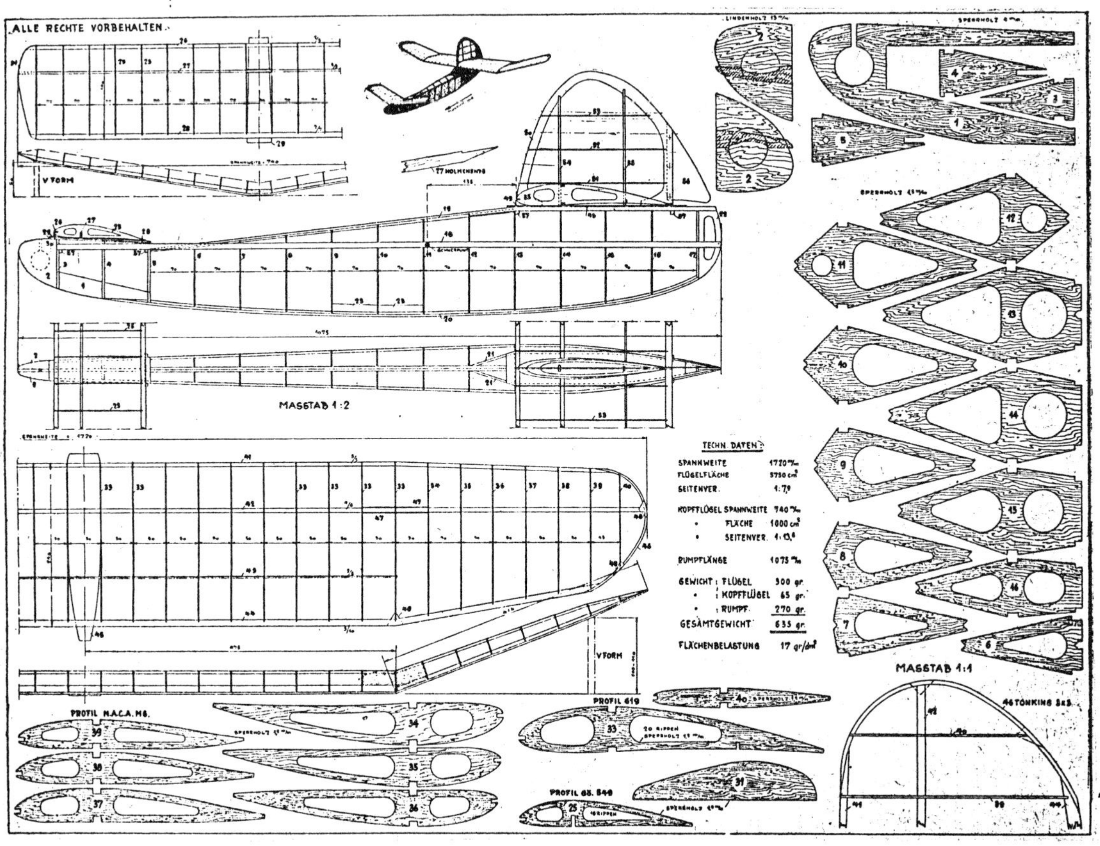

This German canard slope-soarer has a span of about 68in. and although built throughout in ply and hardwood has a comparatively modest flying weight of about 22½ozs.

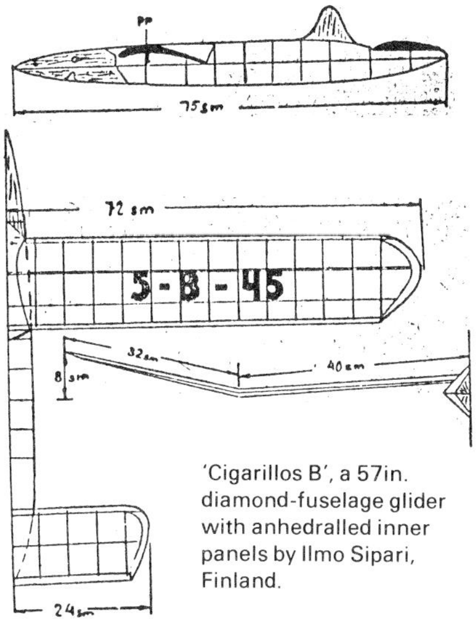

'Cigarillos B', a 57in. diamond-fuselage glider with anhedralled inner panels by Ilmo Sipari, Finland.

Russian stick model dating from 1940, 27½in. span but with a length of 37½in. Not a great deal was heard of aeromodelling activity in the Soviet Union until the 1950s.

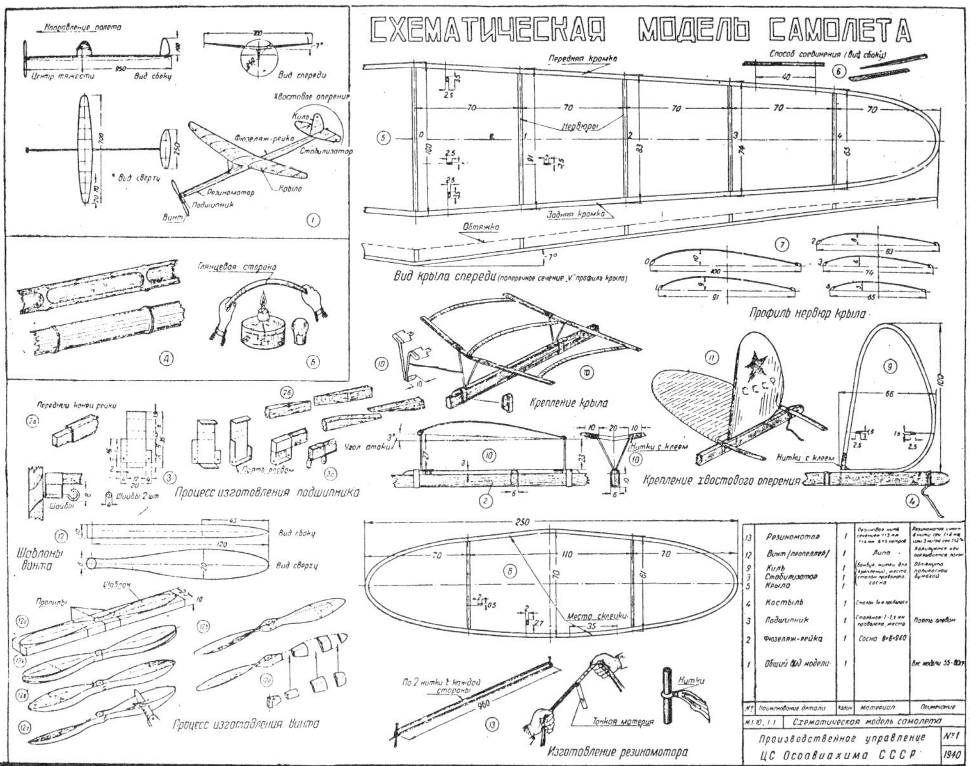

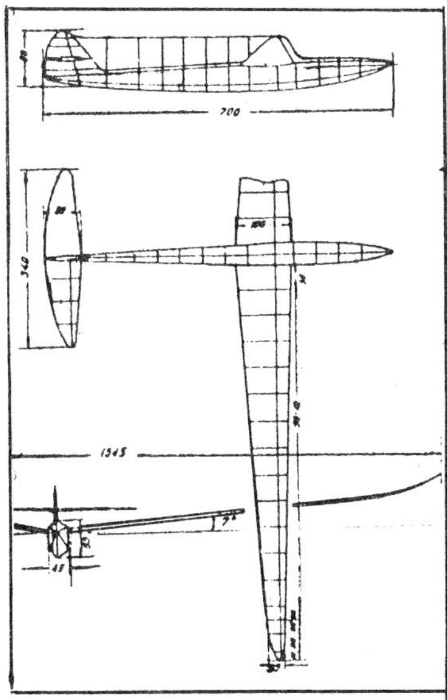

Slope-soarer by Zonov, U.S.S.R., span 61in. Note that a fin and rudder *and* a dorsal fin are both shown.

'Felix III', S3 class sailplane by Rune Andersson, probably ca. 1800mm span.

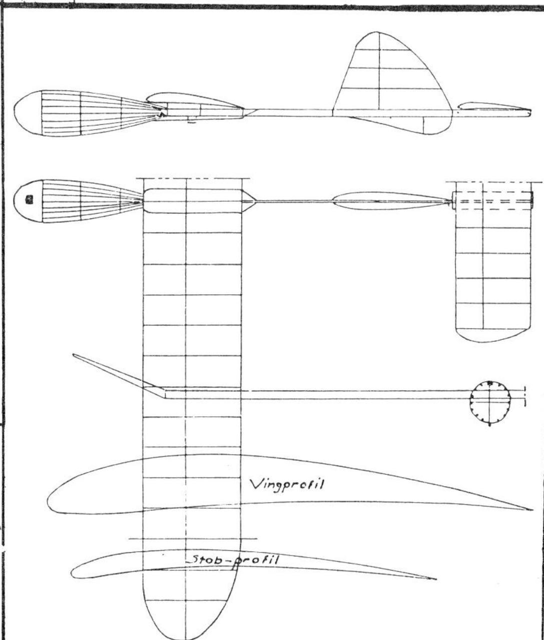

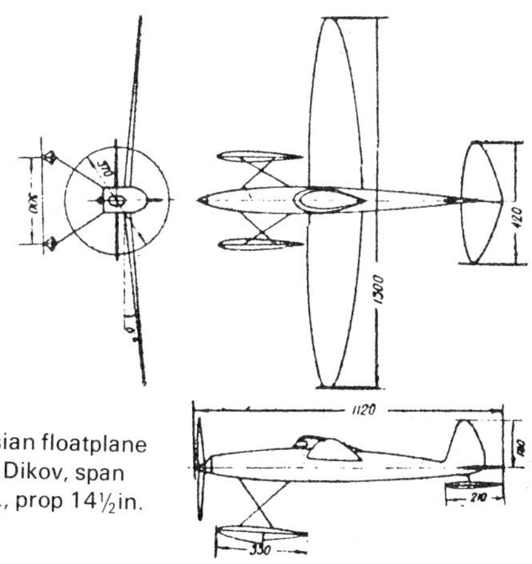

Russian floatplane by J. Dikov, span 51in., prop 14½in. dia.

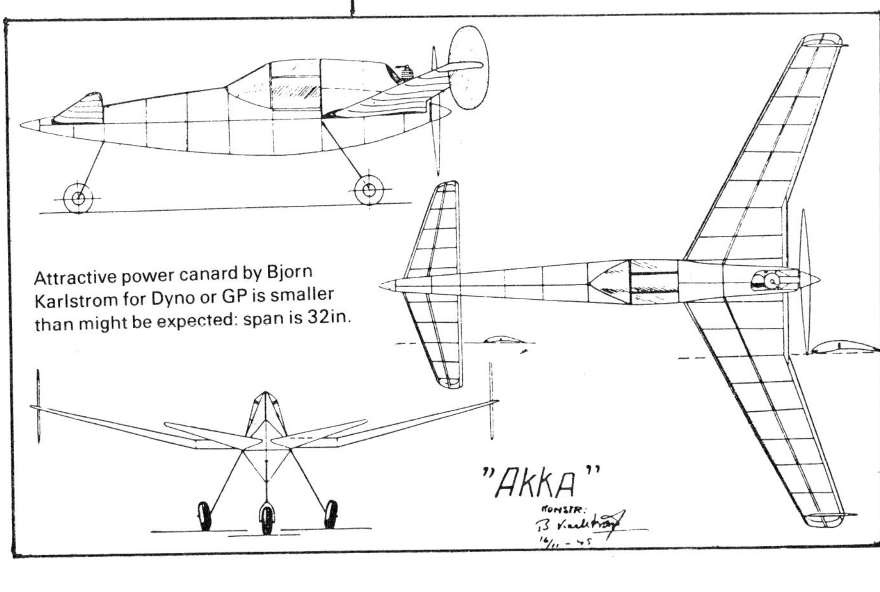

Attractive power canard by Bjorn Karlstrom for Dyno or GP is smaller than might be expected: span is 32in.

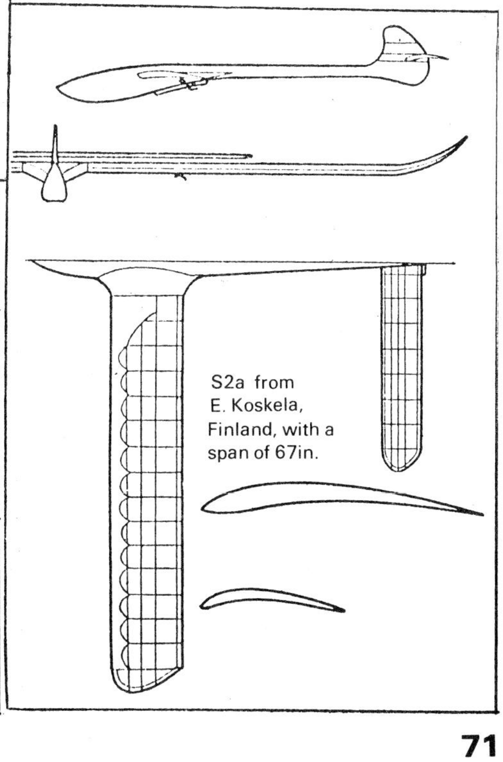

S2a from E. Koskela, Finland, with a span of 67in.

1945-46

Spännvidd 165 cm
Längd 87 cm
Vingyta 25,5 dm²
Stab.-yta 6,4 dm²
Flygvikt 525 g

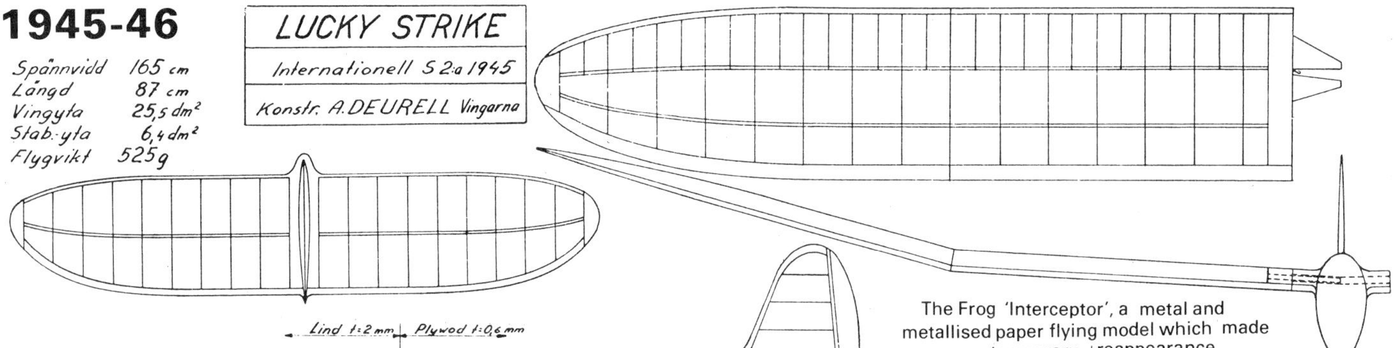

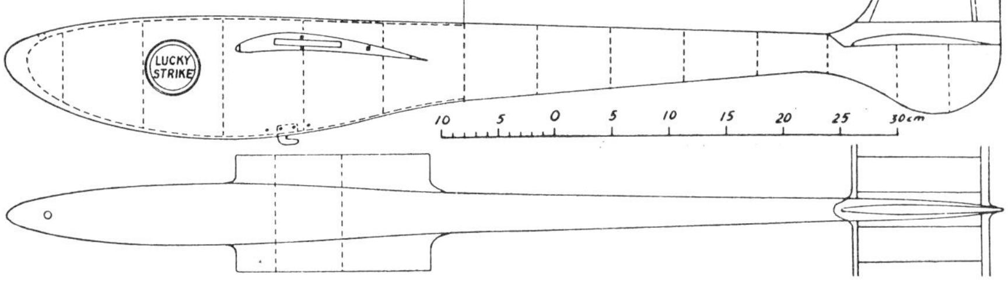

The Frog 'Interceptor', a metal and metallised paper flying model which made a temporary reappearance just after the war.

Possibly one of the earliest published A2s, 'Lucky Strike' appeared in the 1946 Hobby-Boken.

Petrol engine by Laurie Sparey looked like an opposed twin but second cylinder was just the tank.

Eaton Bray, centred on a wartime emergency landing field, was a brave attempt at creating a permanent site for model flying.

November, 1945 AEROMODELLER

OPENING DAY AT EATON BRAY

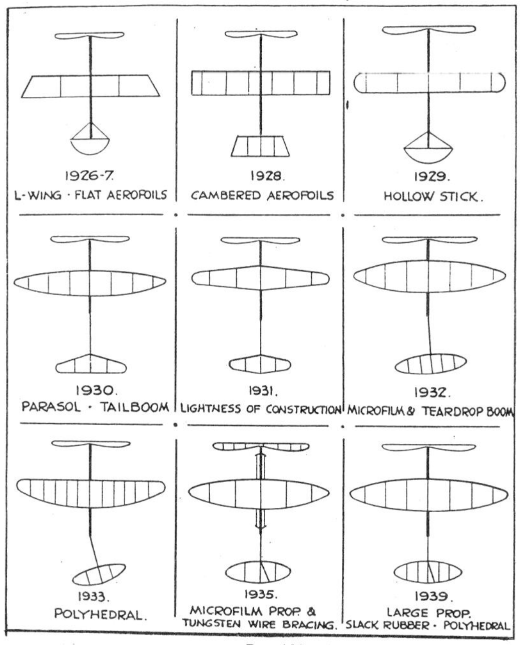

Ron Warring's book *Indoor Flying Models* included this illustration of development of F/F rubber models.

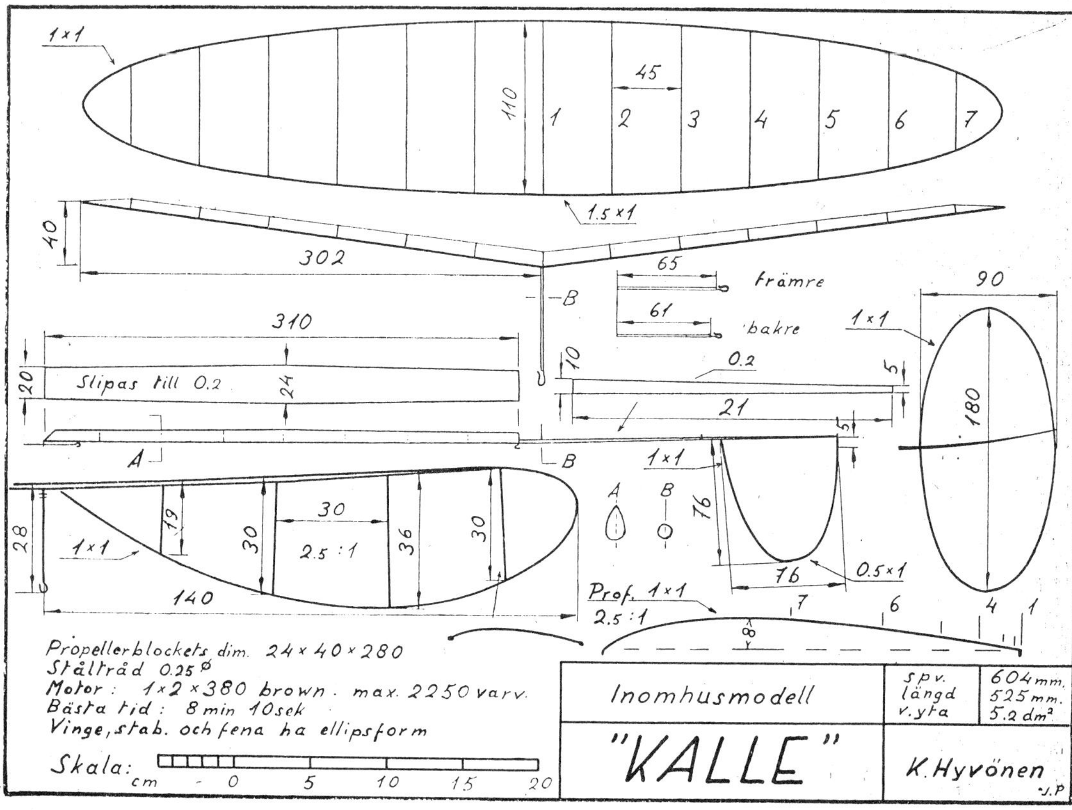

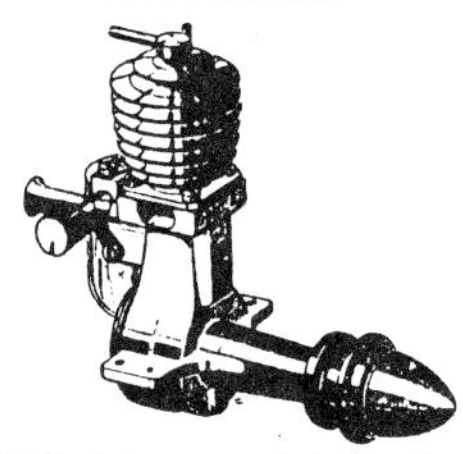

The first commercial diesel seen in quantity was probably the Italian Movo D2, 12 × 17.4mm b. and s., 2cc.

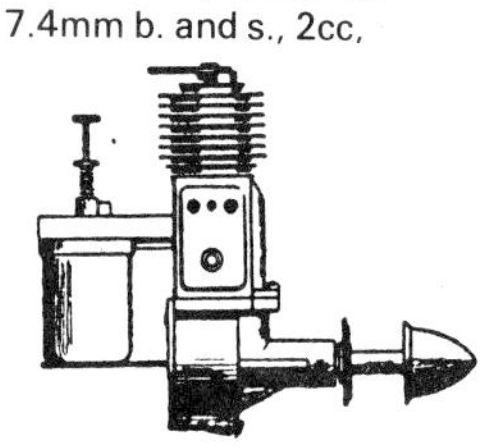

The French were also very early with production diesels. Delmo 2.64cc shown.

Scandinavian microfilm model of 1945 with a best flight of 8:10 but probably capable of longer by experimenting.

1945-46

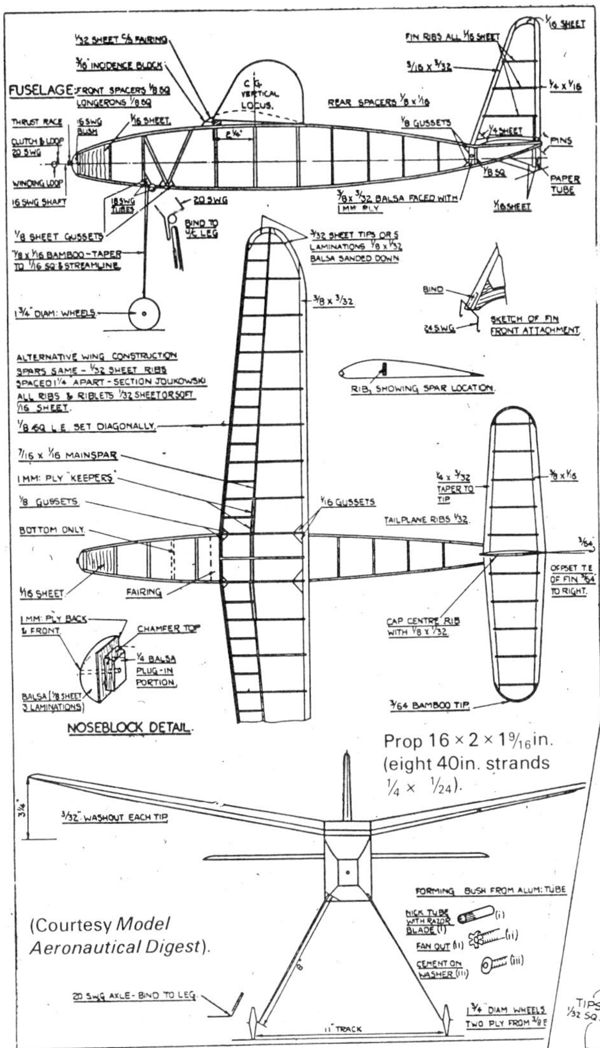

Prop 16 × 2 × $1\frac{9}{16}$in. (eight 40in. strands $\frac{1}{4}$ × $\frac{1}{24}$).

(Courtesy *Model Aeronautical Digest*).

G.W.W. Harris model jet engine. True jet power for models has not yet become commercial, apart from impulse reaction ducts (Dyna-jet etc)

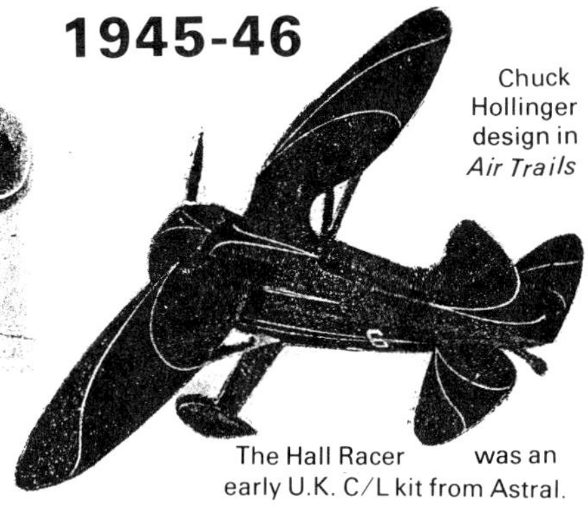

Chuck Hollinger design in *Air Trails*

The Hall Racer was an early U.K. C/L kit from Astral.

Diesel development appears to have been shared by Switzerland, Italy, France, Austria and Germany.

Aeromodeller, December 1945

R.T.P. models were flown in Britain quite a lot during and after the war.

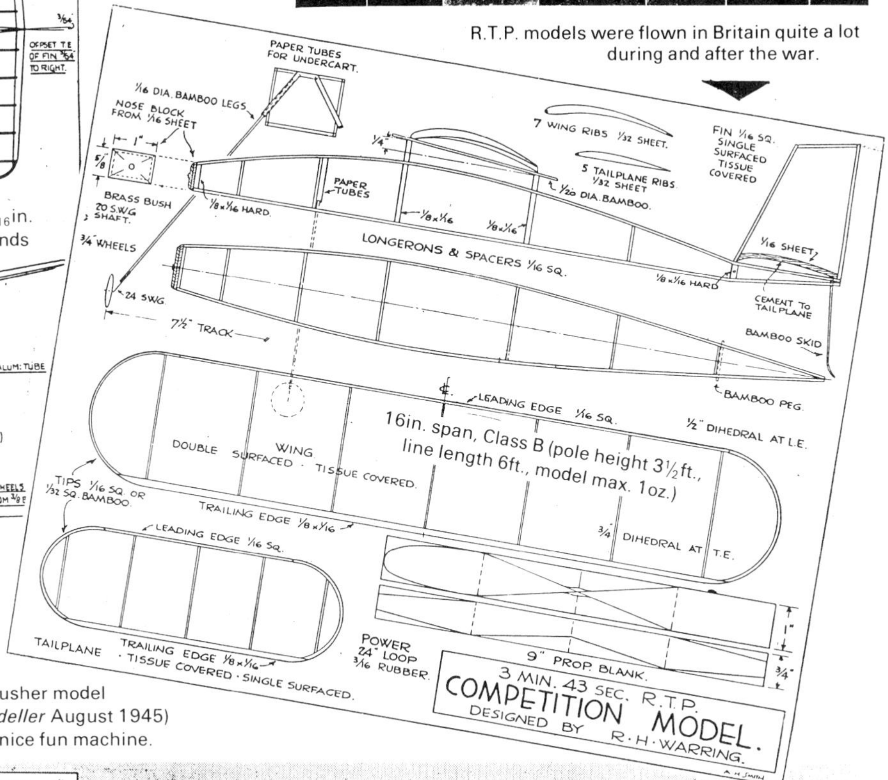

Typical of wartime rubber models is this Warring Lightweight, $36\frac{1}{4}$in. span, 144sq. ins., just over 4oz. flying weight, average flight $2\frac{1}{4}$min.

Early French midget .16cc diesel (*A.M.* Feb. '46)

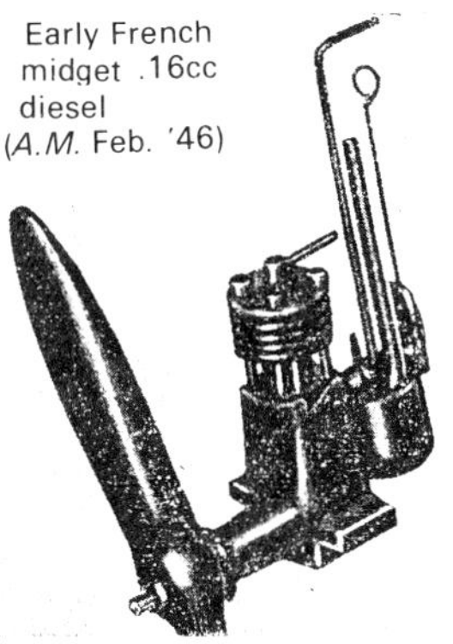

Simple pusher model (*Aeromodeller* August 1945) makes a nice fun machine.

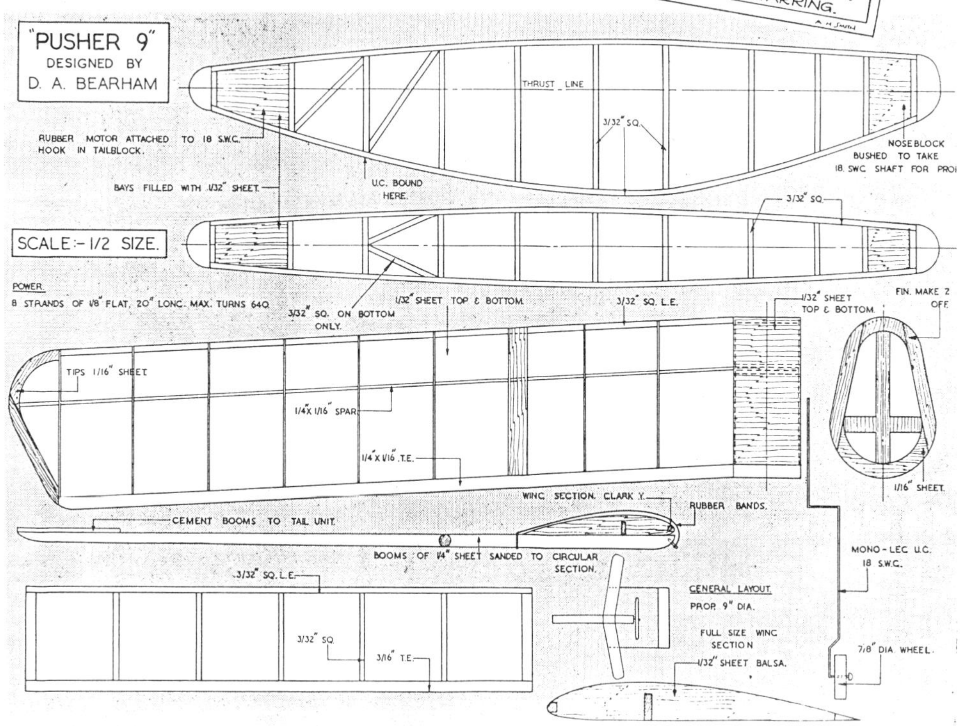

First British diesel on the market (August 1946) and still popular.

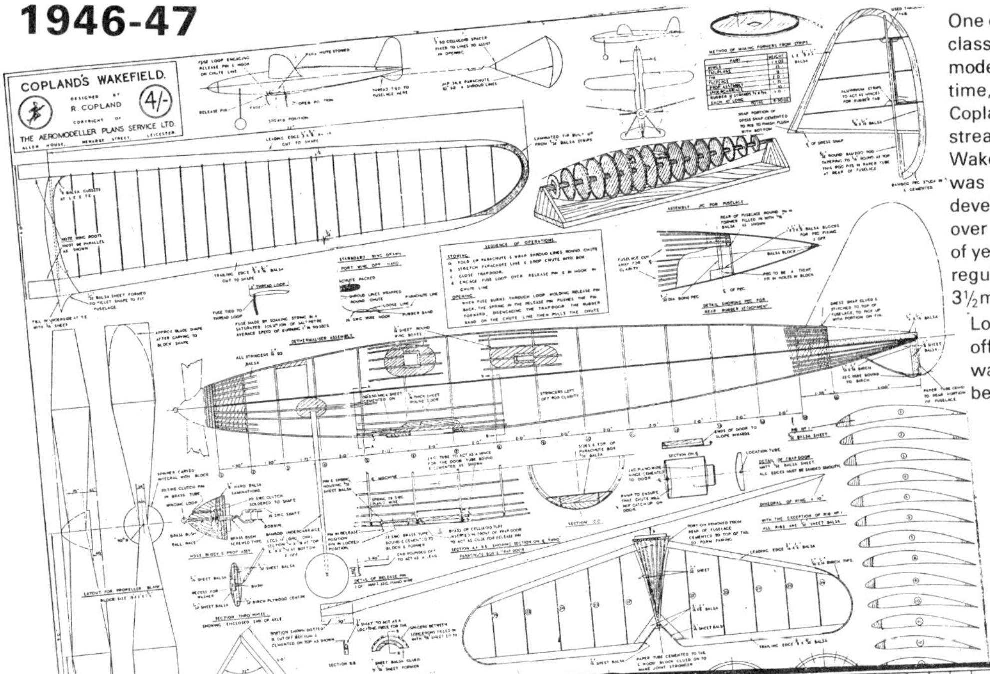

One of the classic models of all time, Bob Copland's streamlined Wakefield was developed over a period of years to a regular 3½min. plus. Longest official flight was a world's best 33:00.

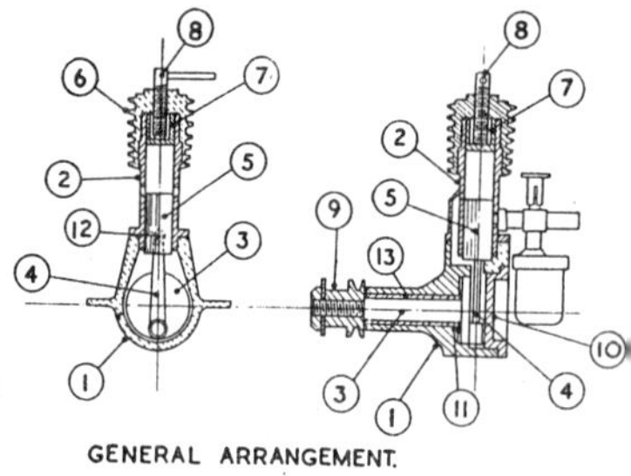

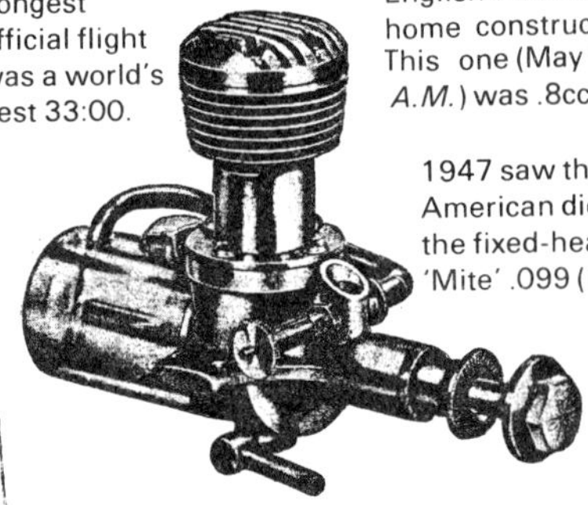

Sparey developed English diesels for home construction. This one (May 1947 *A.M.*) was .8cc

1947 saw the first American diesel, the fixed-head 'Mite' .099 (1.6cc)

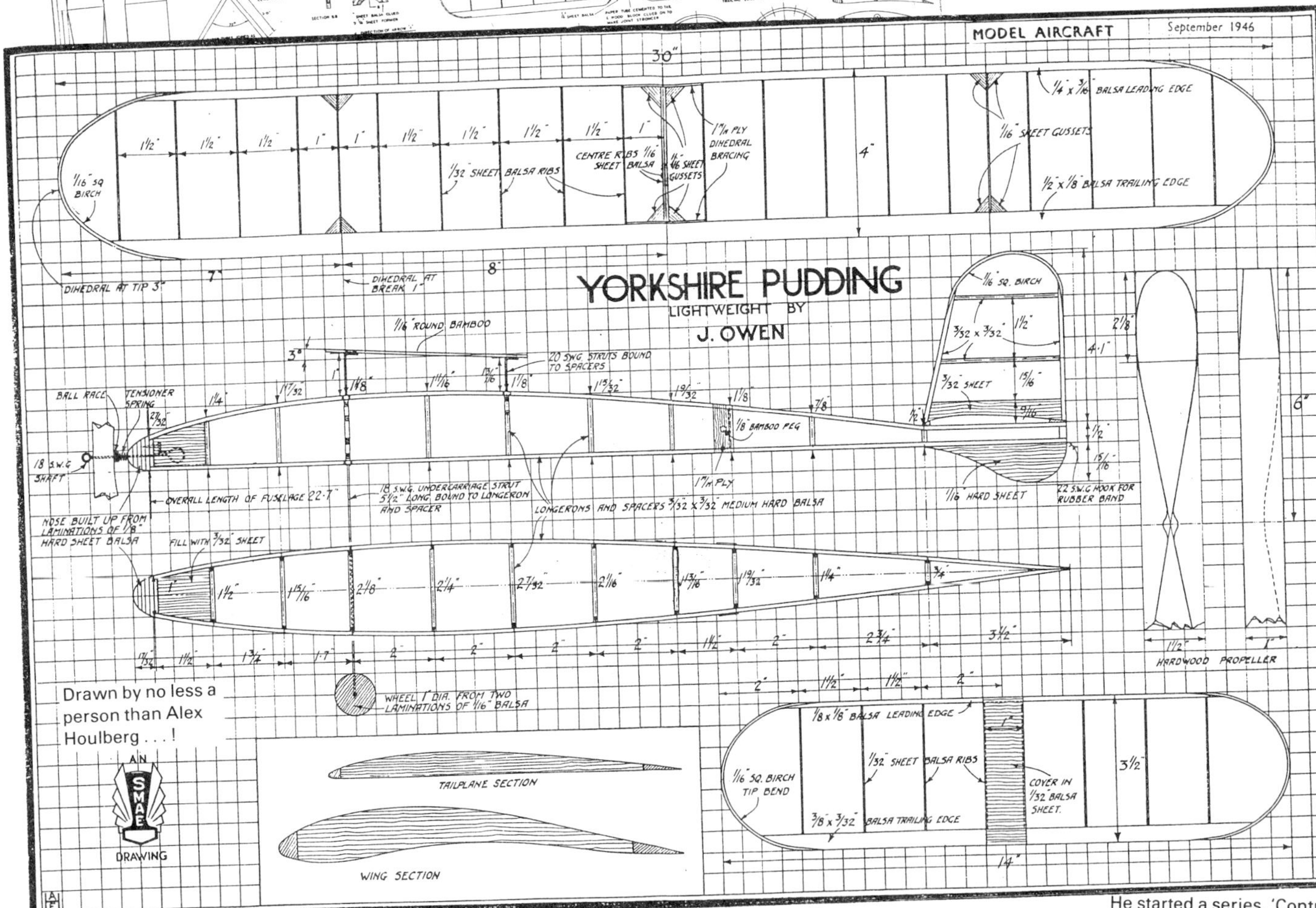

Drawn by no less a person than Alex Houlberg . . . !

MY first control-line model, constructed about 1925, was a rubber-driven round-the-pole model with a controllable elevator. This, I am afraid, was never very successful. As time went on, however, I met with greater success and I feel that readers may be interested in a brief outline of some of my most interesting models. The first success was in 1927, with a compressed air model. This was fitted with a 3-cylinder radial air motor and a Bowden wire-controlled elevator. The compressed air was fed to the model from a rotating gas bracket mounted on the central pylon round which the model revolved. The model flew and controlled quite well, but unfortunately the engine failed to stand up to the constant running and soon wore out. Success number two was finished in 1929 and gave excellent results despite its very unorthodox arrangement. The model was flown round the pole, the central pylon having at its base a ⅛ h.p. Stewart-Turner water-cooled two-stroke petrol engine. The drive from this was a speedometer cable to the rotating top of the central pylon. A flexible line led from the pylon to the model, taking the form of a length of piano wire, 15 feet long, running in a ⅛ in. diameter aluminium tube. This drove the airscrew through Meccano bevel gears, and worked remarkably well—effectively getting round the difficulty of installing so heavy a motor in the actual aircraft. At that time there were no miniature lightweight aircraft engines to be had. The flexible drive gave little trouble and one wire would last for about five hours running time. The elevator control was again by Bowden cable, with a spring return. The throttle and elevator leads from the base of the pylon ran along the ground to the dual control "cockpit" situated just outside the circuit. I still possess a 100 ft. 16 m.m. cine-film of this model in action.

In 1932 I finished a much larger model fitted with a 16 c.c. "Atom-Minor" engine made up from castings. This model had a wing span of 6 feet and weighed 4½ lbs. The low tension supply was led out to the model from the centre pylon, which obviated the necessity for carrying a battery in the plane. This model flew very well and controlled nicely. She did many hours flying before the engine finally wore out, and never once sustained a crash. The wing was only a single spar affair, the leading edge being covered with 1 m.m. ply to take the torsion loads.

There's nothing new! F. B. Thomas had been flying tethered and control-line models since the 1920s and had plenty of evidence

He started a series, 'Control Line Commentary' in May 1947 *Aeromodeller* and his 'Candy II' 38in. trainer appeared in September '47.

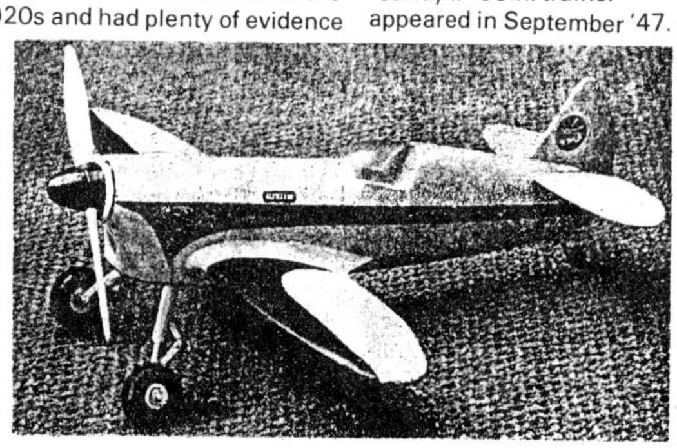

1947

Interest in scale began to grow, though it was eventually radio control which was to boost scale to the most popular category. This cutaway 'Tiger Moth' was by Laurie Bagley

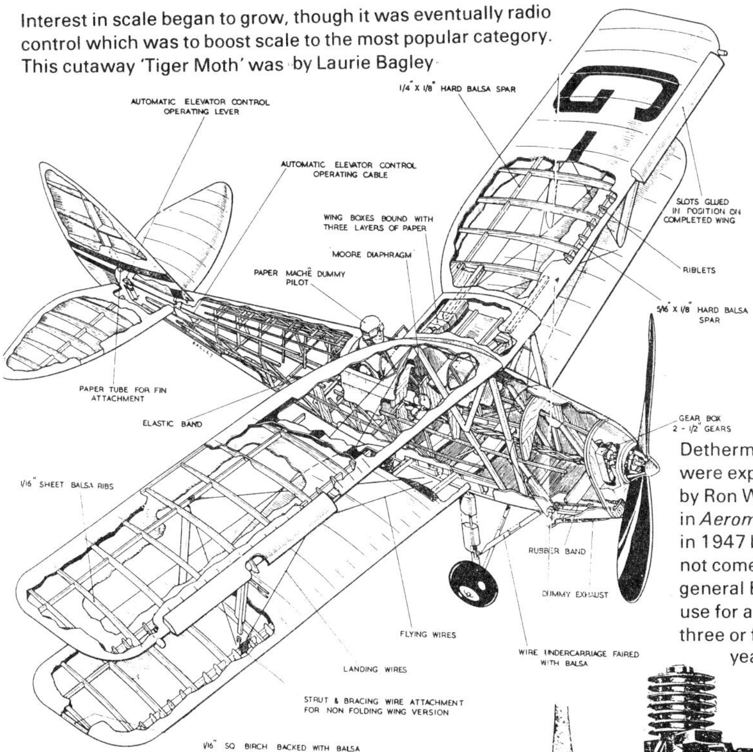

Dethermalisers were explained by Ron Warring in *Aeromodeller* in 1947 but did not come into general British use for another three or four years.

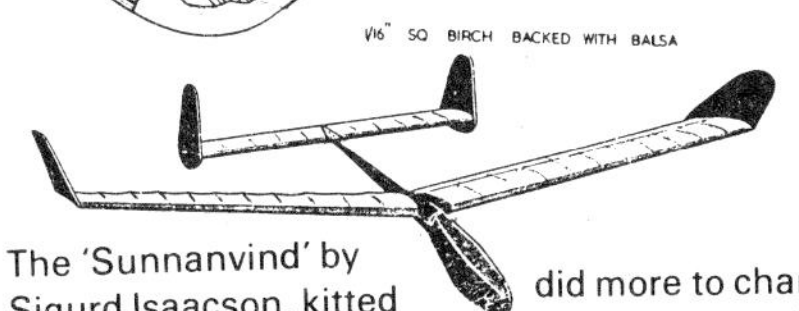

The 'Sunnanvind' by Sigurd Isaacson, kitted by Paramount, probably did more to change attitudes to gliders than any other model.

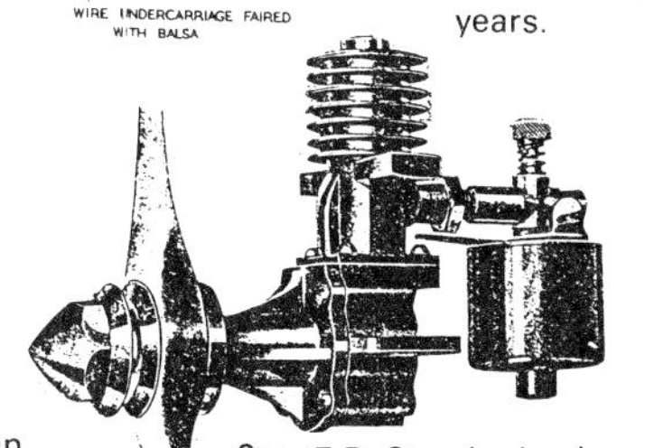

2cc E.D. Standard; coin-slot adjusted compression

The first of the celebrated weeks at Eaton Bray was in 1947.

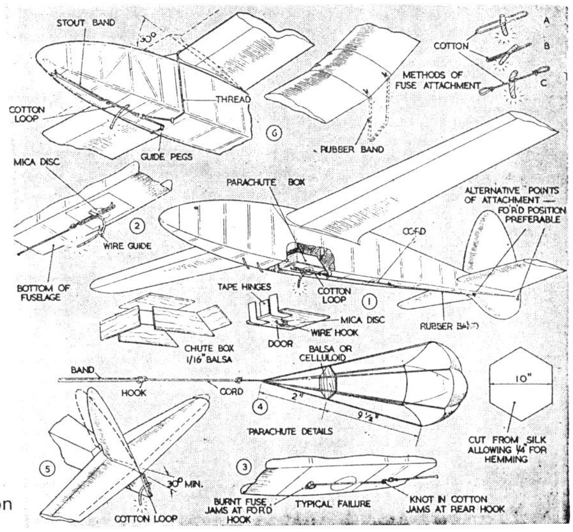

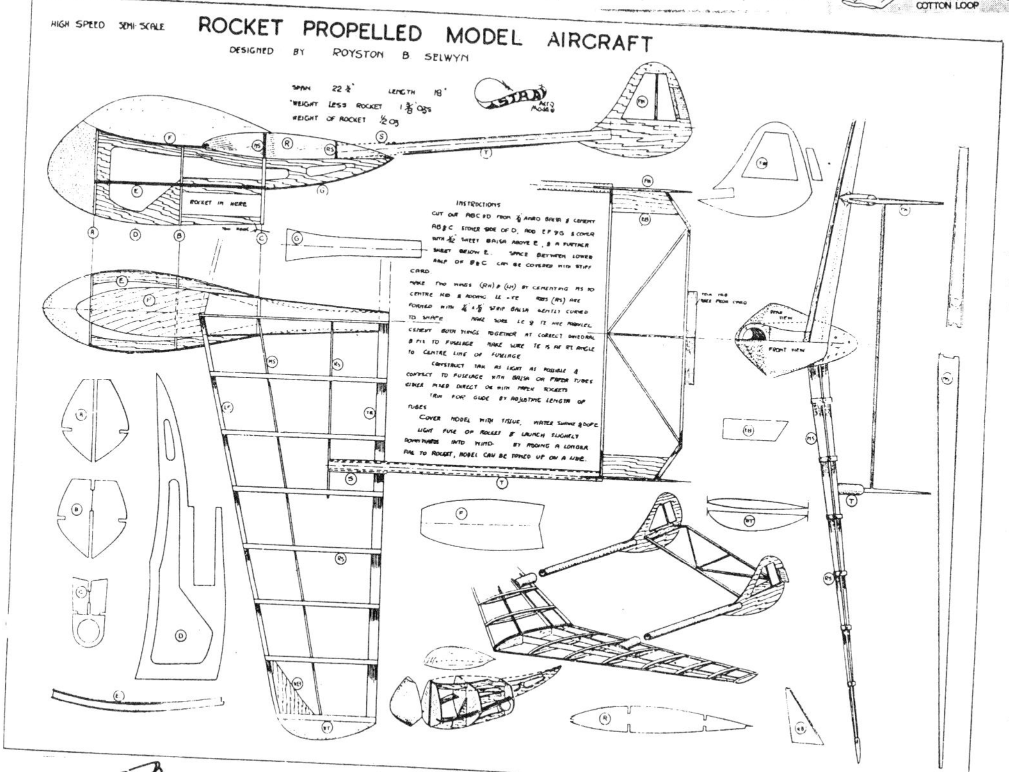

Astral kitted this rocket model in early 1947, with rockets made by one of the famous firework firms. However, it presumably contravened British Home Office regulations — to this day rocket models are not allowed — since it remained on the market for only a few weeks.

While England and Europe swung to the diesel — although a few spark ignition engines were made — in the U.S.A. spark ruled, until in mid-1947 Ray Arden introduced the glowplug, his development of the old hot bulb/hot wire principle. A list of motors advertised in a random 1947 issue of *M.A.N.* is: Aero diesel, Air-O Mighty Midget, Anderson Spitfire, Arden .099 and .199, Atwood .60, Triumph .49/.51and Glo-Devil, Ball .60, Bungay .60, Burgess M5, Buzz .19, .29, .35, .60 and CO_2, Cameron .23, Campus A100 and Bee (CO_2), Cannon 300 and 358, Circle 30, Contestor .60, Deezil diesel, Delong 30, Dennymite, Dooling .61, Drone diesel, Forster .29, .305 and .99, Fox .59, Genie .29, G.H.Q., Glo-Champ .29, Glo-Mite .099, H and H 45, Hassad Sky Devil, Hornet .60, Hurleman, Hurricane, Judco, K&B Torpedo .24, .29, and Glo-Torp, Kes, McCoy .29, .49, .60, Merlin .23, Mite .099 diesel, Mohawk Chief .29, CO_2, Bantam .199, Orwick .64, Pacemaker .59, Phantom P.30, Pierce J and R, Ram .29, Rocket, Rogers .29, Sportsman Junior .36 and Senior .55, Super-Champion JH, Super-Cyclone, Thor B, Vivell .35, .45 and Twin, and Wensen. Plus Dynajet and Minijet, and others not mentioned in that particular issue!

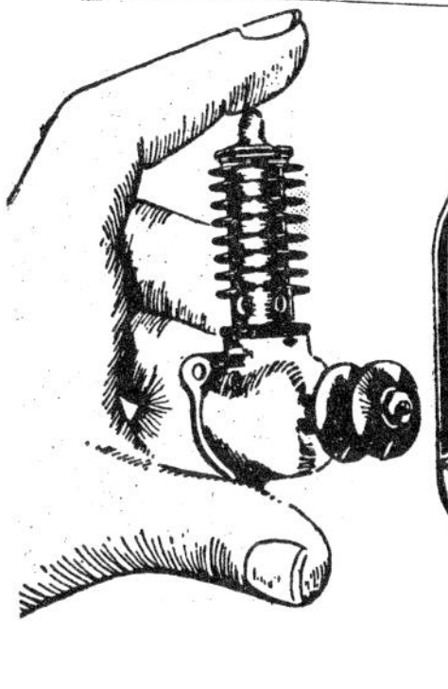

CO_2 SPECIFICATIONS

DISPL.....0178	WEIGHT (Bare engine).....3/4 ozs.
BORE 275	CARTRIDGE (filled).........1 1/4 ozs.
STROKE...300	HOLDER...........................1/2 ozs.
R. P. M. 7,000	WGHT. (Engine accessories) 2 1/2 ozs.

12-22in. span models!

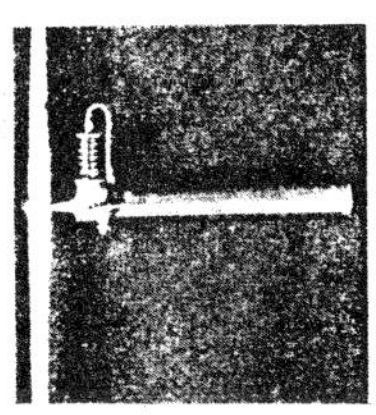

Spring 1947 and Herkimers introduce the O.K. CO_2 motor, claiming 'the first, the only ignitionless expansion type reciprocating engine' and applying for a patent with 14 claims. In December 1947 Bill Brown launched the Campus A100 (right).

Ignitionless, CO_2 operated—weighs 1/4 ounce with propeller and tank—1/8" stroke—1/8" bore—8000 RPM—precision made to tolerances of 1/10,000 of an inch—comes complete with CO_2 charger—No accessories required—8 or more flights per CO_2 cartridge.

Stanzel thumb-controlled single-line system

May 1947

1947

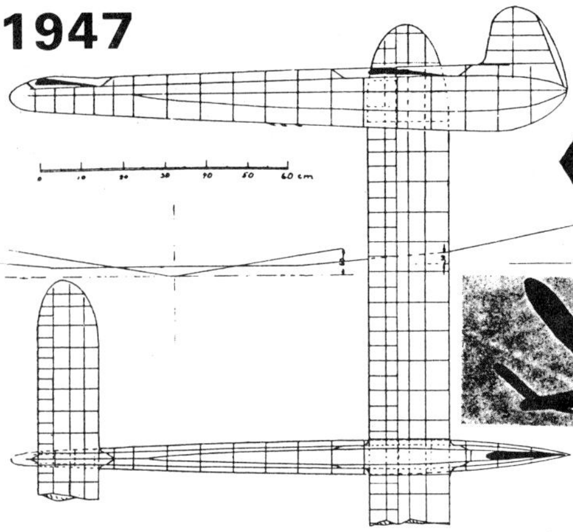

'Kalle-Anka' from Oscar Varg, Denmark. Canard glider, span 2040mm (80in.), 28ozs. both sections 63009.

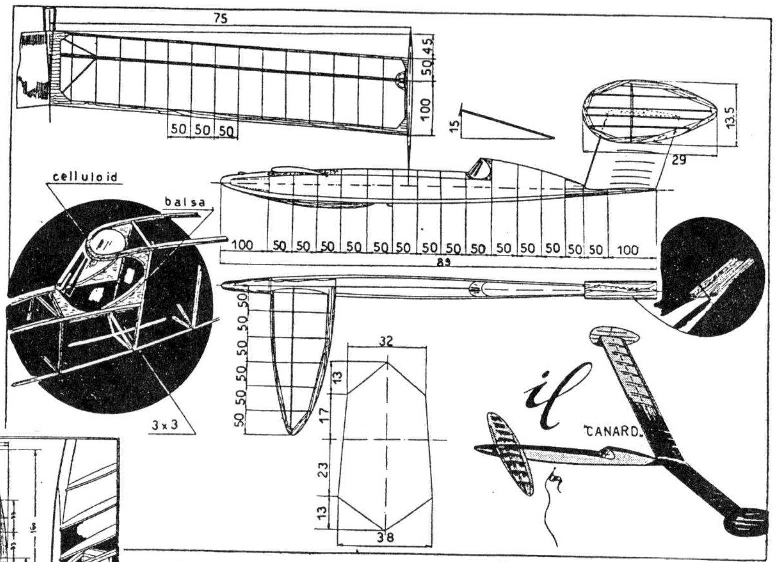

Another canard sailplane, this one from Italy, RA17 by Rappini Adolfo. Span 1500mm (59in.), section Eiffel 400.

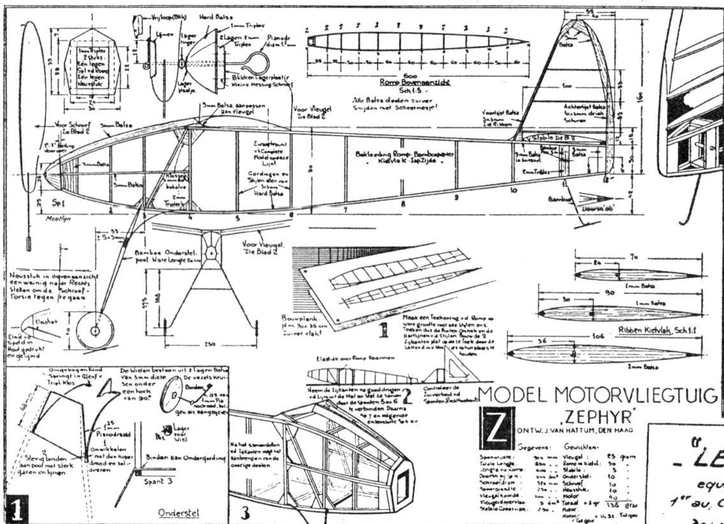

Dutch rubber model 914 × 100mm wing (36 × 3½in.), 4¾ozs. by Juste van Hattum. Tailplane 30p.c. (approx. 13 × 3in.), prop 12½in. dia.

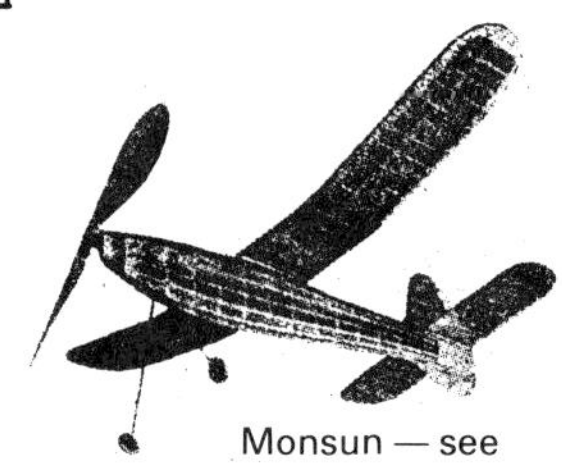

Monsun — see bottom of page

French power model, 79in. span, 42oz., Micron 5cc diesel, 5:30 o.o.s. on 30sec. motor run in contest at Toulouse 7/10/45. By R. Froment.

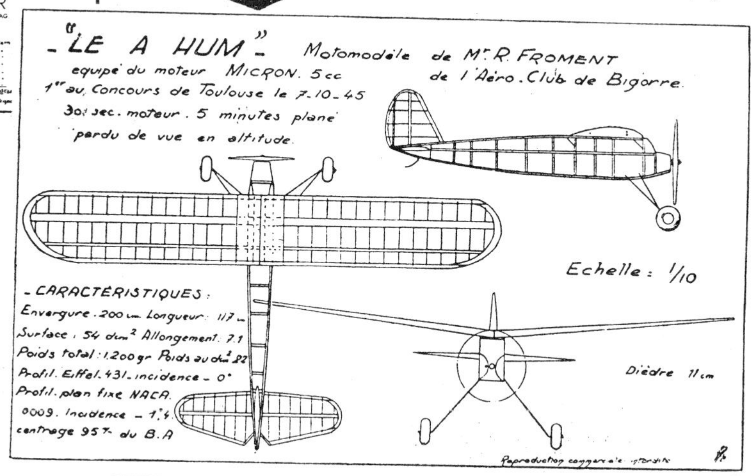

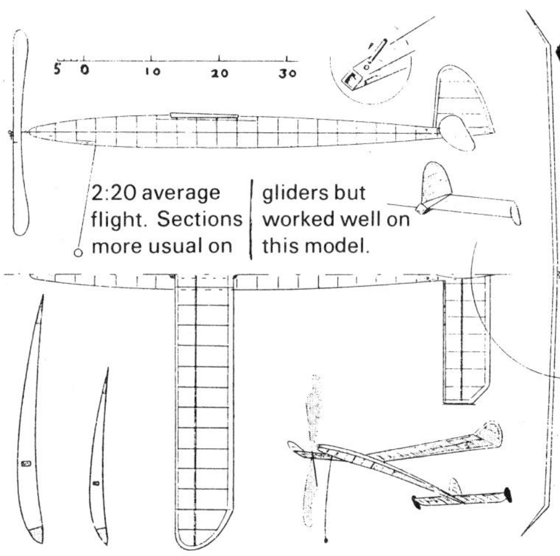

Swedish Class GI rubber model by Lowen. Span 800mm (31½in.), weight 2½ozs., 12in. prop. 2:20 average flight. Sections more usual on gliders but worked well on this model.

Norwegian pylon model by Finn Wahlquist, span 1300mm (51in.) for David-Anderson 3½cc diesel.

Swedish geared Wakefield 'Monsun' by Anders Deurell, built in 1946. Still air performance 3-4mins. An elegant model (see photo above)

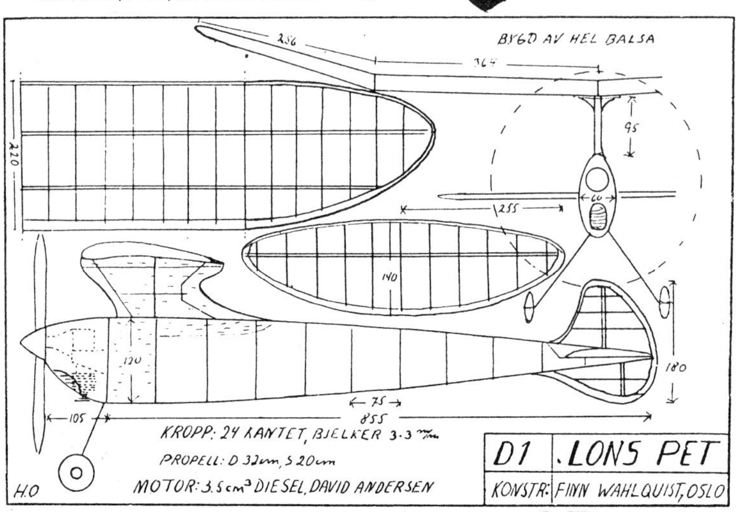

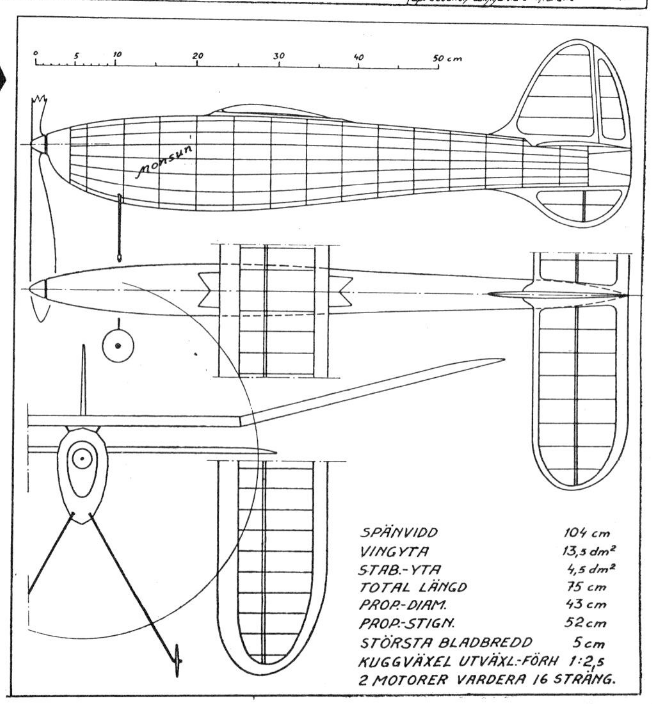

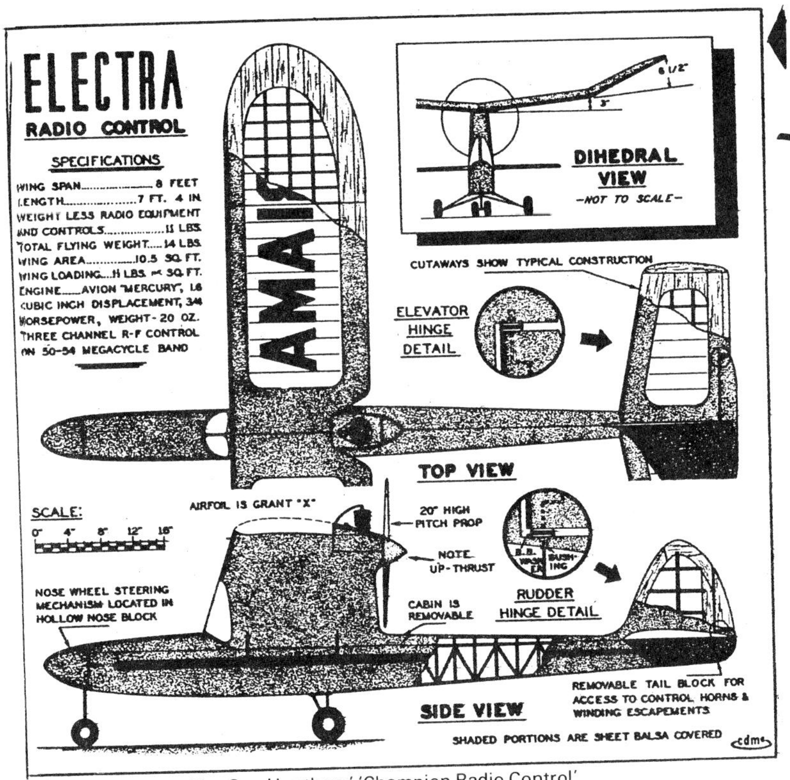

◀ Claud McCullough's design basis for an R/C model was one which seemed to occur to several F/F modellers in the early R/C days.

'Viking 65'

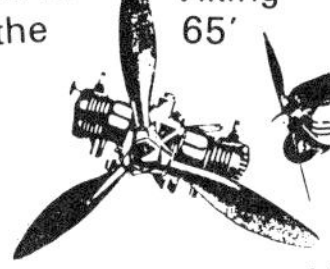

Edco introduced a C/L autogiro in '47, rotor dia. 26ins.

all-metal fuselage

Although it looks like a F/F design, the 45in. prefabbed kit 'Casalaire' was for C/L

The Good brothers' 'Champion Radio Control' receiver, etc. weighed 16ozs. complete; presumably that included batteries.

Ready For Installation . . .

This radio control equipment comes to you completely wired and ready for installation. The receiver is factory fixed-tuned to eliminate difficulties in tuning up for flights. The Good Brothers Radio Control units have been designed around the same basic equipment which won three consecutive National Radio Control Contests.

THRILLING Controlled Flight and You're the Pilot

With "Champion Radio Control" you are the pilot. You have full control of take-offs, turns, spirals, loops and landings in a range of up to two miles. Here is the equipment for all the thrills of model flying that only radio control can give you.

PRICE $59.50 less batteries

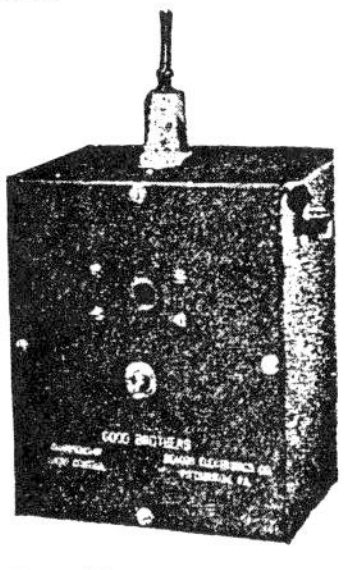

Transmitter Size—Approx. 4" x 5" x 6"
Range—Approx. 2 miles

Flinn's MODEL HOBBIES
909 LIBERTY AVE. PITTSBURGH 22, PA.

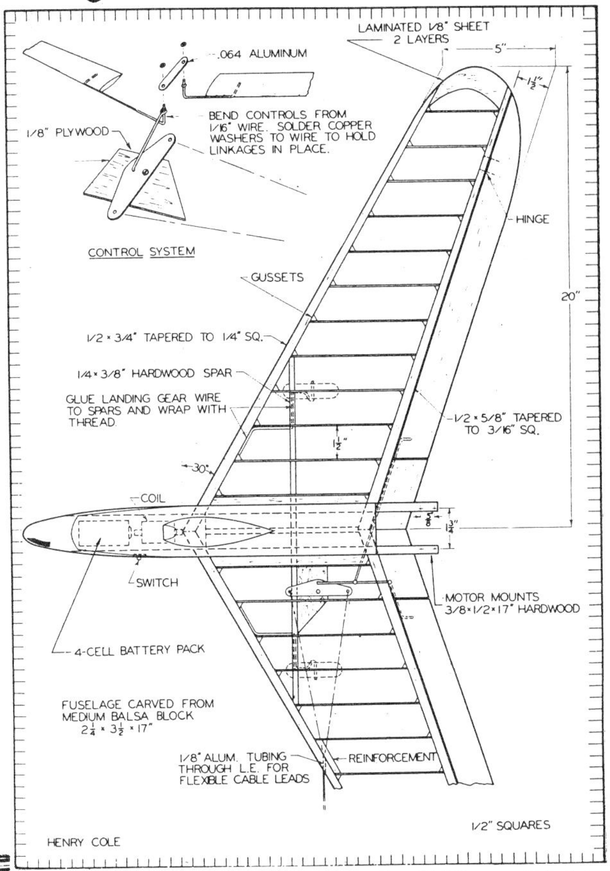

'Arrowhead' by Henry Cole (*Air Trails* October 1947) was a C/L flying wing, a change from the general style of C/L models of the time.

New! Revolutionary!

THE ARDEN GLOW PLUG

ACTUAL SIZE

The first successful application of this engineering principle to small engines.

Designed for Arden Engines

Replaces your spark plug.

Eliminates batteries, coil, condenser, engine timer, wiring . . . and resultant ignition troubles.

Your engine will start easier, and will produce equal or greater power output.

Part of an Arden advt. of early 1948. Just how revolutionary the glow-plug was to prove was perhaps little realised at the time.

Spanish stick model of which little is known, except that it appears to be largely bamboo in construction.

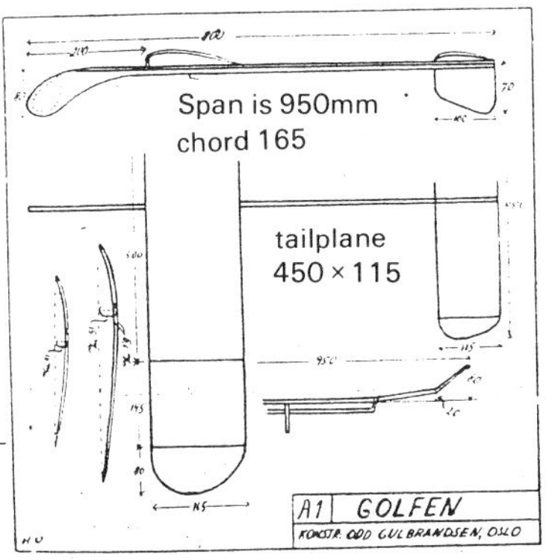

This Norwegian A1 appeared in *Hobby-Boken* 1947/8 and appears to be all-sheet construction. The A1 class received little recognition until 1955.

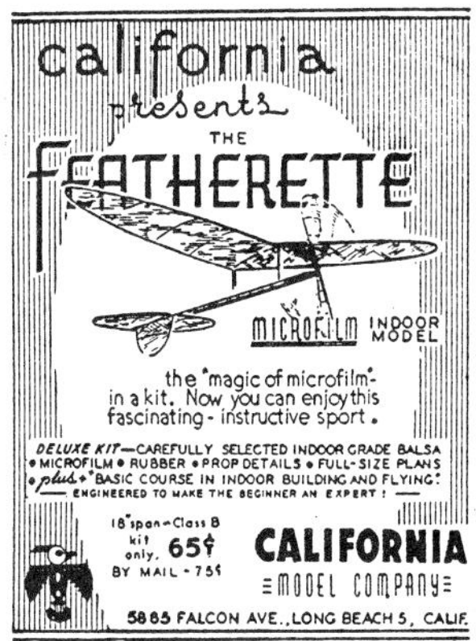

One of the very few attempts to produce a microfilm model kit.

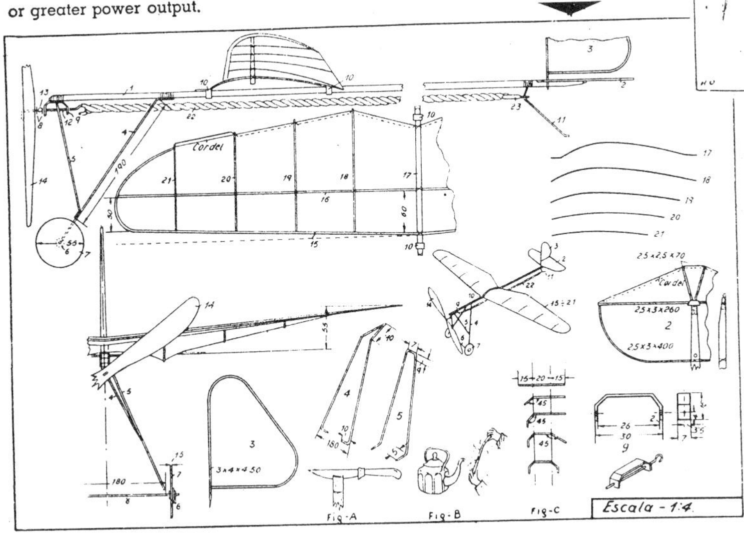

Full-size plans for this neat little flyer appear overleaf. ▶

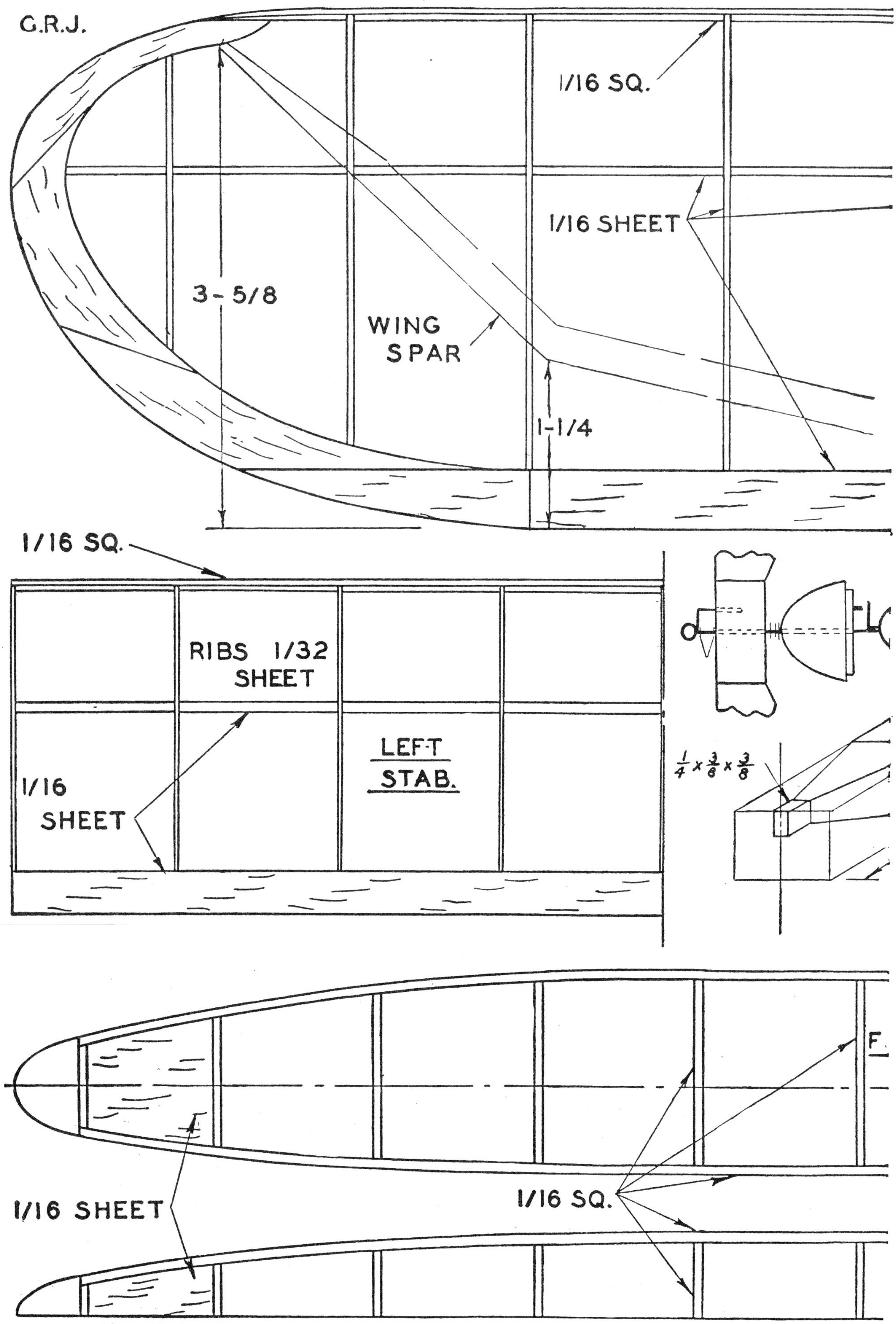
G.R.J.
1/16 SQ.
1/16 SHEET
3 - 5/8
WING
SPAR
1-1/4
1/16 SQ.
RIBS 1/32
SHEET
LEFT
STAB.
1/16
SHEET
1/4 x 3/8 x 3/8
1/16 SHEET
1/16 SQ.

HIGH PERFORMANCE SPORTSTER by C. R. Johnson. (*M.A.N.* August 1948). These plans are full-size, so may be photocopied and pieced together, though it will be necessary to trace wing and tailplane halves. Fuselage is *hard* $\frac{1}{16}$in. sq. balsa. Wing all medium $\frac{1}{16}$in. sheet; glue small block beneath mainspar after covering to give approx. $\frac{3}{32}$in. incidence. Wing held by rubber loop, fins and tailplane cemented in place. Power four strands $\frac{1}{4}$in. flat 20ins. (or more) long.

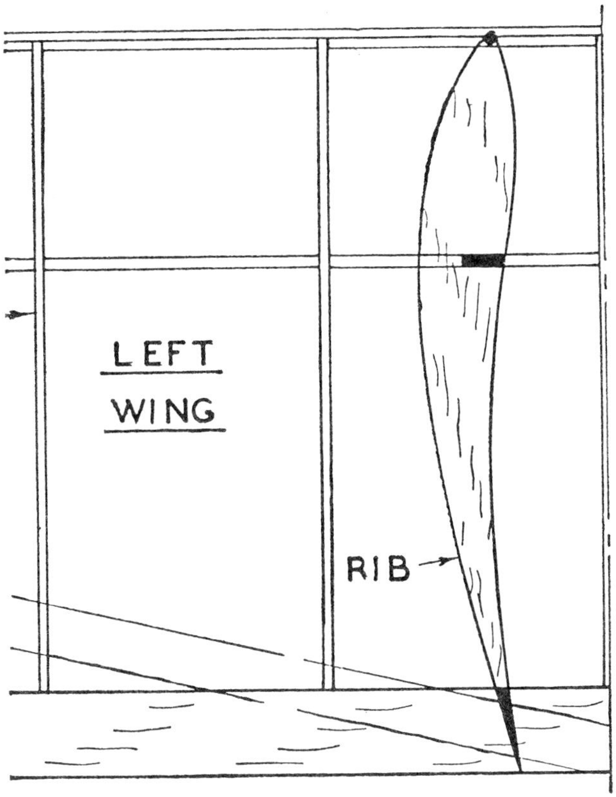

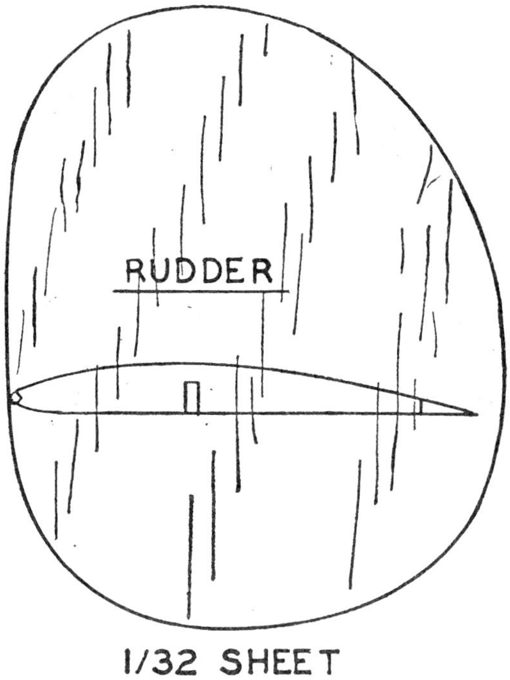

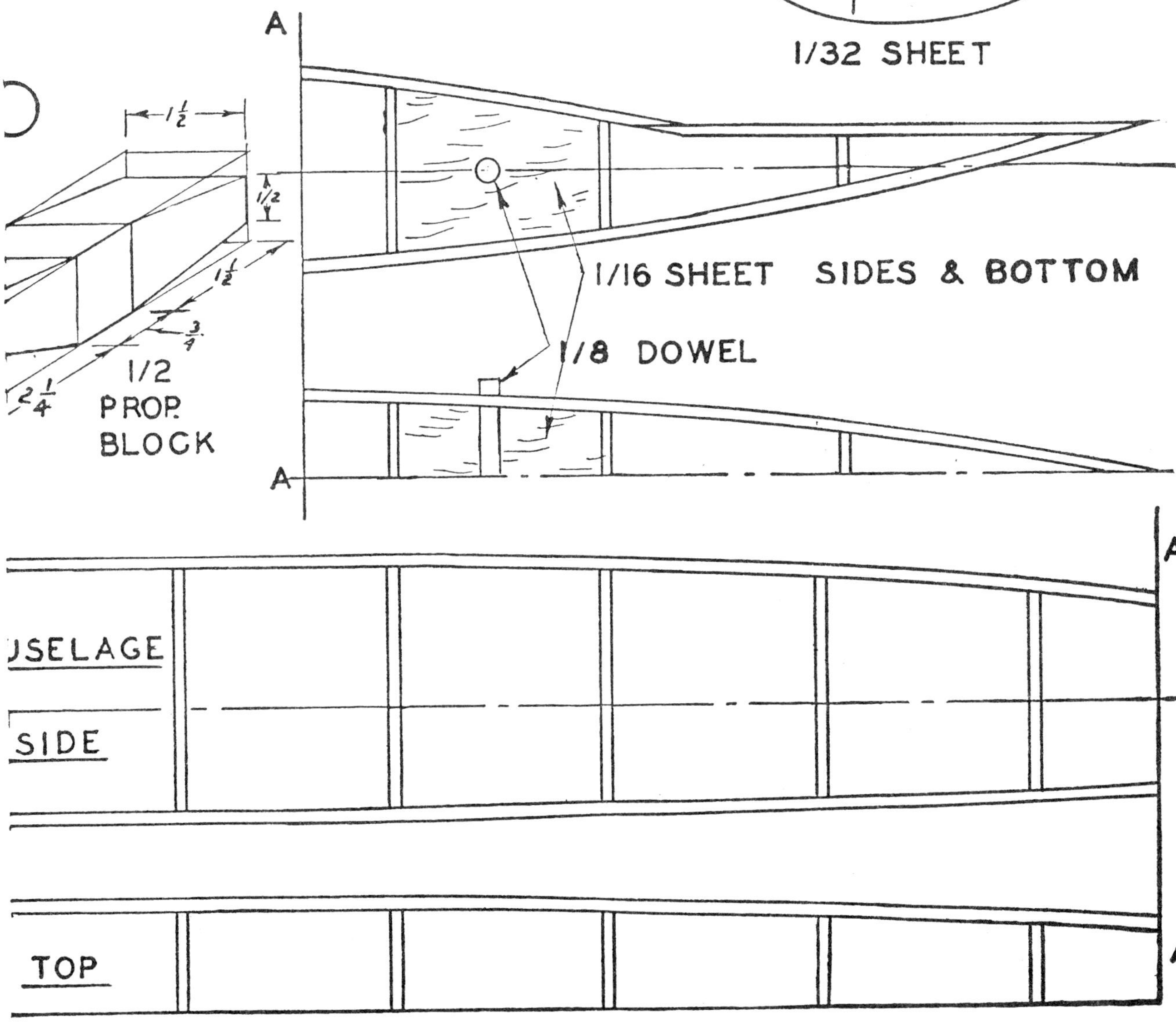

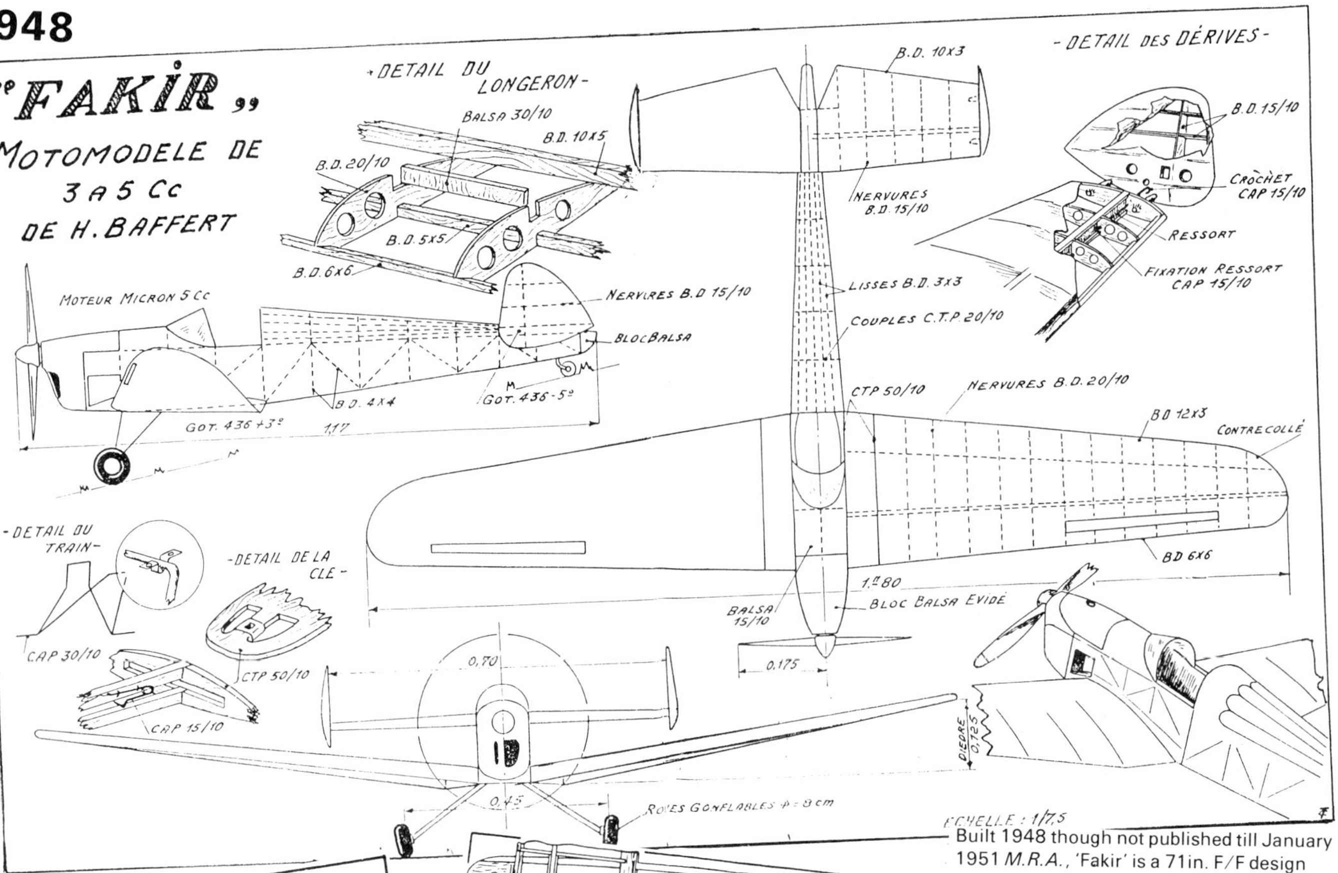

Built 1948 though not published till January 1951 *M.R.A.*, 'Fakir' is a 71in. F/F design which could well suit radio-assist.

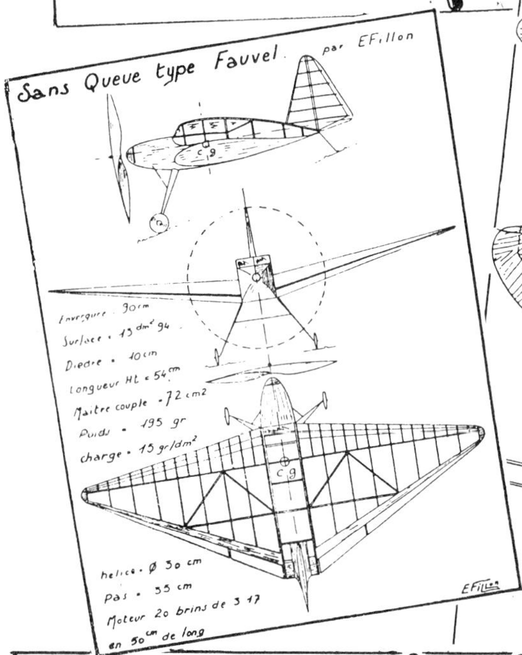

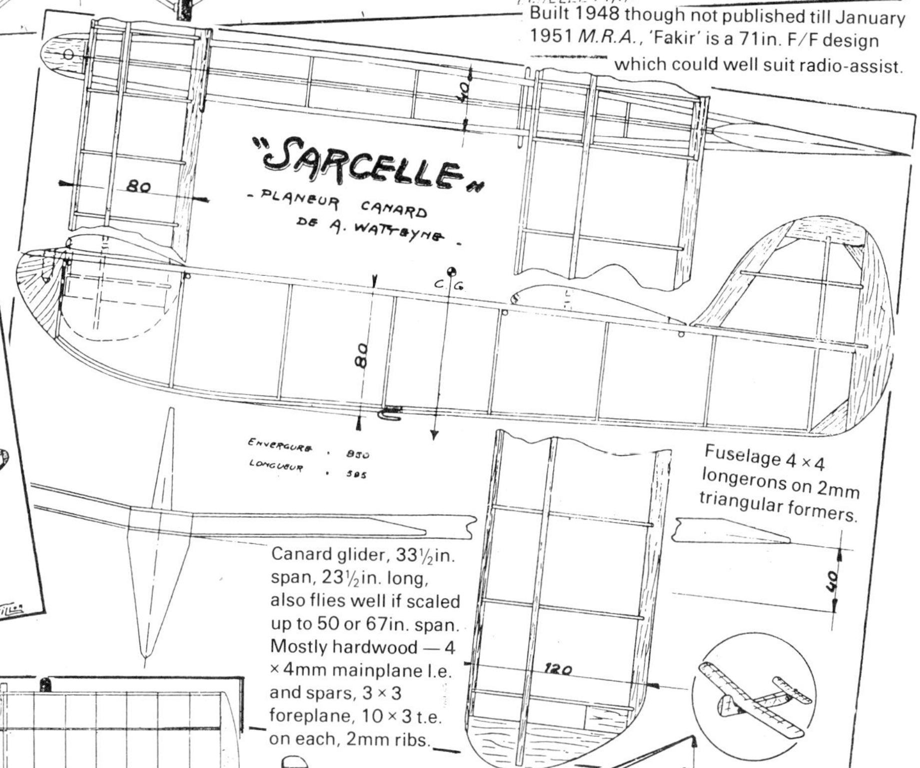

Fuselage 4 × 4 longerons on 2mm triangular formers.

Canard glider, 33½in. span, 23½in. long, also flies well if scaled up to 50 or 67in. span. Mostly hardwood — 4 × 4mm mainplane l.e. and spars, 3 × 3 foreplane, 10 × 3 t.e. on each, 2mm ribs.

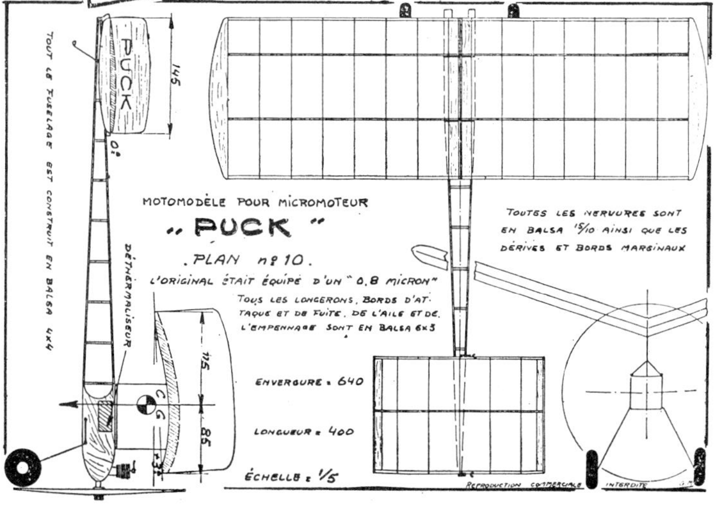

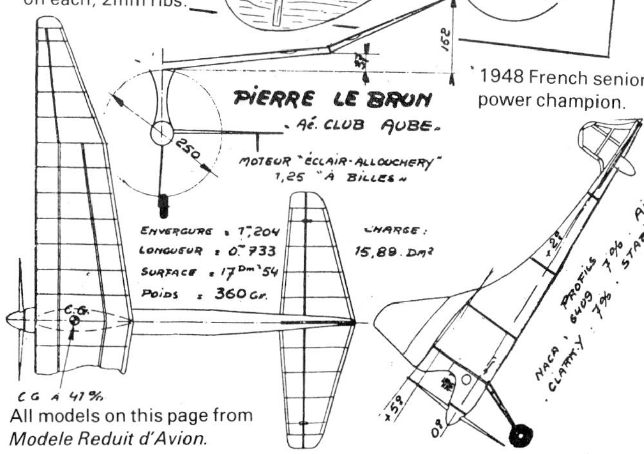

1948 French senior power champion.

All models on this page from *Modele Reduit d'Avion*.

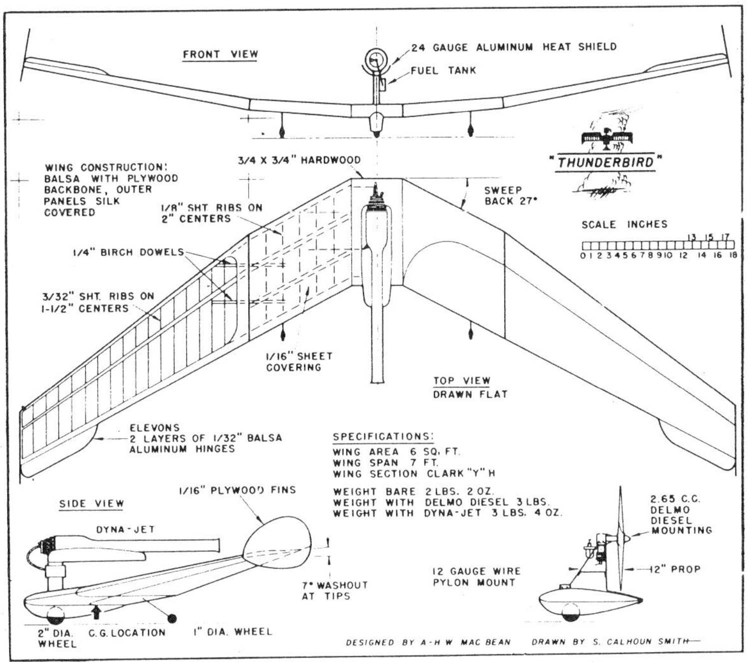

MacBean's 'Thunderbird' flew at Eaton Bray, appeared in both *Aeromodeller* and *Air Trails*.

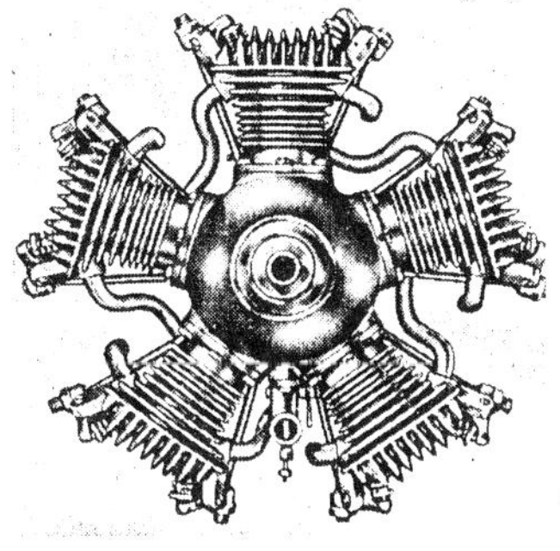

The 5-cylinder Burgess M5 four-stroke totalled 15cc. They are sought-after collectors' items today.

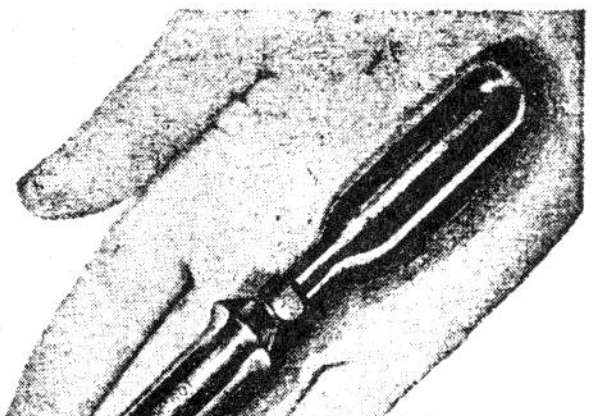

The Minnesota Engine Works sold plans to convert a used

A PRACTICAL

JET ENGINE

CO_2 cylinder to a gasoline-fuelled jet unit, the M.E.W. 601.

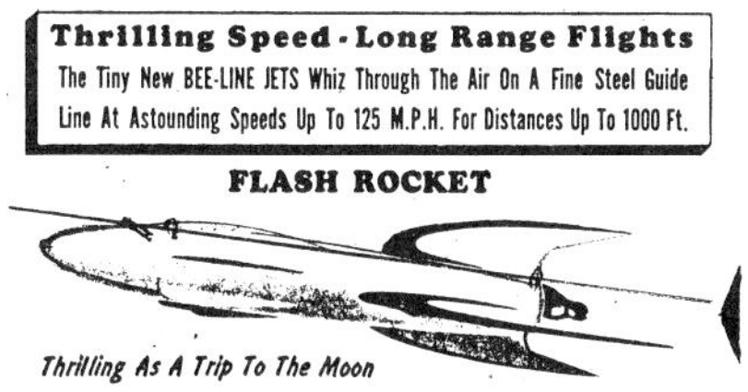

Stanzels came up with aircraft and cars powered by CO_2 capsules which when pierced whooshed the models along a taut wire.

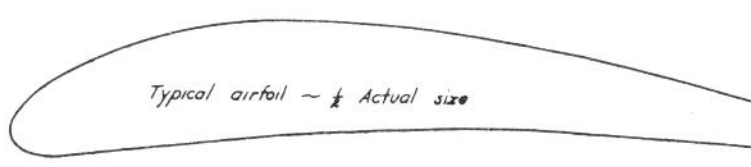

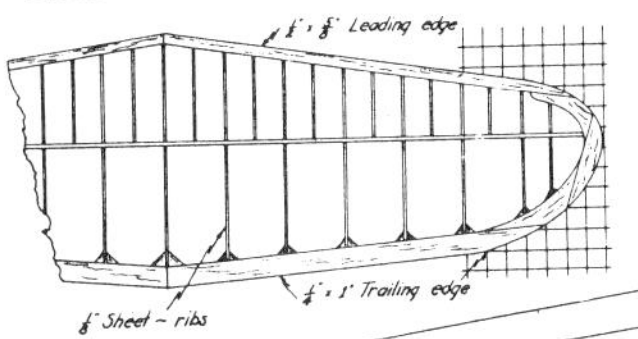

Large Class D model by Dick Everett incorporated R/C rudder for control on glide only, to avoid flyaways.

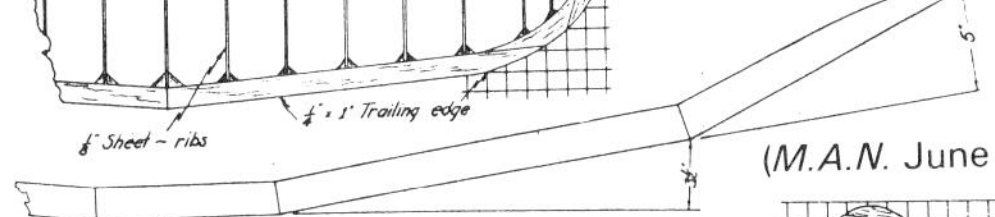

(*M.A.N.* June '48)

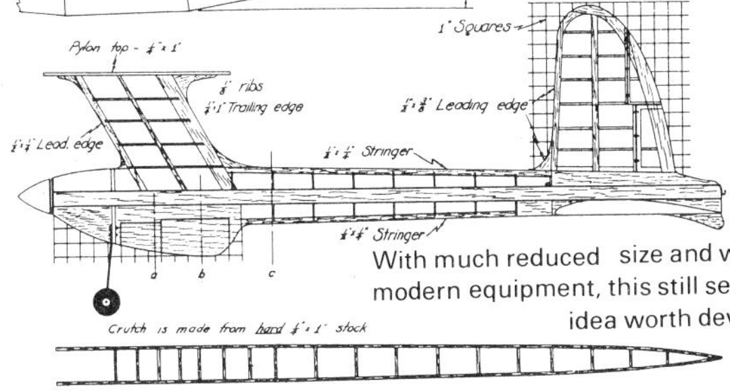

With much reduced size and weight of modern equipment, this still seems an idea worth developing.

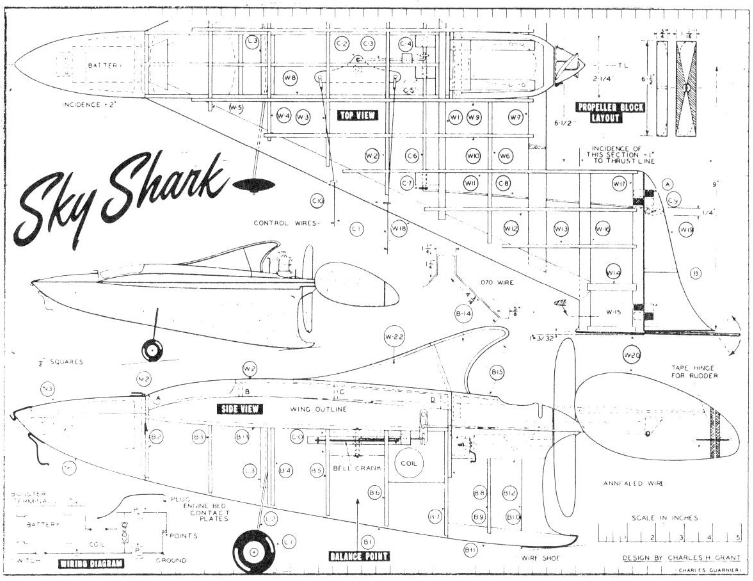

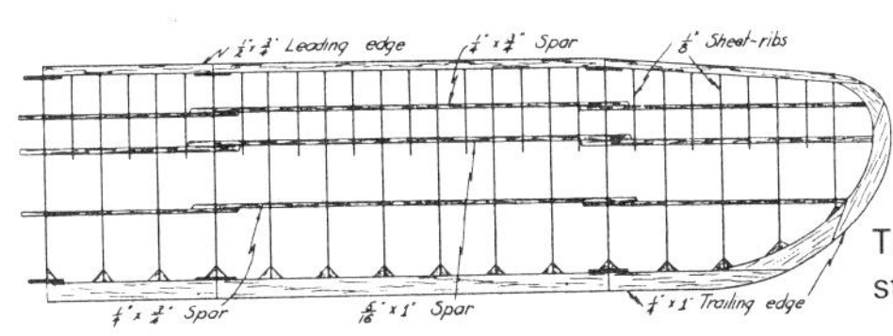

Charles Grant remained very active after retirement from editorship of *M.A.N.* (1932-43) until his death at the age of 93. His 'Sky Shark' (*Air Trails,* March 1948) was certainly a 'different' concept.

This 40in. Class C glider is extremely stable to tow or use on a hi-start catapult.

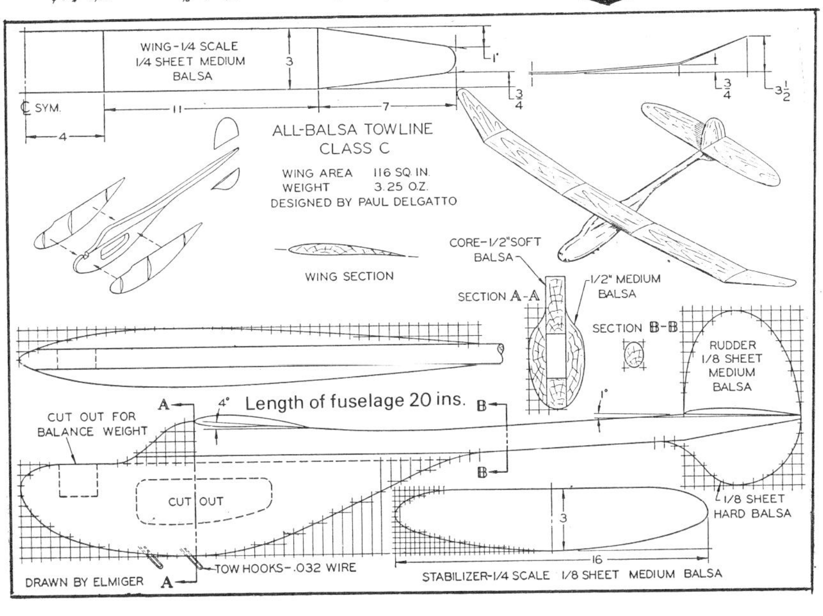

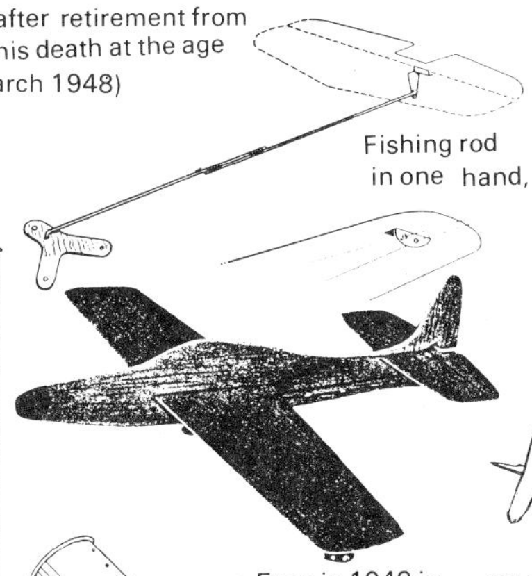

Fishing rod in one hand,

Jim Walker combined whip with U-Control (*A.T.* April '48)

handle in the other

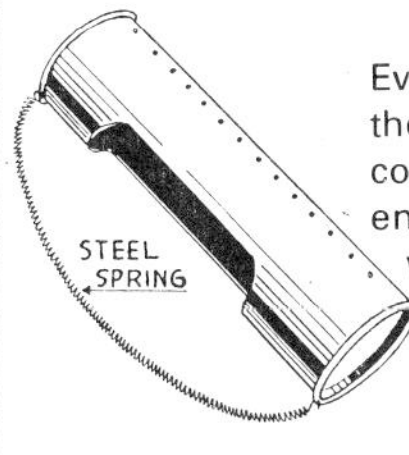

Even in 1948 in the U.S.A. some concern about engine noise was shown.

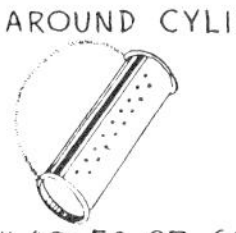

M.A.N. ran an article on simply-made 'mufflers'.

EXPLODED VIEW

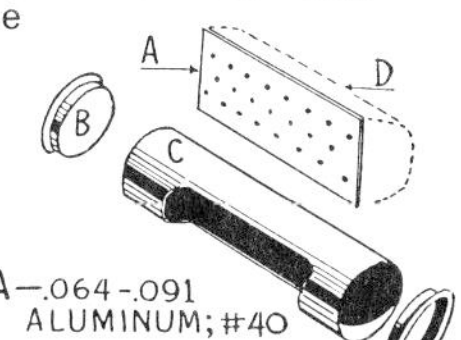

A–.064-.091 ALUMINUM; #40

B–DOOR PULLS.

C–.051-.081 TUBING ¾" I.D.; 2¾" TO 4½" LONG

D–HEAVY STEEL WOOL; PACK BEHIND BAFFLE BEFORE PRESSING IN SECOND DOOR PULL.

1948-49

Dennis Davis' 'San de Hogan' was one of the outstanding contest F/F models of 1948/9.

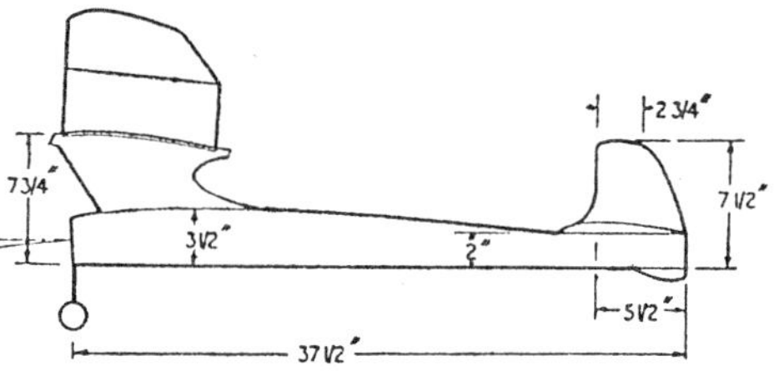

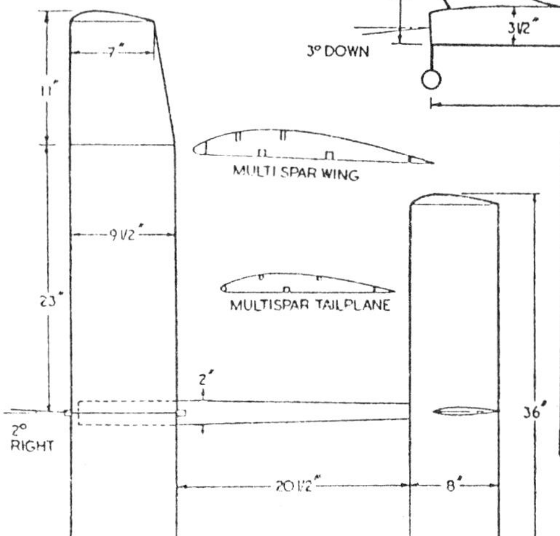

This drawing from a survey of pylon designs in March 1949 *Model Aircraft*.

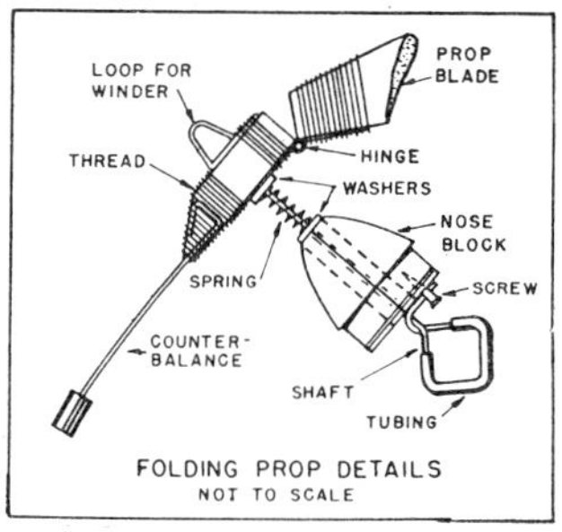

The single-blade folding prop was commonest on contest rubber models at this time, followed by twin-blade folders. Free-wheelers were out of favour but feathering, or feathering/free wheeling, seemed to hold promise.

Model Aircraft August 1949.

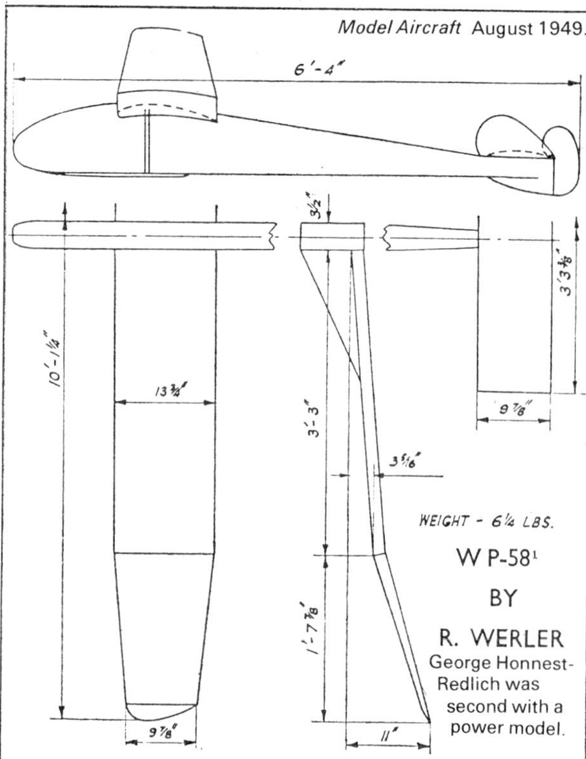

W P-58[1]

BY

R. WERLER

George Honnest-Redlich was second with a power model.

Believe it or not, this was the winner of the first true international radio control contest, held at Bretigny, France, June 12, 1949 with British and French participation.

Below, the Good brothers 'Rudderbug', which appeared in nearly every magazine in 1949 (plans are still available from A.P.S.) was the first widely built, widely successful radio design.

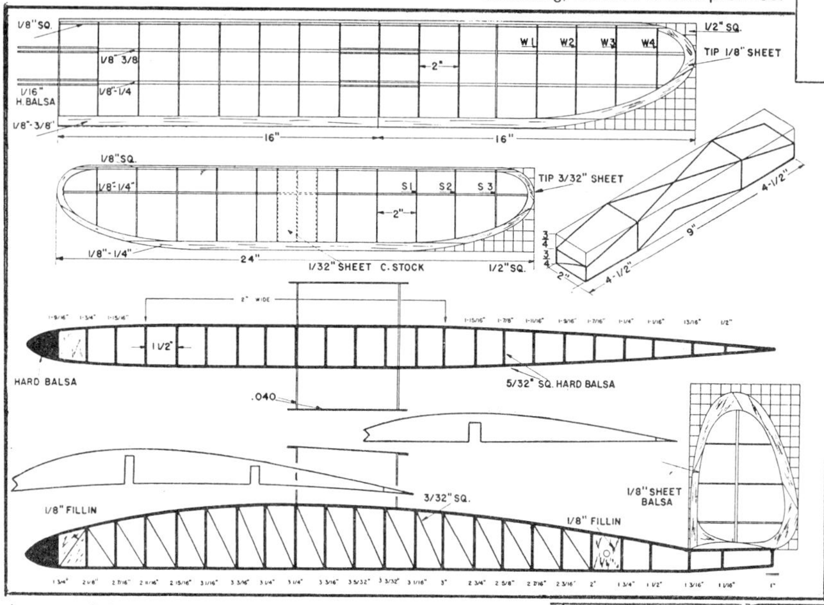

Large and simple rubber models enjoyed quite a vogue at this time. This 48in. Frank Ehling design (*M.A.N.* December 1948) was fairly typical.

The Wakefield Cup was resumed in 1948. Won for Britain by Roy Chesterton, flying Ted Evans' 'Jaguar' design. 6:27.3.

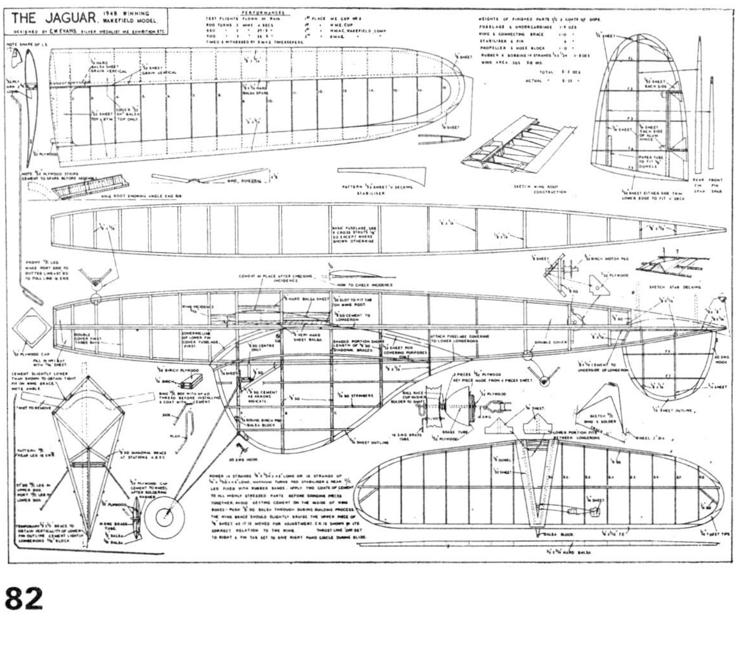

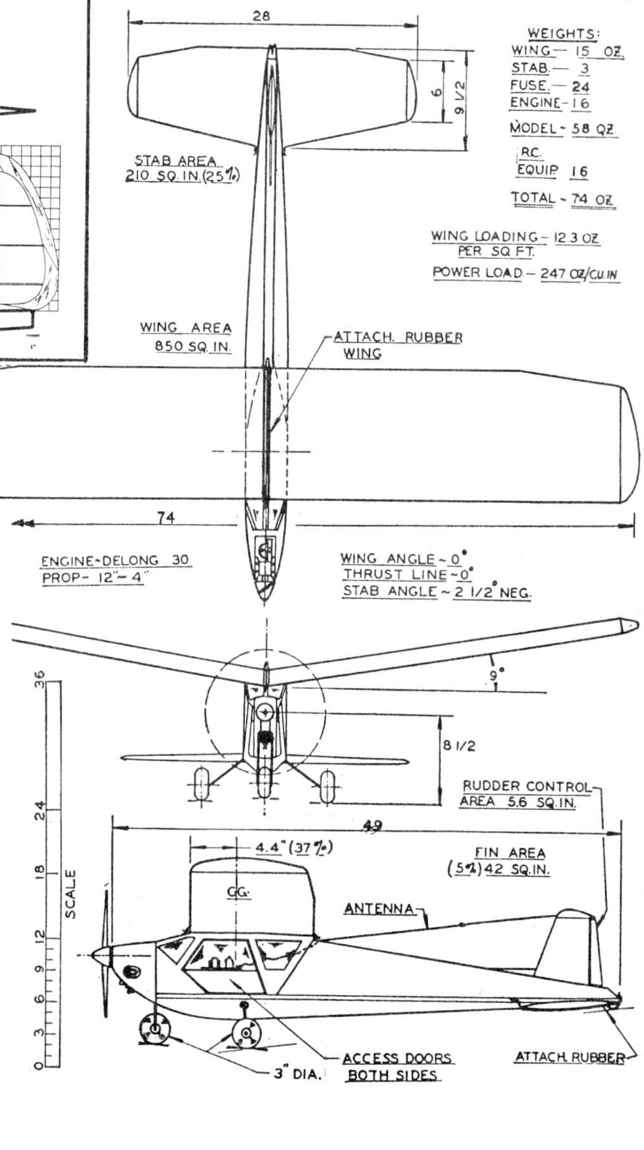

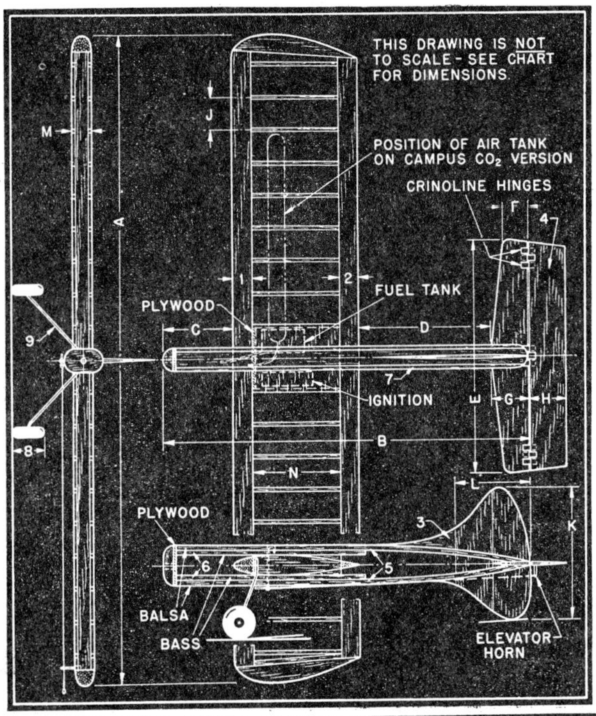

	E-1 A-100	E-2 0-.09	E-3 .09-.199	E-4 .199-.299	E-5 .299-.36	E-6 .36-.49	E-7 .49-.65
1	15/32" SQ.	11/16" SQ.	13/16" SQ.	15/16" SQ.	1" SQ.	1-3/16" SQ.	1-1/4" SQ.
2	7/32"X3/8"	1/4"X1/2"	5/8"X1/4"	5/8"X3/4"	3/8"X1"	1/2"X1"	5/8"X1"
3 & 4	1/16"	3/32"	1/8"	3/16"	3/16"	1/4"	1/4"
5 & 6	3/32"X1/4"	1/8"X5/16"	3/16"X5/16"	1/4"X5/8"	5/16"X7/8"	3/8"X3/4"	3/8"X3/4"
7	1/32"	1/16"	3/32"	1/8"	3/16"	1/4"	1/4"
8	5/8"	3/4"	1"	1-1/2"	2"	2-1/4"	2-1/2"
9	.040"	1/16"	3/32"	3/32"	1/8"	1/8"	1/8"
A	12-1/2"	20"	26"	32"	38"	45"	50"
B	7-1/4"	13"	16-1/2"	21"	24"	29"	32"
C	1-1/2"	2-3/8"	3-1/8"	3-7/8"	4-1/2"	5-3/8"	6"
D	2-3/4"	4-5/8"	5-7/8"	7-3/8"	8-3/4"	10"	11"
E	4-1/2"	7-1/4"	9-1/2"	11-1/2"	13-1/2"	16-1/4"	18"
F	1/2"	7/8"	1"	1-3/8"	1-1/2"	1-7/8"	2"
G-H	3/4"	1-1/8"	1-1/2"	1-7/8"	2-1/4"	2-3/4"	3"
J	1-1/4"	1-1/2"	1-3/4"	2"	2-1/4"	2-1/2"	2-3/4"
K	2-1/2"	3-1/2"	4-1/2"	5-1/2"	6-1/2"	7-1/2"	7-1/2"
L	1-3/4"	2-3/8"	3-1/2"	4-3/8"	5-1/8"	6"	6"
M	1/2"	3/4"	15/16"	1"	1-1/8"	1-3/8"	1-1/2"
N	1-3/4"	2-7/8"	3-5/8"	4-1/2"	5-5/16"	7"	8"
TREAD	3"	6"	8"	9"	10"	11"	12"

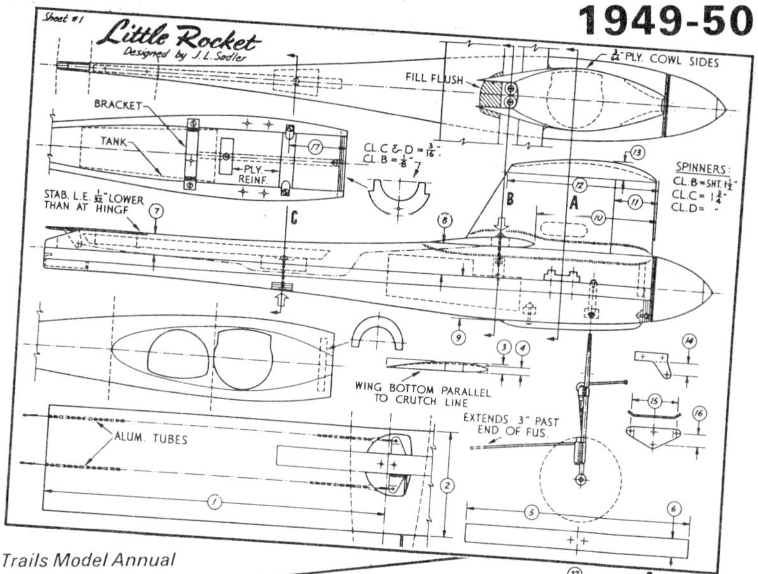

Air Trails Model Annual

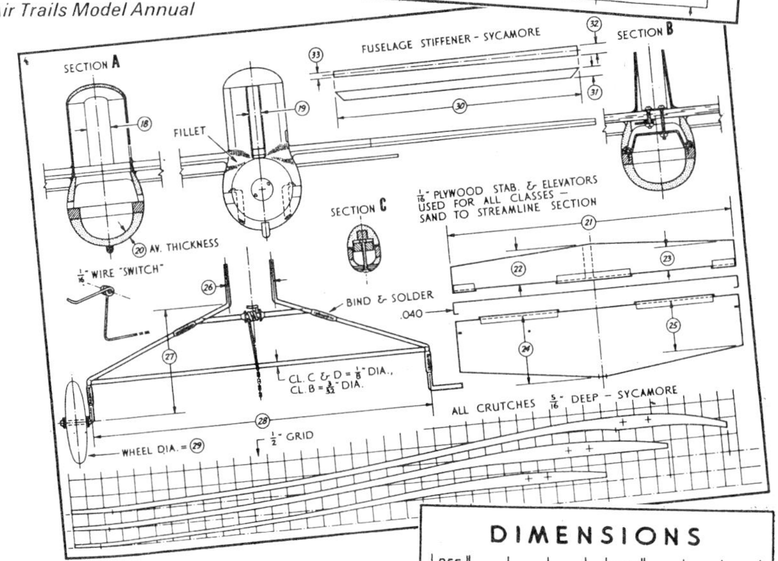

REF. Nº	"B"	"C"	"D"
1	8½"	10"	10"
2	2⅜	3	3
3	7/32	¼	¼
4	5/32	3/16	3/16
5	5	6½	6½
6	½	9/16	9/16
7	7/16	⅝	⅝
8	¾	⅞	1
9	15/16	1 3/16	1 5/16
10	3⅛	3½	3¾
11	1	1¼	1½
12	3¾	4 7/16	4 11/16
13	⅜	½	9/16
14	5/16	⅜	⅜
15	1¼	1⅜	1⅜
16	5/16	⅜	⅜
17	1¼	1⅝	1⅝
18	⅝"	¾"	⅞"
19	5/16	5/16	⅜
20	5/32	3/16	7/32
21	7½	8½	8½
22	¾	1	1
23	½	¾	¾
24	1 11/16	2	2
25	1⅜	1½	1½
26	1⅛	1⅜	1 9/16
27	2¾	3	3
28	8½	10	10
29	2	2⅜	2⅜
30	6½	7¼	9½
31	3/16	¼	¼
32	3/16	¼	¼
33	5/32	3/16	3/16

Control-line had begun to specialise stunt and speed being the first categories. Two designs capable of performing well in different sizes were 'Mr. Ehling's Easy' stunter and, one of the best-known speed models ever, Sadler's 'Little Rocket'.

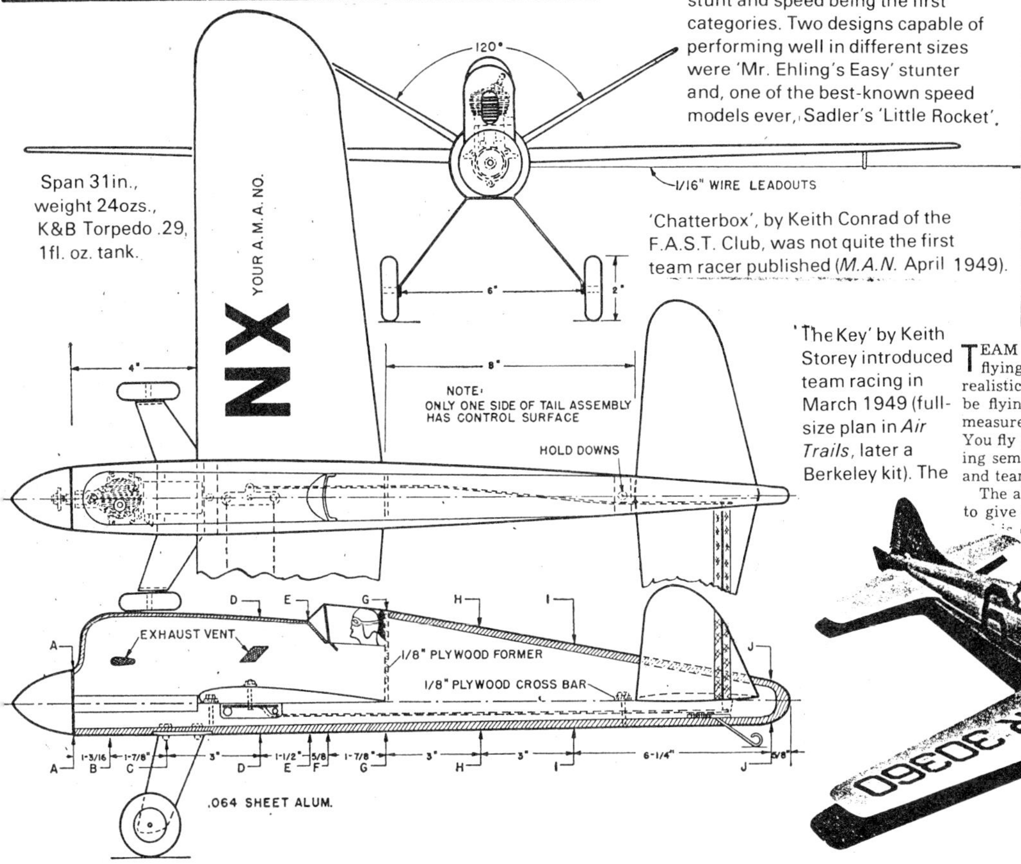

Span 31in., weight 24ozs., K&B Torpedo .29, 1fl. oz. tank.

'Chatterbox', by Keith Conrad of the F.A.S.T. Club, was not quite the first team racer published (*M.A.N.* April 1949).

'The Key' by Keith Storey introduced team racing in March 1949 (full-size plan in *Air Trails*, later a Berkeley kit). The laudable initial aims have not, alas, survived the passage of time unscathed!

TEAM RACING is here! If you want the best in flying fun and thrills, plus actual competition of realistic airplanes, this is your opportunity. You will be flying with three or more fast competitors over measured courses of ½ mile, 1½ miles, and 10 miles. You fly not just made-over speed jobs, but good looking semi-scale models that will give the scale, speed, and team stunt hound a new opening for his hobby.

The aim of Team Racing is fundamental. We want to give the spectator as well as the contestant more ... contest dollar in the way of real competi... ...f competition challenges ... solving new

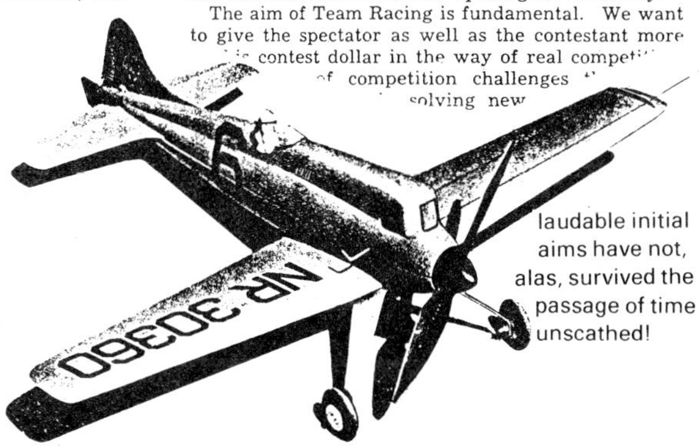

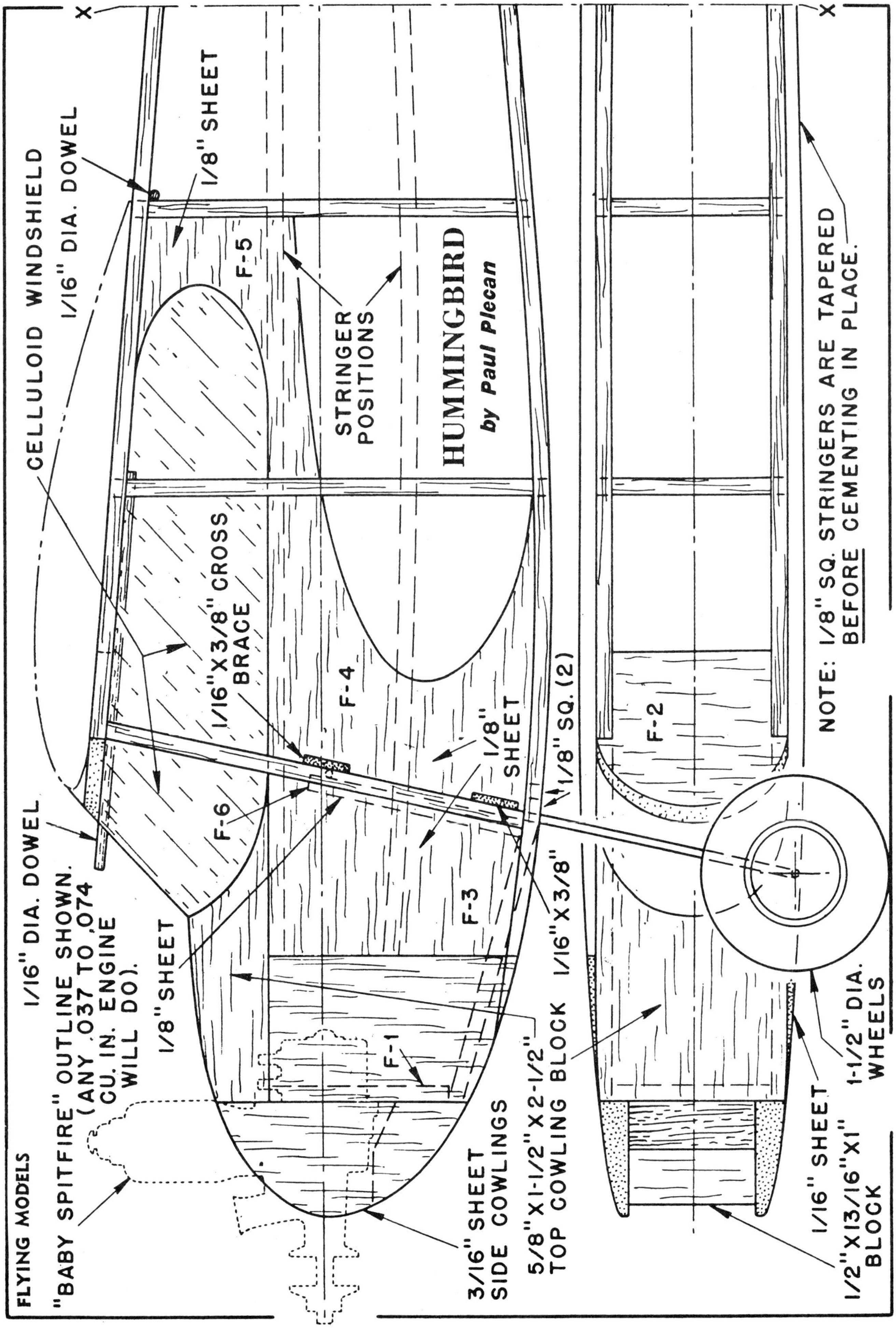
FLYING MODELS
"BABY SPITFIRE" OUTLINE SHOWN. (ANY .037 TO .074 CU. IN. ENGINE WILL DO).
1/16" DIA. DOWEL
1/8" SHEET
CELLULOID WINDSHIELD
1/16" DIA. DOWEL
1/8" SHEET
1/16"X3/8" CROSS BRACE
F-5
F-4
F-6
F-3
F-1
F-2
STRINGER POSITIONS
HUMMINGBIRD
by Paul Plecan
1/8" SHEET
1/8" SQ. (2)
1/16"X3/8"
3/16" SHEET SIDE COWLINGS
5/8"X1-1/2"X2-1/2" TOP COWLING BLOCK
NOTE: 1/8" SQ. STRINGERS ARE TAPERED BEFORE CEMENTING IN PLACE.
1-1/2" DIA. WHEELS
1/16" SHEET
1/2"X13/16"X1" BLOCK

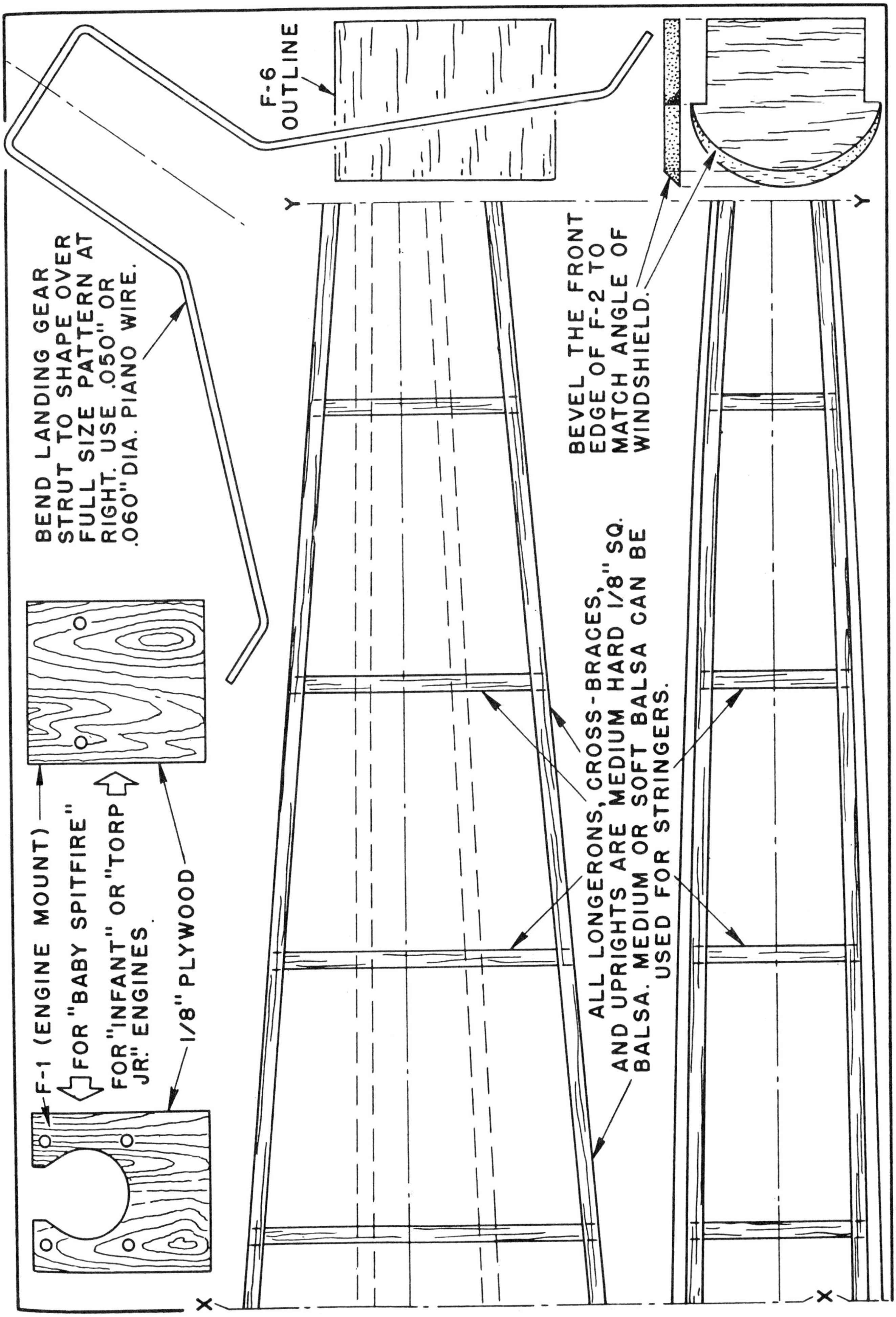
F-6 OUTLINE
BEND LANDING GEAR STRUT TO SHAPE OVER FULL SIZE PATTERN AT RIGHT. USE .050" OR .060" DIA. PIANO WIRE.
BEVEL THE FRONT EDGE OF F-2 TO MATCH ANGLE OF WINDSHIELD.
F-1 (ENGINE MOUNT)
FOR "BABY SPITFIRE"
FOR "INFANT" OR "TORP JR." ENGINES
1/8" PLYWOOD
ALL LONGERONS, CROSS-BRACES, AND UPRIGHTS ARE MEDIUM HARD 1/8" SQ. BALSA. MEDIUM OR SOFT BALSA CAN BE USED FOR STRINGERS.
Y
Y
X
X

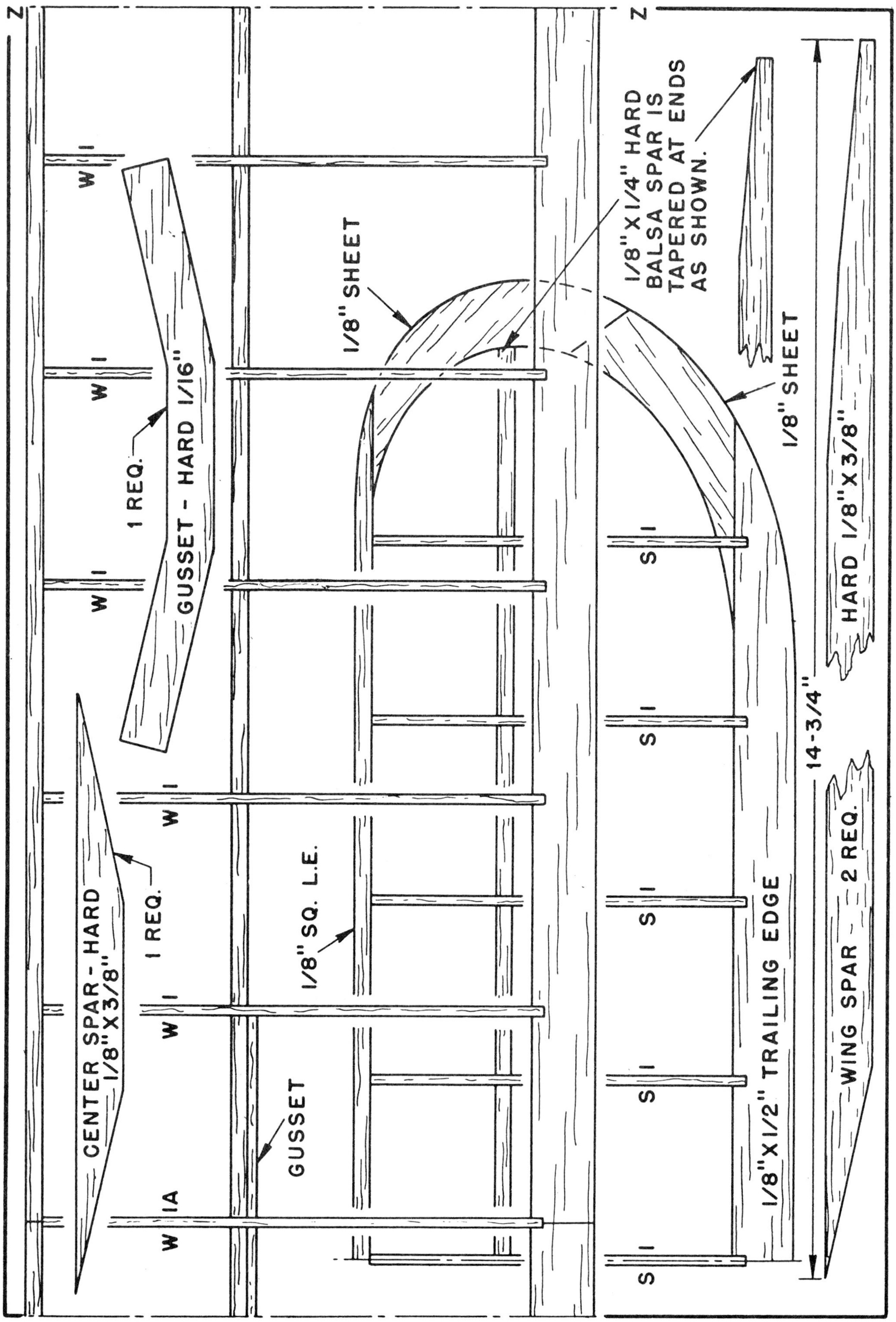

Z
Z
W
W
W
W
W
W IA
S
S
S
S
S
GUSSET - HARD 1/16"
1 REQ.
CENTER SPAR - HARD 1/8" X 3/8"
1 REQ.
GUSSET
1/8" SQ. L.E.
1/8" SHEET
1/8" X 1/4" HARD BALSA SPAR IS TAPERED AT ENDS AS SHOWN.
1/8" SHEET
1/8" X 1/2" TRAILING EDGE
14-3/4"
HARD 1/8" X 3/8"
WING SPAR 2 REQ.

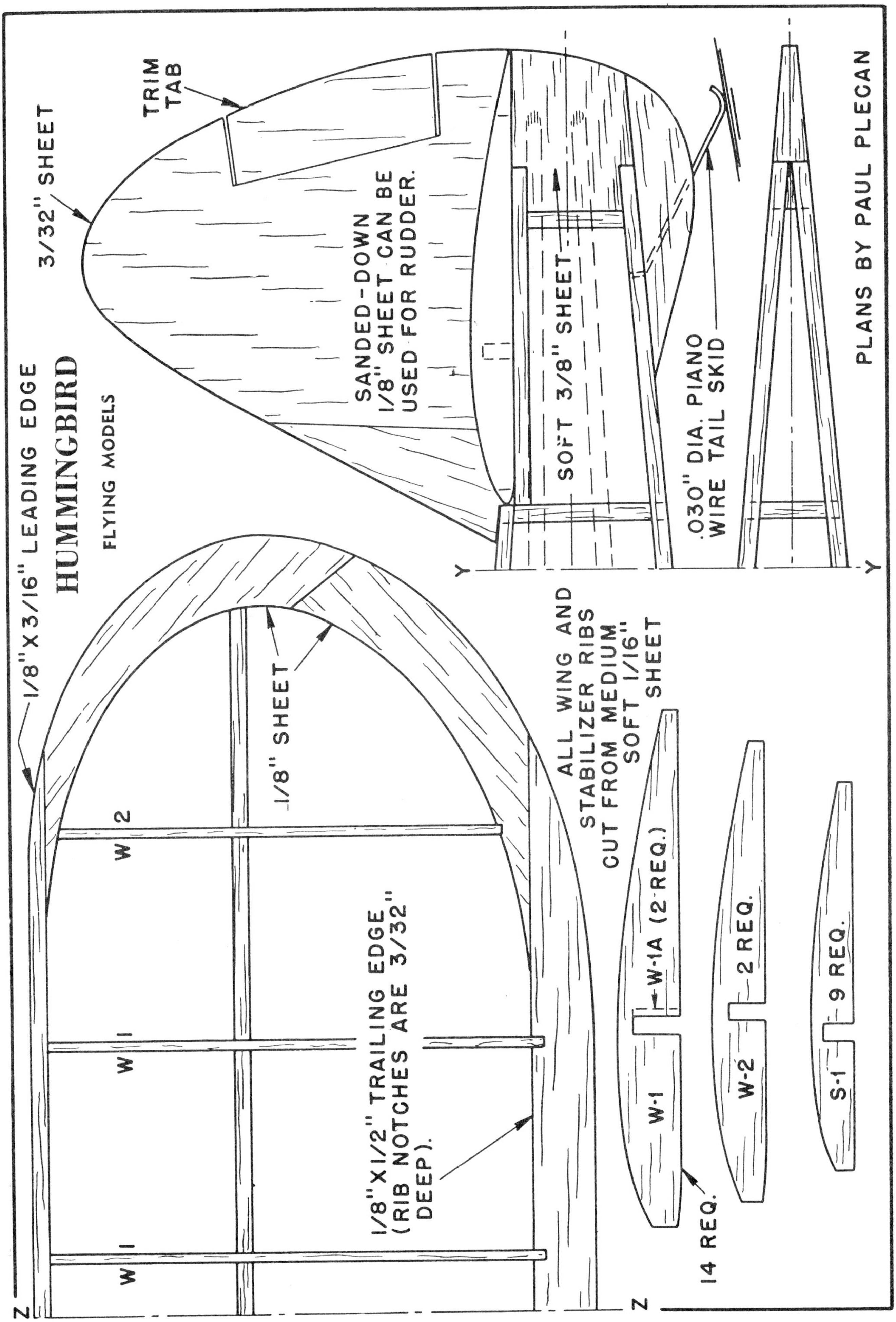
HUMMINGBIRD
FLYING MODELS
1/8" X 3/16" LEADING EDGE
3/32" SHEET
TRIM TAB
SANDED-DOWN 1/8" SHEET CAN BE USED FOR RUDDER.
SOFT 3/8" SHEET
.030" DIA. PIANO WIRE TAIL SKID
PLANS BY PAUL PLECAN
1/8" SHEET
W 2
W 1
W 1
1/8" X 1/2" TRAILING EDGE (RIB NOTCHES ARE 3/32" DEEP).
ALL WING AND STABILIZER RIBS CUT FROM MEDIUM SOFT 1/16" SHEET
W-1A (2 REQ.)
W-1
14 REQ.
W-2
2 REQ.
S-1
9 REQ.
Y
Y
Z
Z

1949 -50

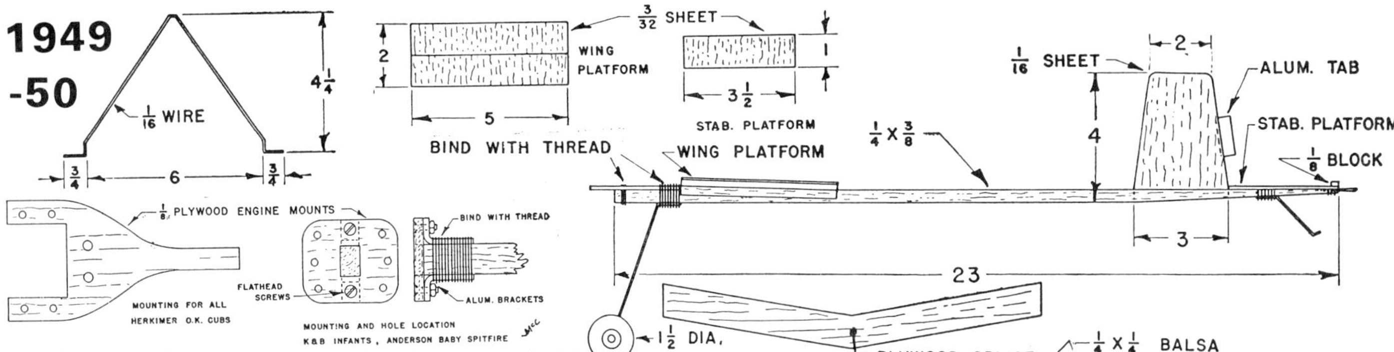

The advent of small motors encouraged J. L. McLarty to experiment with a beefed-up conventional 'Baby ROG' stick model with flat-plate wing and tail (*M.A.N.* April 1950). It worked perfectly happily!

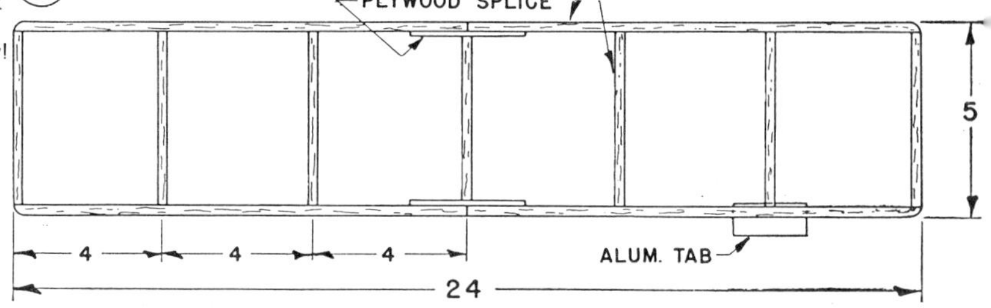

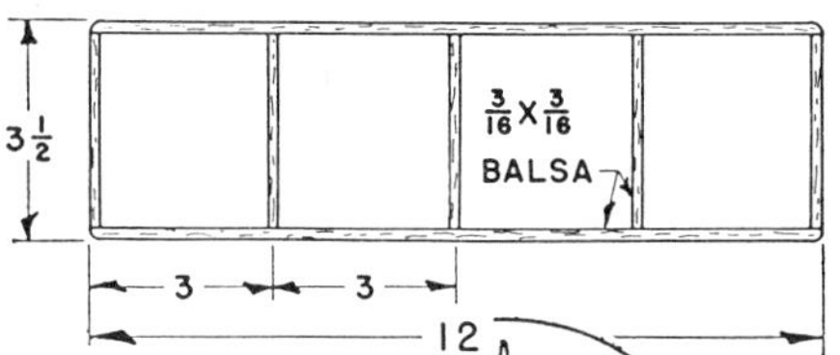

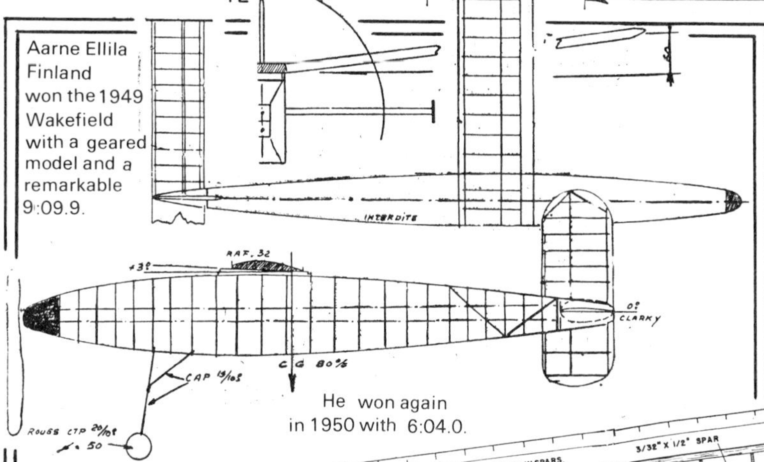

Aarne Ellila Finland won the 1949 Wakefield with a geared model and a remarkable 9:09.9.

He won again in 1950 with 6:04.0.

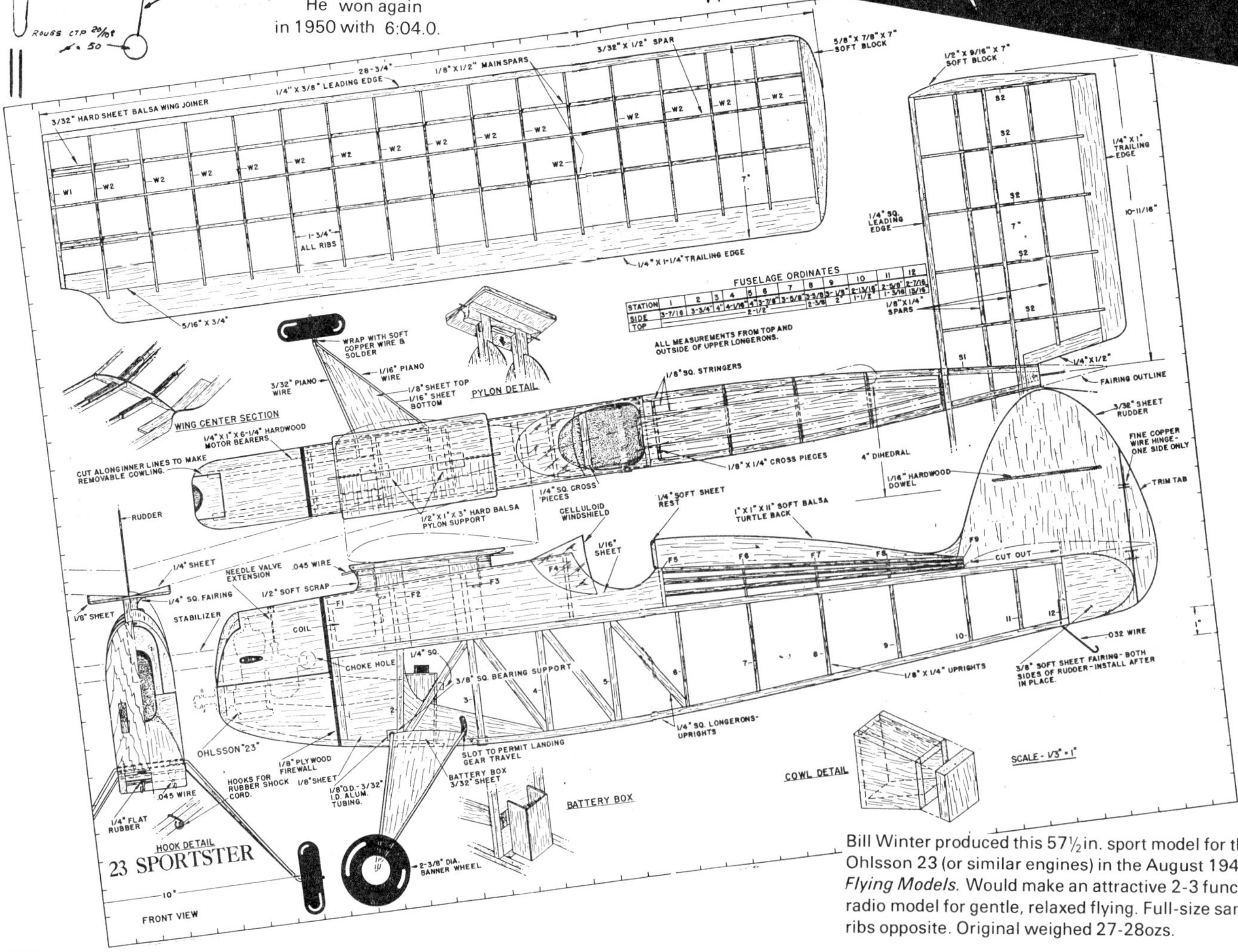

Bill Winter produced this 57½in. sport model for the Ohlsson 23 (or similar engines) in the August 1949 *Flying Models*. Would make an attractive 2-3 function radio model for gentle, relaxed flying. Full-size sample ribs opposite. Original weighed 27-28ozs.

1949-50

23 SPORTSTER

Full-size specimen ribs for model opposite.

The forerunner of plastic kits, advertised in January 1950.

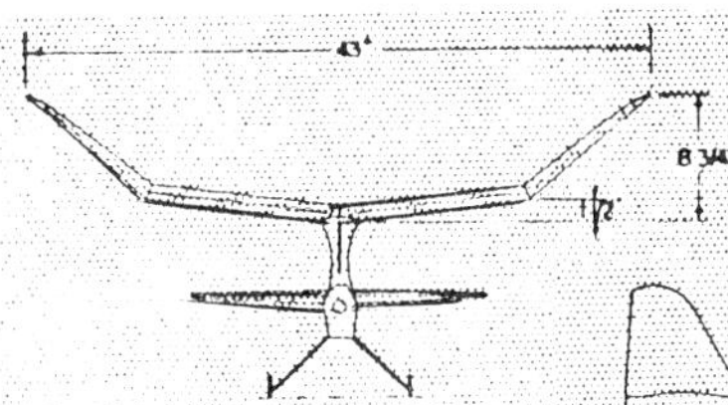

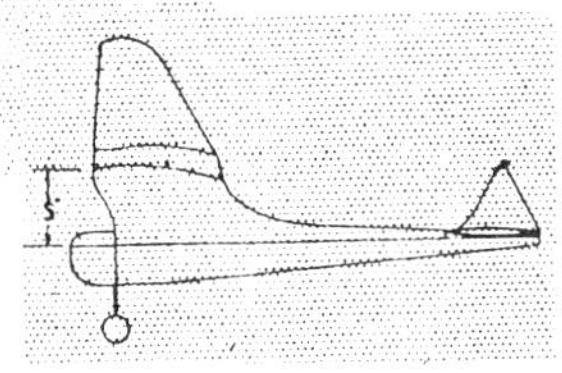

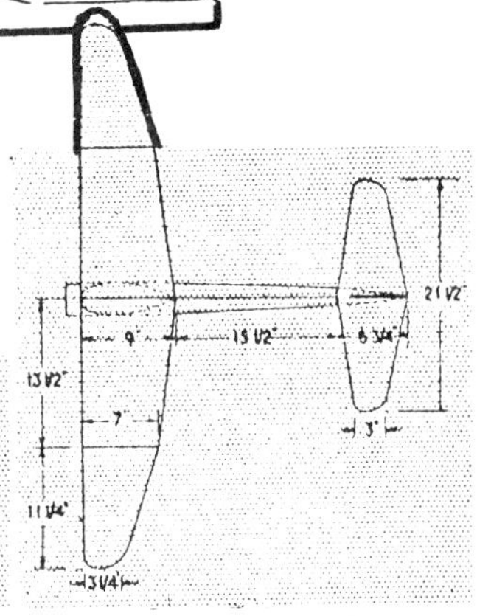

Leon Shulman's 'Banshee' was a top model in 1948/9.

The 'Banshee' was kitted in both America and Britain.

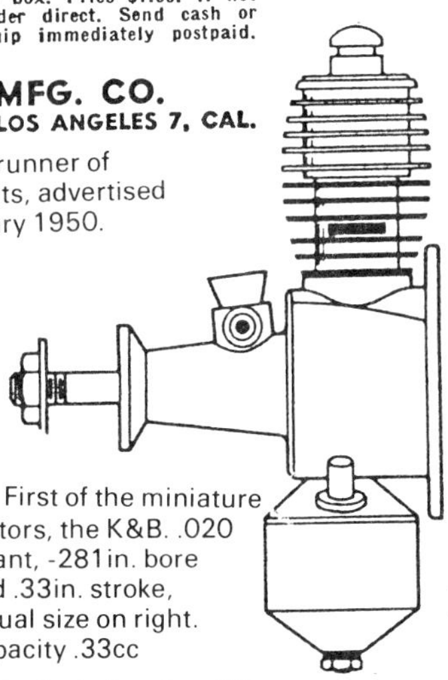

First of the miniature motors, the K&B. .020 Infant, -281in. bore and .33in. stroke, actual size on right. Capacity .33cc

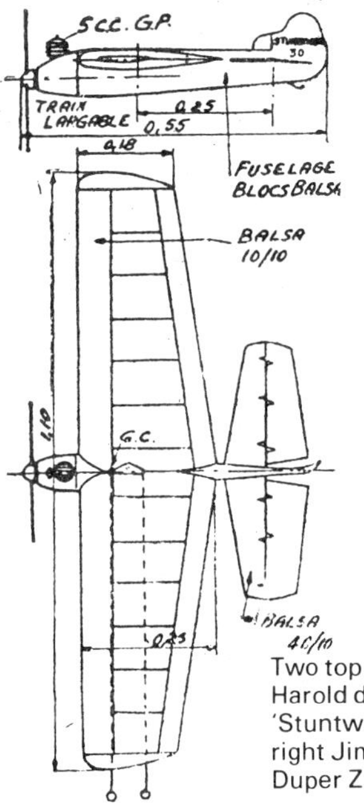

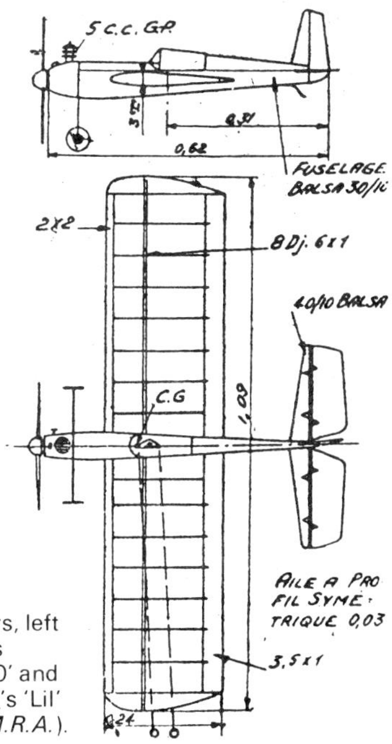

Two top stunters, left Harold de Bolt's 'Stuntwagon 30' and right Jim Saftig's 'Lil' Duper Zilch'. (*M.R.A.*).

Insignia for nine different countries using the Curtiss Hawk, full-size plans for which follow.

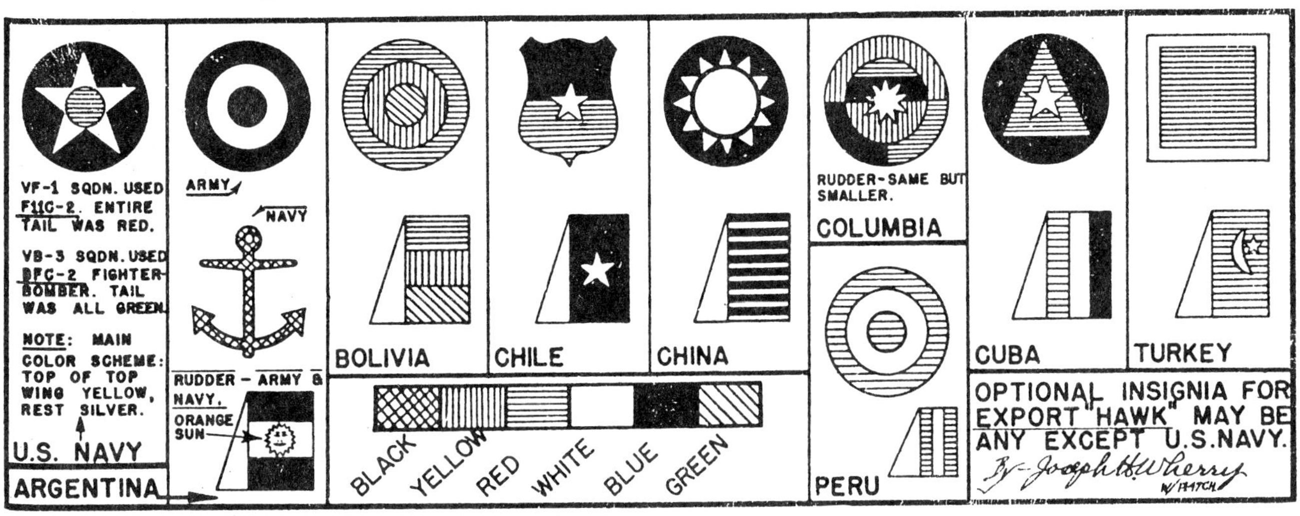

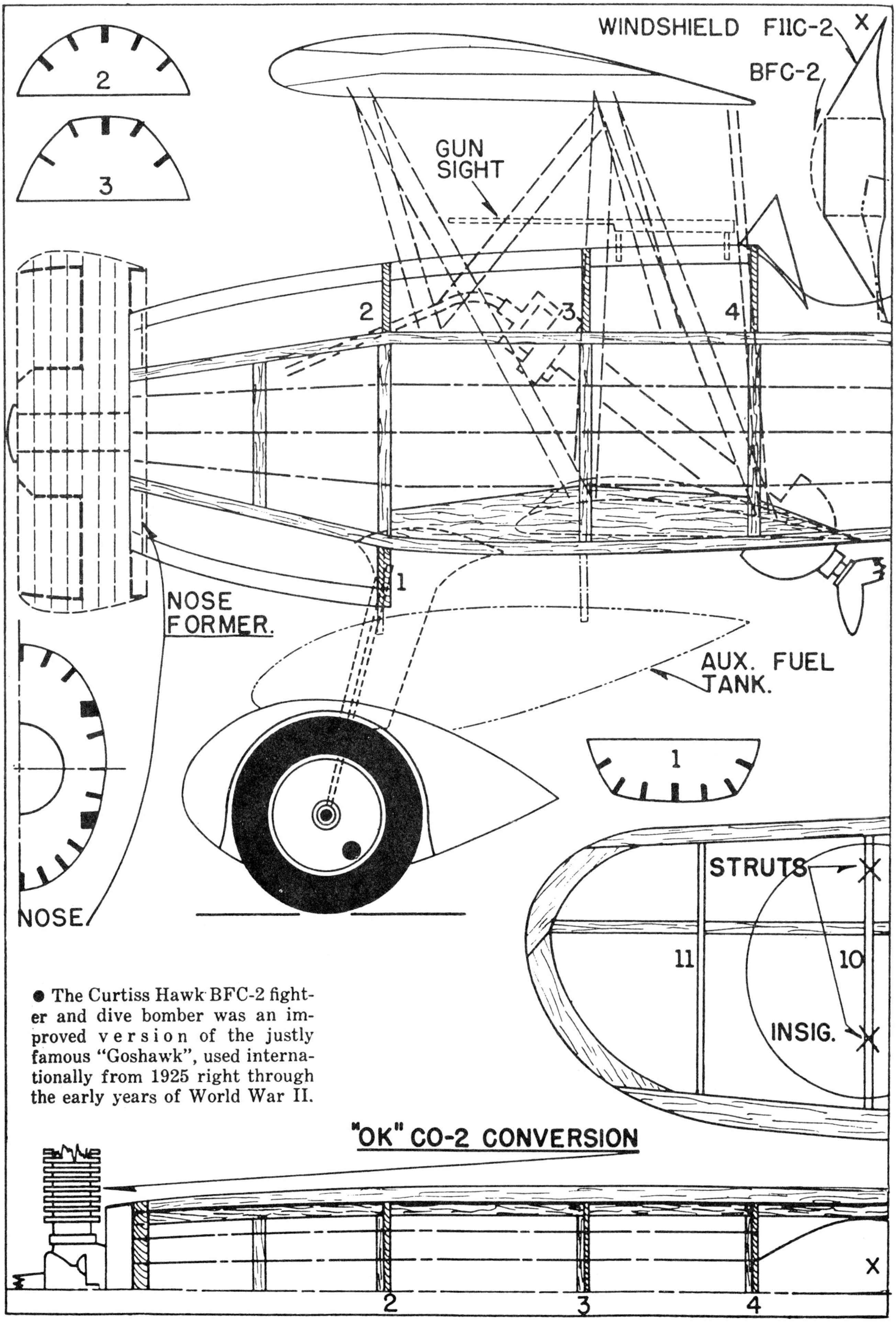

● The Curtiss Hawk BFC-2 fighter and dive bomber was an improved version of the justly famous "Goshawk", used internationally from 1925 right through the early years of World War II.

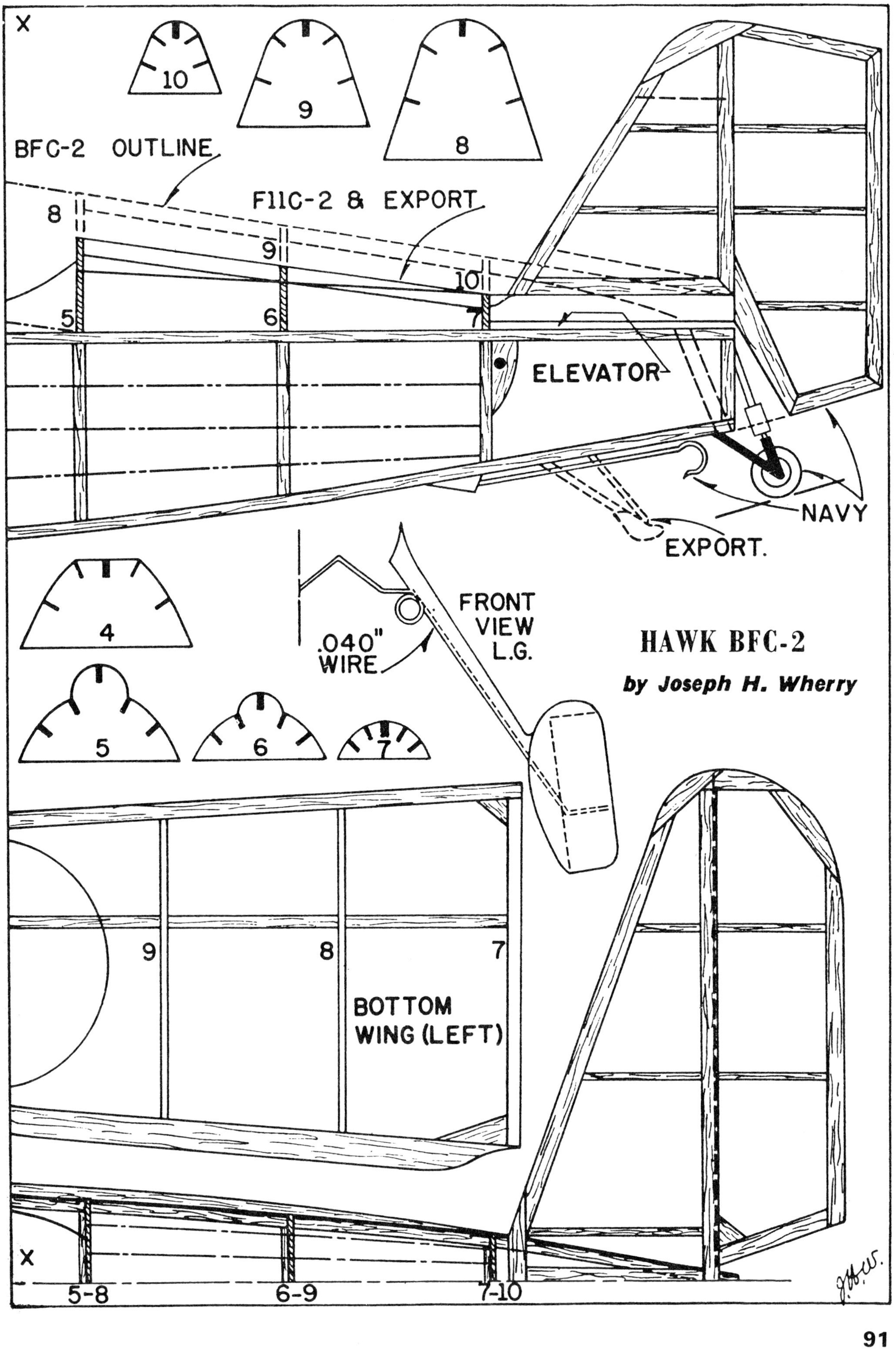
X
10
9
8
BFC-2 OUTLINE
F11C-2 & EXPORT
8
9
10
5
6
7
ELEVATOR
NAVY
EXPORT.
4
5
6
7
FRONT
VIEW
L.G.
.040"
WIRE
HAWK BFC-2
by Joseph H. Wherry
9
8
7
BOTTOM
WING (LEFT)
X
5-8
6-9
7-10

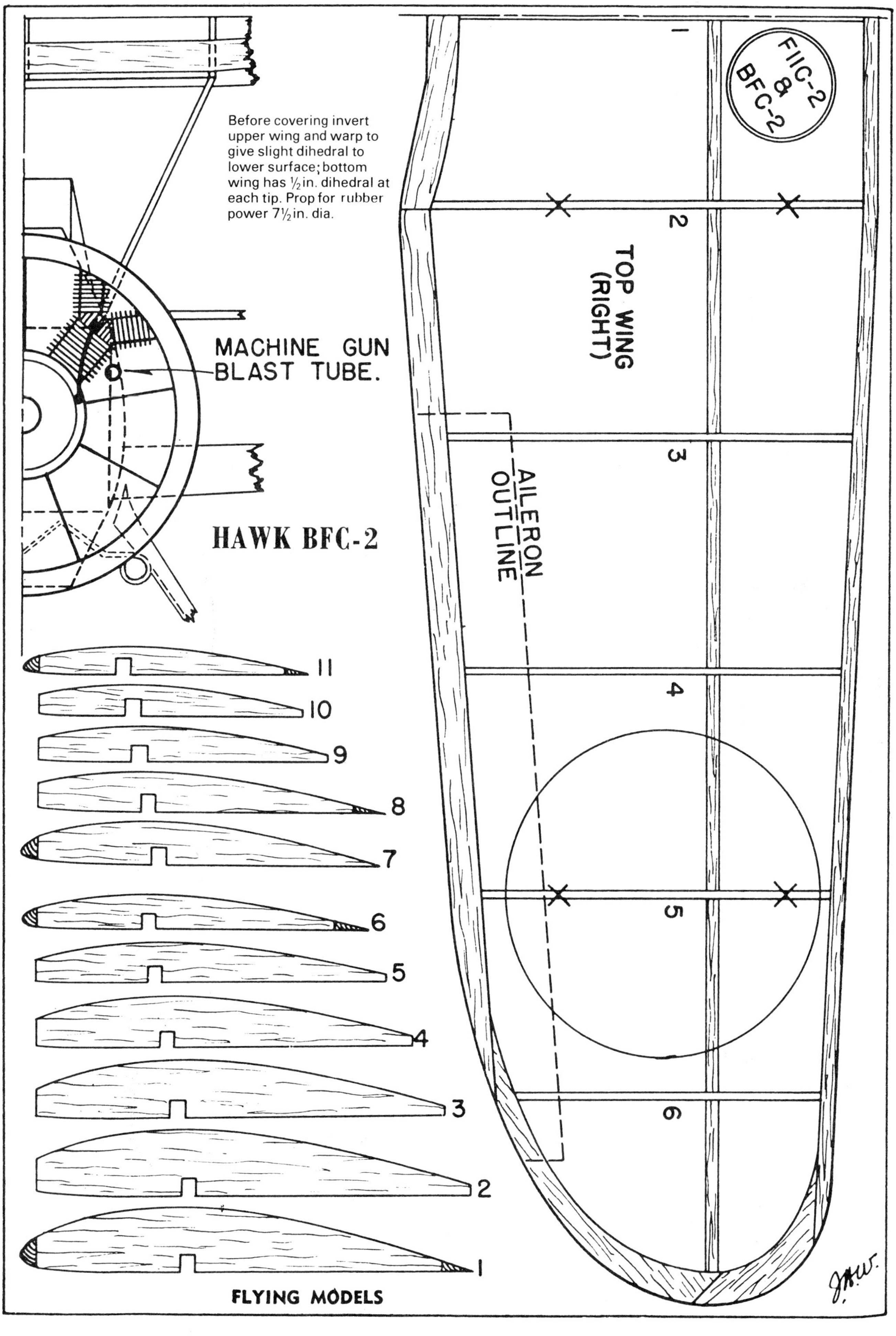
Before covering invert upper wing and warp to give slight dihedral to lower surface; bottom wing has ½in. dihedral at each tip. Prop for rubber power 7½in. dia.
MACHINE GUN BLAST TUBE.
HAWK BFC-2
FIIC-2 & BFC-2
TOP WING (RIGHT)
AILERON OUTLINE
1
2
3
4
5
6
7
8
9
10
11
FLYING MODELS

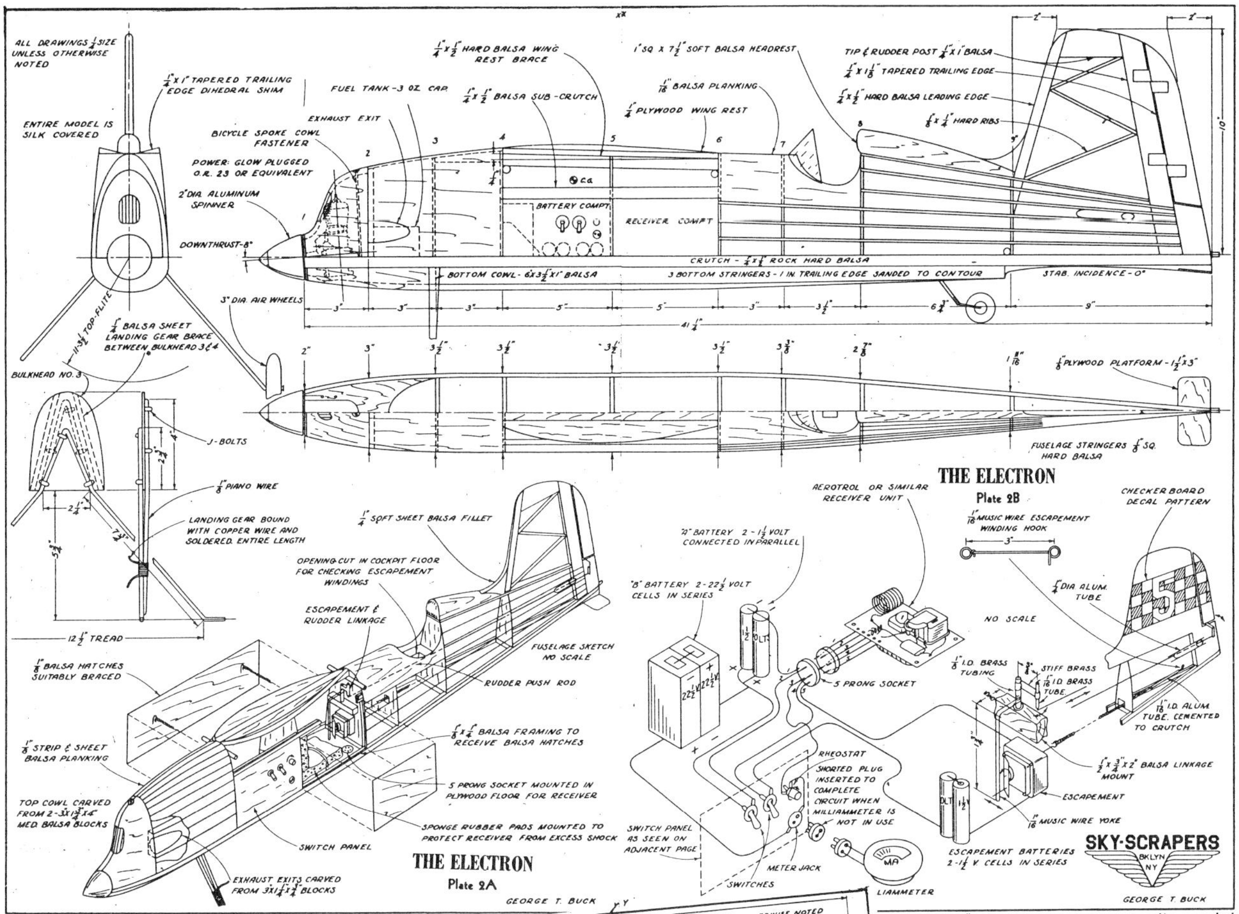

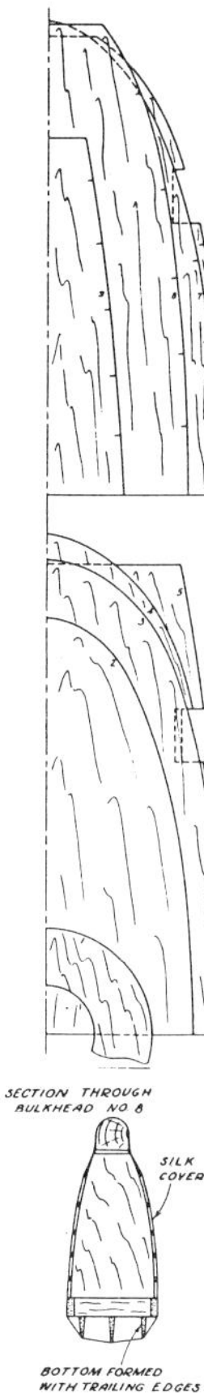

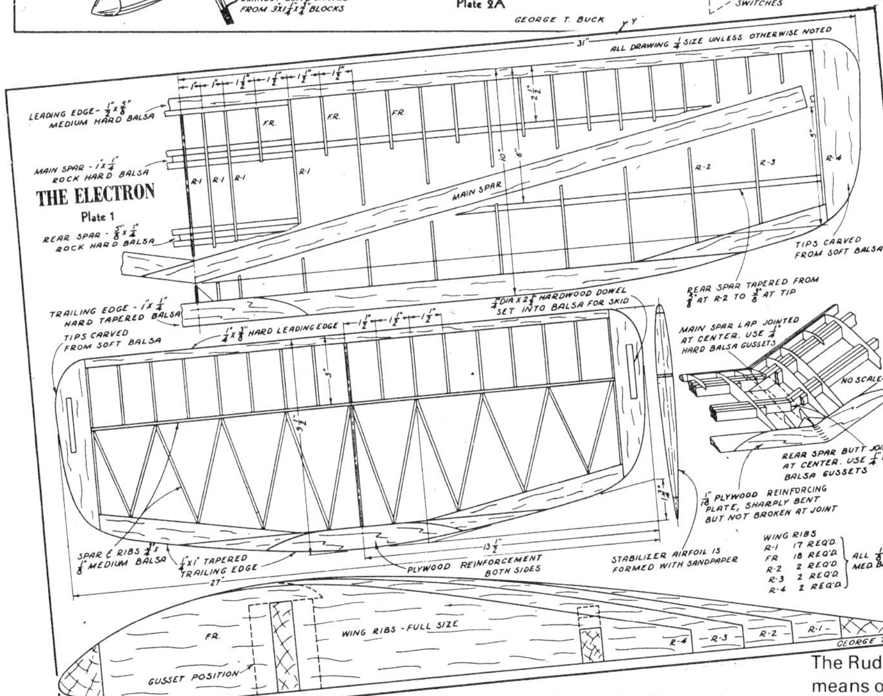

'The Electron' was a 60in. radio model by Norman Rosenstock, published in December 1950 *Flying Models*. Power was an Ohlsson 23 or similar and the original model weighed no less than 5½lbs.

In mid-1950 a single transmitter to operate on 27mc (MHz nowadays!) was licensed in the U.S.A. for the AMA to evaluate the potential of this waveband for model control. This same frequency was in use by radio enthusiasts in Britain

Americans needed an amateur radio licence, which involved a written test and the ability to take Morse at 13w.p.m.!

The Rudevator was a clever means of extending the controls available beyond the simple rudder-only control which was all that most of the early equipment would provide.

A typical valve (tube) receiver operated a sensitive relay, closing its contacts and thus triggering the rubber-powered escapement (above) applying left or right rudder. Memory was involved — which rudder comes next?

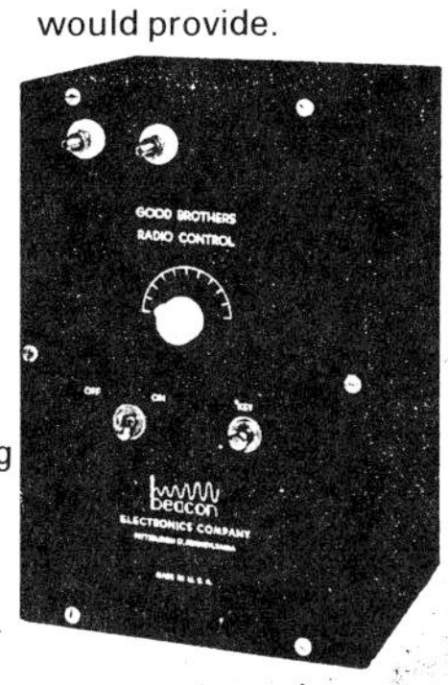

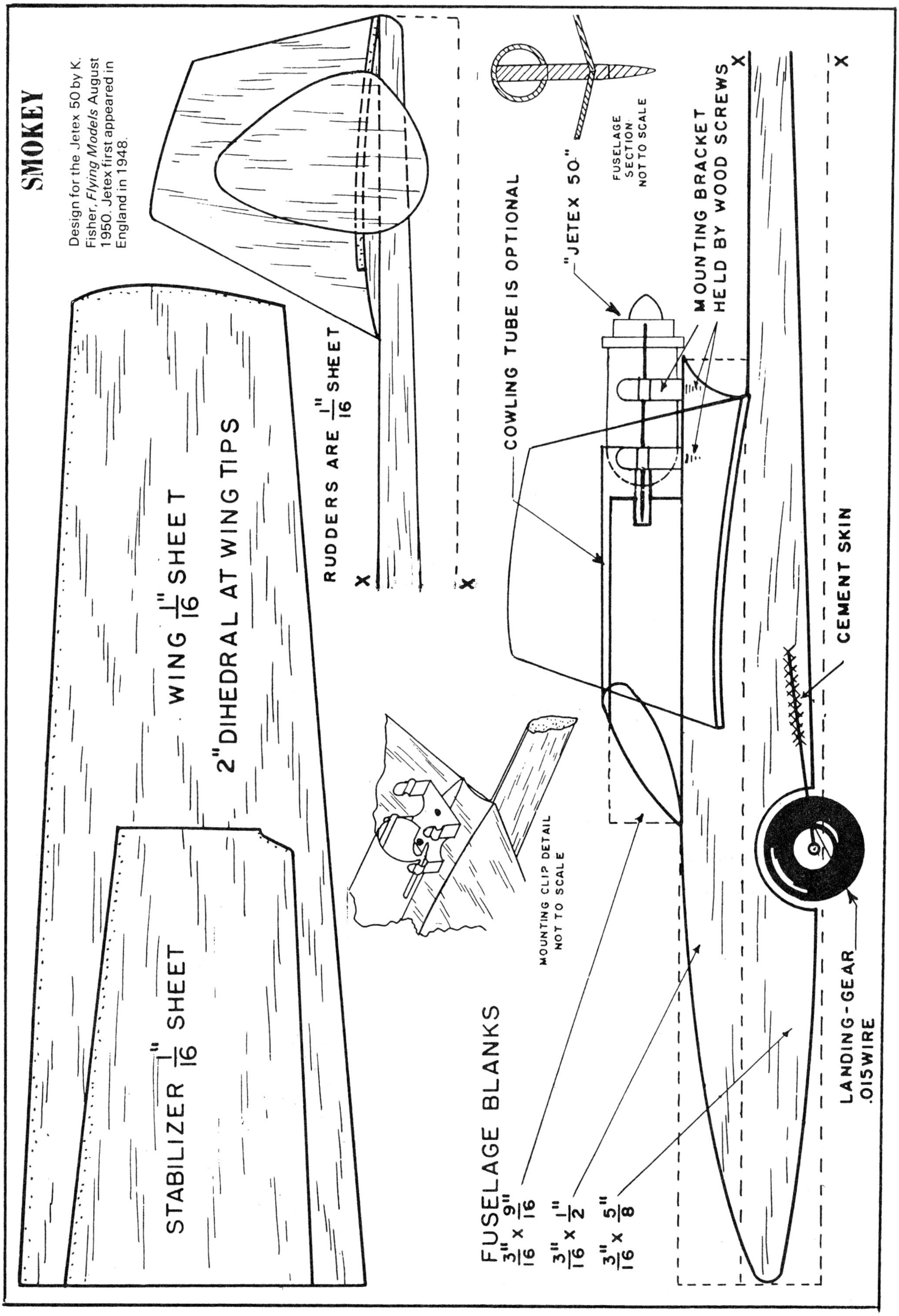

Design for the Jetex 50 by K. Fisher, *Flying Models* August 1950. Jetex first appeared in England in 1948.

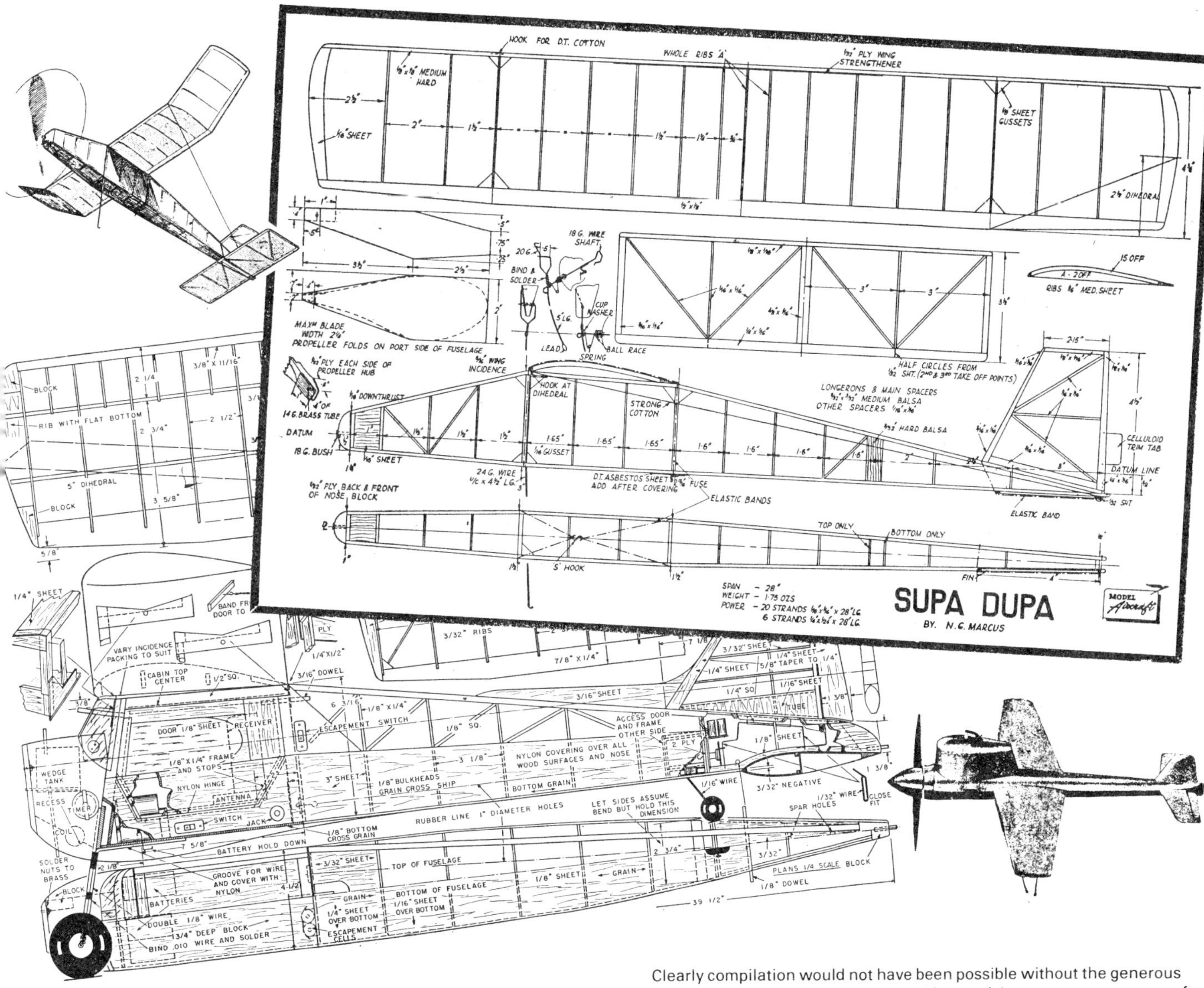

Conclusion

As well as ending the half-century, 1950 could also be regarded as something of a watershed year, a time when the explosion of post-World War 2 activity began to level off but when future development was also clearly signalled. People had stopped making 'solid' or 'shelf' models, but plastic kits were just over the horizon; control-line flying had become well established in stunt and speed, though team-racing had only just started. Gliders had attracted a growing following and were about to enter a 'large phase' — 10-12ft. span — before settling to the A2 size. Jetex power attracted a following and lightweight rubber models were popular. Power models continued to develop as engines grew more powerful. The Wakefield remained the supreme model and the major international competition category, though soon to be rivalled by other classes, and radio control as a commercial proposition was just beginning.

Looking back at the types of models and the way we flew them in competitions tempts the suggestion that 1950 was virtually the end of the age of innocence and that since then things have become a great deal more sophisticated, not only in constructional techniques but in tactical flying and the like. Since aeromodelling continues to enjoy world-wide support its attractions, even with complications,obviously still exist, though it is noticeable that the average age of model flyers is gradually increasing in most countries and that the vintage movement, concerned with building and flying pre-1950 models, has been making rapid progress over the last few years. If you have enjoyed this book, you are very probably a potential vintage flyer; if you have enjoyed it but with reservations in respect of omissions, you are quite likely to be an actual vintage flyer, or perhaps an enthusiast who would be flying if he could find the time?

Clearly compilation would not have been possible without the generous cooperation of publishers, copyright holders and, in some cases, owners of copies of early magazines and books. In particular we have leaned heavily on *Model Airplane News* for the American scene in the late '30s and much of the '40s, for a number of reasons. It was the most model-orientated of the U.S. publications, the others carrying a fair amount of full-size and fictional material, and it came nearest in those days to an official journal of the organising body. For almost all of the period it was printed on good quality white paper, which helps with reproduction, but more importantly its method of presentation was more condensed; it would have been a pleasure to include more drawings from, for example, *Flying Aces*, but on average they extended over five or six pages and would have rapidly filled this book without giving quite the variety sought. A further factor favouring *M.A.N.* was the availability of a virtually unbroken run of copies from the mid-1930s on, partly as a result of an agreement for use of each other's material, if desired, between *Aeromodeller* and *M.A.N.*

As well as excerpts from our own titles such as *Model Engineer, Model Aircraft, Models for Flying*, etc., we gratefully acknowledge the following publications and individuals:

Airborne (Australia)
Air Trails (U.S.A.)
American Boy, The (U.S.A.)
Flugmodell (Germany)
Flugsport (Germany)
Flying Aces (U.S.A.)
Flying Models (U.S.A.)
Hobby-Boken (Sweden)
Model Aeronautical Digest (G.B.)
Model Airplane News (U.S.A.)
Model Aviation series (G.B.)
Modele Reduit d'Avion (France)
Modelltechnik (Germany)
Plane Model Winners (U.S.A.)
Practical Mechanics (G.B.)

John Blake
Ken McDonough
Ron Moulton
Geoff Nicholls
Frank Zaic

CONTENTS

INTRODUCTION

ONE OF THE MOST FASCINATING aspects of London bus operation is the appearance of a type of bus operating a given route that it does not normally work. In busmens' and bus enthusiasts' parlance this is an unscheduled working, a strange visitor or an odd man out. 'Odd Men Out' was the title of the very first transport book I compiled, published by Capital Transport detailing such workings in 1977. Along with a second edition published in 1981, this book is long out of print, as also is 'The Odd Spot' on the same theme, which I compiled for the North London Transport Society in the 2000s. So I felt the time was right to look at this interesting subject again in this new volume.

During the years when I first took a serious interest in London Transport bus operations, the late 1950s and early 1960s, the allocation of bus types to the routes they operated was very rigid in LT's Central Area and in any case, the fleet had achieved a very high degree of standardisation by 1956, when all double-deckers, except for the 76 low-height RLHs and the first RM, were of the RT family. Thus the scope for seeing 'strange visitors' on routes they did not usually operate was almost non-existent. All that could usually be seen was an RTL working on loan to a garage that had RTs, and vice-versa, or maybe one of the rapidly dwindling number of TD-class single-deckers standing in for an RF.

Things began to change once London's trolleybuses had been replaced between 1959 and 1962. Over a thousand Routemasters entered service to replace the them, and since some trolleybus depots and bus garages which had been near one another were closed in the interests of economy as the Trolleybus Conversion Programme progressed, RMs began to be housed in the same garages as RTs, RTLs or RTWs. This led to RT-types sometimes working former trolleybus routes normally allocated RMs, usually on a scheduled basis where one of two RTs of RTLs would have a couple of turns on RM routes in rush hours. At first, the converse, RMs working routes normally RT, RTL or RTW-operated was less common, though as time went by, they were actually allocated to such routes at weekends when many of them would otherwise have stood idle.

However, owing to the busmens' unions being very stringent as regards type-to-route allocations (largely because at first staff were paid more to operate an RM than an RT owing to the former's higher seating capacity), the appearance of completely unscheduled types, for instance an RM on a route always RT or RTL operated, was very rare indeed.

Fortunately, between 1965 and 1978, I worked in Central London and usually commuted from North London by bus. I always carried a camera with me, and in addition, frequently travelled around London at weekends and on days off. Therefore I was often 'in the right place at the right time' when very rare workings appeared.

In the early 1980s, when more different types of bus were operating for London Transport, the strict union guidelines were relaxed somewhat, resulting in such extreme cases as single-deck Leyland Nationals working on routes usually double-deck operated, for instance the 271, when the need arose.

This collection of photographs is not intended as a complete tabulation of all unscheduled and unusual, seldom seen workings - that has been done before in my 'Odd Men Out' volumes, but rather shows a cross-section of them from the 1960s up to the mid-1990s, when London Buses was privatised, and in any case many routes tendered out to other operators thanks to government policy. Many of the pictures show very rare workings indeed; also most have not been published before. They are presented in route number order, and then by date.

This volume generally does not cover special 'last day' workings of odd types on routes' last days of crew operation, or their last days of all. That too has been done before, in my book 'Farwell Performance', which I produced for NLTS in 2006.

I hope readers will find the selection of photographs I have chosen here interesting and enjoyable; many of them show aspects of life in London in the second half of the last century other than just the buses themselves, thus adding to the nostalgia. May I thank my publisher Ken Carr for encouraging to compile this volume, also Colin Clarke who spent about six years scanning all my original negatives! Thanks are also due to the P.S.V. Circle and the London Omnibus Traction Society, on whose records of London bus workings many of my records, from which comments herein are derived, were originally based.

Jim Blake
Palmers Green
20th July 2017

This book is dedicated to my two wonderful daughters, Margaret and Felicity.

Main Cover Photo: One of the most unusual of all strange visitors to appear in the period covered by this book was West Ham Garage's RTL502, substituting for an RML on route 249 on 21st August 1968. This view sees it at Chingford, Royal Forest Hotel terminus. By this time, such a working was very rare indeed and just over three weeks later, on 7th September 1968, all RTLs in the East End perished. Route 249 was also withdrawn that day, and the long-serving bus terminus abolished too!

Rear Cover Photo: Route 217B was never scheduled for RM operation, having been introduced with RTs in April 1976 and converted to DMS OPO in August 1977. However odd RMs from Ponders End Garage's small allocation for route 279 occasionally strayed onto it, especially towards the end of crew operation. On 16th August 1977, RM2210 has a crew change on the Great Cambridge Road near its junction with Southbury Road just four days before OPO conversion. Route 217B was subsequently renumbered 317, and still runs today.

1 - 14

Right: Until the end of 1966, Willesden Garage had a small allocation of RTLs on route 1 on Saturdays, which ran alongside the route's main allocation of New Cross RTs. RTL1583 (AC), one of the last RTLs to be overhauled a few months previously, is seen in Waterloo Road on 30th July 1966 - the day England won the World Cup!

Below: Route 1, by now operated by RTs from Catford and New Cross Garages, converted to RM operation in March 1975, but RTs from the former continued to stand in for RMs on it until the very end of RT operation, thus gaining the route the distinction of being the very last to be RT-operated through the West End. On 23rd August 1978, just a couple of days before then, RT4210 (TL) heads south from Elephant & Castle.

Route 1 was never scheduled for RML operation, but in its last couple of years of RM working, odd examples from New Cross Garage's share of route 37 found their way onto it. Here, RML2662 (NX) crosses Waterloo Bridge on 20th October 1985. It is working short to Surrey Docks Station, possibly as a result of a staff cut.

Left: RMs replaced RTLs on routes 2 and 2A at Stockwell Garage shortly before those based there for other routes (the 77A/C, 168 and 181) were replaced by RTs in the summer of 1967. RTs then put in odd appearances as substitutes for RMs. On 27th February 1968, RT206 (SW) crosses Vauxhall Bridge on a short working to Victoria.

Below: Appearances of RTs substituting for RMs on route 2A were somewhat rarer. However RT3973 (SW) has found its way on to the route on 30th August 1973, it arrives at its West Norwood, Rosendale terminus. Route 2A was withdrawn in January 1974.

Route 2B had been shared between Norwood RTs and Stockwell RTLs prior to its conversion to RM in the autumn of 1966. Appearances of Stockwell RTs on it after the RTLs had gone were much rarer than on the 2 and 2A. On 5th April 1973, RT2629 (SW) is working the 2B at Vauxhall Station. This RT was later preserved and I was a co-owner of it in the 1980s.

RMs were allocated to route 3 late in 1964, replacing RTs from Norwood Garage and RTLs from Chalk Farm. After the latter had been ousted by RTs on their other routes, these RTs occasionally appeared on the 3 as RM substitutes. Complete with upper-case via blinds, RT10-bodied RT4273 (CF) passes Kennington Park on 7th June 1968.

Chalk Farm's Sunday allocation on route 3 was converted to crew DM operation when the 24 received these contraptions in October 1975, though their actual appearance was erratic. However, on 25th July 1976, DM1142 (CF) has made it to Crystal Palace and is turning into College Road, Dulwich on its way back to the garage. RMLs returned to the 24 in April 1979, rendering these workings defunct.

At various times, RMLs from Chalk Farm Garage were allocated to route 3 at weekends, though their actual appearance was also erratic. On 16th January 1983, RML2380 (CF) heads away from Crystal Palace through Gypsy Hill for the long journey to Camden Town.

In the last few years of its crew operation, RMLs from Camberwell Garage appeared on route 3 substituting for RMs. On 27th June 1992, former London Country RML2411 (Q) is approaching Herne Hill Station. It has been very unusually curtailed at Millbank owing to the Gay Pride march which ended in nearby Brockwell Park causing severe traffic congestion.

Route 4 in its present form was introduced on 4th September 1971 with RTs, running from Tufnell Park to Farringdon Street on Mondays to Saturdays, with a Monday to Friday extension to Waterloo. On 6th July 1972, RM1065 (HT) leaves the latter point to where RMs were not scheduled to work the 4 at this time, as they were only allocated on Saturdays. The route was converted to crew DM in June 1975, and it was not until February 1981 when RMs replaced them that these ran to Waterloo on a scheduled basis.

Above: In its last few months of crew operation, route 4 also had unscheduled RML workings. On 29th February 1984, RML2735 (HT) is about to pass Islington Town Hall in Upper Street.

Left: Route 6 converted from RTW to RM operation in the spring of 1965, and then to RML in the summer of 1966. However, until their demise from Hackney Garage on 7th September 1968, RTLs made rare appearances on the route. On 19th December 1967, RTL1321 (H) is seen in Fleet Street heading for home.

Route 6A converted from RTW to RM and then RML operation along with the 6. Until the route's withdrawal on 7th September 1968, RTLs occasionally substituted for Routemasters. RTL1592 (H) stands at Cornwall Road bus park, Waterloo on 31st January 1967. Hackney's RTLs were withdrawn on the same day as the 6A itself.

RMs replaced RTLs on route 7 in the autumn of 1963, some two years before the remaining RTLs at Middle Row Garage working routes 187 and 187A were replaced by RTs. Until they too were replaced by RMs early in 1975, RTs substituted for RMs on route 7. On 14th March 1973, RT3143 (X) overtakes three Routemasters at Oxford Circus.

Following the closure of the cramped, antiquated North Kensington, Middle Row Garage in August 1981, crew-operated DMs allocated to the replacement Westbourne Park Garage for route 18 sometimes appeared substituting for RMs on route 7. DM2579 (X) negotiates the junction of Westbourne Grove and Chepstow Road on 5th September 1981. Its conductor, complete with Gibson ticket machine, may be clearly seen standing next to the driver.

Route 8 was an RTW stronghold until they were replaced by RMs early in 1965. After that, RTLs from Bow Garage's large allocation for routes 10, 25 and 26 sometimes substituted for RMs until their own demise. On 15th March 1967, RTL342 (BW) heads along Proctor Street, Holborn in the morning rush hour.

A very rare working indeed on 26th February 1968 is Saunders-bodied RT3123 (BW) on loan from Seven Kings Garage to Bow, subbing for an RM on route 8 in Threadneedle Street. Bow did receive RTs to replace their RTLs just over six months later, and these put in the odd appearance on the route until 1972.

Route 8A had also been RTW-operated prior to conversion to RM early in 1965, after which RTLs based at Bow Garage occasionally substituted for RMs until their replacement by RTs on 7th September 1968. On 27th April 1967, RTL1315 (BW) crosses the Great Eastern main line in Shoreditch High Street.

In the summer of 1963, route 9 converted from RTL to RM operation from Dalston and Mortlake Garages. The change removed RTLs from the latter, but they remained at Dalston for routes 47 and 78 until June 1968, so still made rare appearances on the 9 as substitutions. On 19th January 1967, RTL1313 (D) heads west at Aldwych.

For five years after route 9's conversion to RM operation, the route had a long Sunday extension to Becontree Heath, which replaced weekday route 23. Barking RTs had an allocation on it, running alongside the RMs. On 21st April 1968, Saunders-bodied RT2219 (BK) is seen at Plaistow, Greengate on its long journey to Mortlake. The extension was withdrawn upon the first stage of LT's Reshaping Plan on 7th September 1968.

RTs based at Dalston Garage for routes 47 and 78 occasionally substituted for RMs on their share of route 9 until withdrawn from the garage in January 1975 - the 9 had been RTL operated prior to receiving RMs. On 13th August 1973, RT4765 (D) is seen down the 'Dilly. It is one of 34 RTs reacquired from London Country in September 1972 to help with shortages of the type. The RT following is also substituting for an RM, on route 38.

RMs were only allocated to route 10 (Victoria - Abridge) at weekends, and its Sunday extension to the latter point ceased in September 1968. However, in its last year or two of crew operation, RMs allocated to Bow Garage for routes 8, 8A and 25 sometimes found their way onto the route during the week owing to a shortage of RTs, and therefore, worked that section of route. On 27th September 1972, a month before the 10's conversion to DMS, RM2097 (BW) heads for Abridge, passing Wanstead Station, on one of the 10's handful of evening rush-hour workings.

London's famous 11 route was RTW-operated for many years prior to conversion to RM early in 1966. Until replaced by RTs in June 1968, RTLs allocated to Dalston Garage for routes 47 and 78 put in appearances on the 11, as illustrated by RT10-bodied RTL1427 at Parliament Square on 8th November 1967. Only 27 RTLs ever carried roofbox bodies, 23 of them having been overhauled in 1964/65, as this one was. It is now preserved.

RTs from both Dalston and Riverside Garages worked the 11 as RM substitutes after the RTWs had gone. RT4452 (R) from the latter is at Charing Cross on 18th August 1966.

Long before RMLs were officially allocated to the 11 following the opening of the new Ash Grove Garage in 1981, at least three of them worked the route from Riverside Garage on Saturday, 13th December 1975 for some reason, possibly because on that day RMs replaced RTs on route 72 at Riverside, and there may not have been enough of them to go round on the first day. The RMLs were allocated there for route 74B, which ran only on Mondays to Fridays. This view shows RML2605 (R) setting off from Brook Green, Hammersmith for Aldwych.

The last RT-type to work route 11 in normal service was RT3871 (D), which was sent to Dalston Garage for a few days in May 1977 to help out with a shortage of RMs. On 2nd May 1977, it calls at St-Martin's-In-The-Fields on its way from Victoria to Liverpool Street. By coincidence, forty years later this RT survives in the London Bus Company's heritage fleet, and has specially worked the 11 again in recent years. The 1930s London Transport shelter on the right is also of note.

Long before route 12's protracted conversion from RT to RM operation during 1973, RMs from Shepherd's Bush Garage had a small Saturday allocation on the route. Their RM297 (S) rounds a snowy Marble Arch on 8th February 1969.

For a time after their appearance on route 45 in February 1975, crew DMs operated by Walworth Garage occasionally substituted for RMs on route 12. DM1130 (WL) loads up at Shepherd's Bush Green on 29th February 1976. Subsequently, these DMs theoretically had a small Sunday allocation on the 12 prior to the 45 regaining RMs early in 1981.

Conversely, Shepherd's Bush had some extra RTs during the spares shortage of 1976 to help their RMs out. One of them, RT2992 (S), has terminated at Charing Cross on 1st October 1976, and is about to head back to Harlesden.

Route 12 also had visits from RMLs long before they were allocated to the route daily early in 1987. This resulted from route 88's conversion to RML in February 1975, from which some of Shepherds Bush Garage's share had Saturday workings on route 12. One of these, RML2687 (S) is in Cockspur Street on 13th December 1975.

More than four years after route 12's long-protracted conversion from RT to RM operation was completed, RT4053 (WL) is one of a handful of RTs also kept as spares to cover for RM shortages at Walworth Garage, and will in fact be the very last such allocation. The flags are out for the Queen's Silver Jubilee as it works an evening rush hour trip to Park Royal down the 'Dilly on 28th September 1977. As may just be discerned, it is overheating.

During the four years or so that the MD class Metro-Scanias were based at Peckham Garage for the 36 group of routes, they were scheduled to work route 12 on Sundays from October 1977, alongside the RMs that did so from its other garages. With Christmas decorations as a backdrop, MD20 (PM) is in Regent Street on 16th December 1979. These workings ceased four weeks later upon the 36 group's reversion to RM operation.

Following the relaxation of union rules regarding type-to-route workings, some DMs and DMSs were reclassified D and equipped to work as either OPO or crew buses. A flap beneath the windscreen told passengers whether to pay the driver or the conductor. DM2540 (S) substitutes for an RM on route 12 at Elephant & Castle on 24th August 1981. In the end, RMs and RMLs survived on route 12 until November 2004, when they were replaced by bendi-buses!

Another instance of an OPO vehicle being used in crew mode to substitute for a Routemaster is M567 (FY), shown passing Finchley Road Station on 2nd September 1982. At this period, route 13 was operated by RMLs from Finchley Garage. Metrobuses subsequently worked the 13 on Sundays, but as OPO vehicles. This one correctly shows 'pay conductor'. They were never allocated to the 13 daily in the end, since it retained RMs until October 2005, nearly 43 years after the Routemasters were introduced onto it!

Above: An oddity to appear on route 13 shortly after it was tendered out to BTS and worked by refurbished RMLs leased to them by London Buses was RM104, which had actually been purchased by BTS for use as a spare. As its re-registration suggests, it had initially been sold for use in the Glasgow area to Clydeside Scottish. This view shows it about to turn from Pall Mall into Waterloo Place on 3rd July 1994 - the driver seems to have forgotten to change his blind for the return journey to Golders Green!

Above: Route 14 was converted from RT operation (Holloway Garage) and RTW (Putney) to RM operation in the autumn of 1963, then to RML operation in the winter/spring of 1966/67. Odd appearances by RT-types continued to be made after this, as shown by RT2562 (J) in the company of new RML2604, also a Holloway bus, at Hyde Park Corner on 11th July 1967. By coincidence, this RML (numerically speaking, anyway!) worked route 14 on its last day of crew operation almost exactly 38 years later!

Right: RTs reappeared on route 14 in the autumn of 1976, from both Holloway (by now the former trolleybus depot) and Putney Garages, when allocated to cover shortages of RMs and RMLs. On 21st October 1976, a very shabby RT3257 (HT) which had not been repainted for more than nine years runs down the 'Dilly in pouring rain. RTs from this garage worked the 14 as late as October 1977.

15 - 21

Route 15 also converted from a mixed allocation of mainly RTWs plus a few RTs to RM operation in the spring of 1964, and then to RML in the summer of 1966. However, a few RTs from its Upton Park allocation still had scheduled duties on it, and in later years also worked as substitutes for the RMLs, as well as from its weekend RM allocation at Middle Row. On 21st February 1968, in pouring rain, Saunders-bodied RT1525 (U) approaches Blackwell Tunnel and appears to be running empty!

More exotic was the use of former Green Line coach RMCs on route 15. A batch of these had been resurrected from training duties (for which they had been re-acquired from London Country some ten years before) to work new Docklands Express route X15, with which they interworked the 15. On their first day, 6th March 1989, gleaming RMC1490 (U) is seen on diversion along Mincing Lane in the City.

Even more bizarre was the use of two former British Airways forward-entrance RMAs in normal service on route 15! RMA8 (U) is seen at Trafalgar Square on 28th January 1994. These too had originally been refurbished for Docklands Express route X15, and had originally seen London bus service on route 175 in 1975/76.

Upton Park's RMC's also appeared on the short-lived route 15B, when it followed the 15's old routeing via Bank rather than its later one via Tower Hill. On 9th August 1989, RMC1456 (U) passes Mansion House Station.

Route 16 was one of the first to convert from RT to RM operation, doing so at the end of 1962. Shortly after this, the remaining RTs at its garage, Cricklewood, were exchanged with RTLs displaced by RMs elsewhere. However, RTLs occasionally worked the route after this, such as RTL431 (W) at Victoria Bus Station - the Red Arrow XMS beside it proves this is more than three years after the 16 gained RMs. In fact the date is 7th June 1967.

In turn, once Cricklewood lost its last RTLs (on routes 142 and 240) in October 1968 and regained RTs to replace them, these too appeared on route 16. RT938 (W), whose driver has forgotten to change the destination blind, leaves Victoria in the evening rush hour on 28th March 1969, with an XMS in pursuit. RTs finally left Cricklewood in January 1971.

Route 16 converted from RM to crew-operated DM's in the autumn of 1973. During their disastrous tenure of the route, a trial batch of five Metrobuses was placed on the route for evaluation in the summer of 1976. One of these, M4 (W), departs from Victoria on 21st April 1980, shortly before route 16 (and the associated 16A) converted to RML operation, largely using vehicles re-acquired from London Country. However, the Metrobus proved a success with London Transport and successors, and in fact examples replaced the RMLs on the 16 and 16A in November 1987; the type remained at London's service until the early 2000s.

On the same day as the previous picture, M2 (W) is at Marble Arch on the 16A. This route originally ran from Victoria to the new Brent Cross Shopping Centre, effectively being a short working of the 16 that previously terminated at Cricklewood Garage extended and diverted to the new terminus. Subsequently, in April 1981, the 16A was diverted at its southern end to run to Oxford Circus, rather than Victoria, as the next picture shows.

Above: Route 16A never had a full RM allocation, starting life with crew DMs, then converting to RML in May 1980 and to Metrobus in November 1987. However, odd RMs from route 266's allocation at Cricklewood appeared on occasion. On 28th April 1984 RM74 (W) has just departed from Brent Cross Shopping Centre and pulls away from the Tilling Road stop alongside the North Circular Road.

Above: Trolleybus replacement route 17 was one of several working from Holloway Garage which converted from RM to crew DM operation in January 1975. After that, RM's seldom appeared on the 17, but on 10th April 1975, RM2120 (HT) did, it crosses the junction of Grays Inn Road and Theobald's Road in Holborn. Most of the other crew DM routes at Holloway reverted to Routemaster operation in 1981/82, but by then the 17 had been withdrawn as part of a series of route changes in October 1978. When it was reintroduced in August 1985, it was operated by OPO Metrobuses.

Right: During the time period covered by this book, there were relatively few instances where RTWs appeared as oddities, since most were used on busy routes serving the City and West End. An exception however, was a small Saturday allocation of them at Willesden Garage for route 18, using those operated on route 176 during the week. This in a way was a remnant of the former 18B route which had used them prior to being replaced in January 1962 with the long daily extension of the 18, also replacing the 662 trolleybus, with the result that the 18 at its fullest form ran from London Bridge Station to Aldenham LT Works - albeit in overlapping sections! On 22nd May 1965, RTW481 (AC) arrives at Paddington Green, a few weeks before RTLs replaced them on these workings.

Representing the Willesden Saturday RTL allocation on the 18, RTL626 (AC) has turned short at Craven Park on 24th August 1968, a fortnight before this allocation ceased. This RTL was itself an oddity, having been numerically one of the 450 Metro-Cammell-bodied RTLs (RTL551-1000) whose bodies could not be interchanged with standard Park Royal or Weymann ones upon overhaul. Yet it was outshopped with a standard body in January 1962, apparently by mistake, then overhauled again in 1965! RTLs based at Stonebridge Park Garage for route 112 also occasionally substituted for RMs on route 18.

In April 1979, route 18 exchanged its RMLs with route 24's crew DMs, apparently as a result of requests by crews for doored buses following mass-muggings of conductors and passengers in the Harlesden area. However, in the spring of 1980, a few RMLs returned to Stonebridge Park and worked route 18; two newly-overhauled examples, RML2480 and RML2487 (SE) are in the garage yard on 10th May 1980. They moved on to convert routes 16 and 16A from DM to RML operation a few days later.

Above: In the summer of 1983, comparative trials were carried out on route 18 (then worked by crew-operated DMs) between a handful of new Metrobuses and Leyland B15 Titans. These were based at the new Westbourne Park Garage which had replaced Middle Row and Stonebridge Park two years earlier. On 23rd July 1983, T794 (X) passes the Jubilee Clock, Harlesden. Two years later, the Titans moved away leaving the 18 operated by Metrobuses, now working in OPO mode. It had in fact had them for a year or so in 1980/81 working from Stonebridge Park Garage before it was closed.

Centre: On 9th March 1975, RML2647 (HT) works what should have been a Sunday allocation of RMLs on route 19, and is in St. Paul's Road, Highbury. However, in practice, RMLs were seldom used, with the weekday allocation of RMs sufficing instead. By the time route 19 had gained a full RML allocation, the route had long since ceased to run to Tooting Bec Station, to where this one is heading.

Bottom: Route 19 was very late, for a busy Central London route, in gaining RMs, which it did not do until August 1972. However, for three years after that RTs from Holloway Garage still substituted for them until finally ousted from their share of route 172. On 23rd July 1975, RT689 (HT) crosses Islington Angel junction, as RM2124 waits to turn right into Pentonville Road on the 30, a route which had gained RMs in the summer of 1964. A batch of RTs returned to Holloway in the autumn of 1976 to cover for RM shortages, and often appeared on route 19. The impressive Blue Coat Boy pub, and the shops behind it, were all demolished in the early 1990s to make way for a new office complex embodying the enlarged Angel tube station.

Some of the last workings of the unscheduled RTs used for RM cover during 1976/77 were at Holloway Garage on route 19, a route I have known since early childhood, when the post-war RTs were new. On 12th November 1977, RT1069 (HT) arrives at Highbury Barn terminus, where three RMs already occupy the stand. This particular RT had been new to the Country Area, being overhauled in red in 1965.

By virtue of their being used mainly on routes where single-deckers were needed owing to low bridges or weight restrictions, RFs were not often 'strange visitors' on routes they were not scheduled to work. However, on 25th September 1973, RF685 (L) has been pressed into service on busy route 20, then the preserve of the troublesome MB class. It is seen stuck in heavy rush hour homegoing traffic at Waterworks Corner, Woodford. At the time, a small number of RFs were based at Loughton Garage for route 254. This one was one of a batch of Country Area RFs transferred to the Central Area in January 1969 and subsequently repainted red, then overhauled again. It was the highest-numbered RF ever to run in red livery.

Route 21 converted from RT to RM operation on 1st March 1975, ironically the day after the Moorgate tube disaster which caused disruption to their 'final' day of operation. Needless to say, however, RTs continued to operate the route from Sidcup Garage for more than two years afterwards, substituting for RMs. On 25th March 1977, a very shabby RT219 (SP) stands at Moorgate, Finsbury Square terminus. Its front in-house LT adverts leave no doubt that this is long after RM conversion; those for cigarettes on the side now seem to belong to another world!

22 - 30

Route 22 converted from RTL to RM operation in the summer of 1967, when it was operated by Hackney and Putney Garages. Clapton took over the former's share on 7th September 1968 and subsequently RTs from both garages could be seen on the route substituting for RMs. On 2nd March 1971, I was turned to stone at Holborn Station on my way home from work to find the last surviving roofbox RT at London's service, Saunders-bodied RT1903 (B) working the 22 in the evening rush hour. This RT had survived all the others because it had been fitted with saloon heaters by mistake upon its last overhaul in 1966. However, just a day or two after this picture was taken, it suffered rear-end damage when working route 31, so was withdrawn.

On 21st November 1987, route 22 was officially converted to RML operation from Putney Garage, but cut back from Homerton to Tottenham Court Road Station. However, for a few days beforehand, RMLs from both Ash Grove and Putney Garages strayed onto the 22 and worked through to Homerton. Thus on 12th November 1987, RML2502 (AF) is shown turning from Clifden Road into Median Road on the one-way section between Homerton and Lower Clapton. New route 22B replaced the 22 here, later being renumbered 242.

The original route 23, which linked Becontree Heath with the City and West End, had a very complicated set of vehicle and garage allocations in the 1960s and 1970s, involving Barking, Upton Park, West Ham and Poplar. The latter had first begun to operate it following stage five of the Trolleybus Conversion Programme, using new RMs, in November 1959. However, by 21st August 1968, when RTL533 (PR) was at Aldgate Minories Bus Station, RTLs from that garage had scheduled rush hour workings on it. RTs replaced these on 7th September 1968, when the route was introduced on Sundays but withdrawn west of Aldgate. On that same day, all RTLs in the East End perished!

Despite converting to RM in the spring of 1964, route 23 had scheduled RT workings from Barking Garage on weekdays until its cut-back on 7th September 1968. On the same day as the previous picture, RT2393 (BK) escorts an RTL on the 25 through Holborn in the evening rush hour. It is about to pass what is now The Penderels Oak pub, on the right, has been one of my favourite 'watering holes' since Wetherspoons converted it to a pub in 2003. Barking RTs continued to work the 23 on Sundays, from its reintroduction that day on 7th September 1968 until the spring of 1974.

RMLs from Upton Park Garage's large allocation for the 15 could be found working route 23 alongside Barking RTs on Sundays in 1970. RML2597 (U) approaches the junction of Commercial Road and Burdett Road on 5th July that year. From 1978, RMLs from both West Ham and Upton Park worked the route daily.

One of the shortest-lived allocations to route 23 during the years this book covers was that of Titans from Barking Garage on Sundays, between January and April 1981. They worked alongside DMSs from Poplar, that type having worked from Barking on Sundays since the 23 converted to one-man operation that day in the spring of 1974. On Easter Monday, 20th April 1981, the last day of this arrangement; T180 (BK) passes Gardiners Corner, scene of the so-called Cable Street riot of 1936. The following Saturday, the 23 was recast to run between East Ham and Ladbroke Grove, largely duplicating the 15. It was actually renumbered 15 in 1985 running via Tower Hill with the 15's routeing via Bank renumbered 15A. Seven years later, the 15 was withdrawn west of Paddington, with a 'new' route 23 replacing it between there and Ladbroke Grove, paralleling it as far as Mansion House and then continuing to Liverpool Street. As this book goes to press, the 23 is due to be drastically revised again.

Above: Route 24 had been a bastion of RTW operation until RMs took over in the autumn of 1963. Occasionally RTLs worked it instead, such as Metro-Cammell-bodied RTL780 (CF) at Euston Square Station on 9th November 1963. They also appeared after RM conversion.

Right: Between November 1965 and May 1966, the experimental XA-class Leyland Atlanteans were allocated to route 24. They were prone to frequent breakdowns, and RMs used on Chalk Farm Garage's other routes usually covered for them. However, on rare occasions, RTs which had recently replaced RTLs and RTWs at the garage on routes 31, 39 and 45 also appeared. RT2467 (CF)passes a traffic policeman at Parliament Square on 21st April 1966.

The 24 was also chosen as a 'test-bed' for the prototype Leyland B15 Titan, at the time when it was operated by crew DMs from Chalk Farm Garage. Here it has turned short at Trafalgar Square on 19th July 1976. The Titan did indeed become 'London's Bus of Tomorrow', more than 1,100 of them were produced for London Transport and they shared with 1,400-odd Metrobuses the honour of replacing the awful DMSs! Some Titans survived at London's service until the early 2000s.

Above: Route 26 was introduced in July 1970, running from Victoria to New Barnet (later Barnet, Chesterfield Road) replacing parts of routes 2, 2B and 260. It was always RM-operated, but in August 1976, two RTs were sent to Finchley Garage to cover for RM shortages, and operated the route for over a year. They were noteworthy for their odd blind displays, as shown by RT2143 (FY), which had also been fully repainted, in Park Lane on 5th August 1976.

Right: Following the allocation of a number of RMLs to Finchley Garage for route 13 in October 1978, these were freely intermixed with route 26's scheduled RMs. RML2492 (FY) is in Lyonsdown Road, New Barnet shortly before the 26 converted to OPO, on 7th September 1980.

Right: Another instance of RTWs being oddities was their use on Riverside Garage's share of route 27 on Saturdays, running alongside RTs from Holloway and Turnham Green Garages. RTW429 (R) calls at Camden Town Station on the long journey from Archway Station to Teddington on 28th September 1965.

Below: Similarly, RMLs derived from Riverside's weekday route 74B were used on route 27 on Sundays in the late 1960s, before its daily conversion from RT to RM, and early 1970s. RML2655 (R) approaches Kentish Town Station on 23rd February 1969. From January 1974, Holloway RMLs also ran on Sundays.

RTs sent to Holloway Garage in the autumn of 1976 as unscheduled spares to cover for RMs occasionally worked on route 27 for a year or so. This busy scene in Richmond town centre on 5th March 1977 finds a very shabby RT2052 (HT), which had not been repainted for almost ten years, heading for Archway Station.

Also in 1977, a very odd vehicle to be found working route 27 was former Leeds Corporation Roe-bodied Daimler 7517UA, being used by Dennis as a 'test-bed' to develop new types of buses. Painted in all-red livery, it was based at Turnham Green Garage owing to its proximity to Chiswick Works. Here, it is well-loaded in Chiswick High Road on 28th September 1977, working short to Camden Town. Of note is the emergency exit on the offside behind the driver's cab.

A very unusual working on the 27 on 5th October 1983 is that of brand new Metrobus M909 (V) working as a crew vehicle, complete with 'pay conductor' flap is arriving at Archway Station. It has recently replaced DMSs on OPO routes at Stamford Brook Garage, which had replaced Turnham Green in May 1980. Crew-operated Ms were scheduled to work the 27 on Sundays from Fulwell Garage alongside RMs in 1981/82. Just over three years after this picture was taken, route 27 converted to OPO Metrobuses daily.

Above: In its last few years of crew operation, RMLs from Holloway Garage occasionally appeared on route 27 during the week as well as when scheduled on Sundays. On a very wet 21st October 1986, a week before its OPO conversion, RML2295 (HT) calls at Gunnersbury Station.

Left: Wandsworth was another garage to receive unscheduled RTs for RM cover in the latter part of 1976, and they were used on route 28. On 18th September that year, RT799 (WD) leaves RM1550 to pick up all the passengers on the first stop out of Golders Green Station!

On Christmas Day 1978, both DMs and DMSs from Westbourne Park Garage were used on route 28 instead of the scheduled RMs. DMS2345 (X) turns from Kensington Church Street into Kensington High Street. The 28 was never destined to have a daily allocation of DM-types, instead converting from RM to midibus early in 1989.

For a few weeks prior to its conversion from RM to midibus operation in the spring of 1989, RMLs based at Westbourne Park Garage for the 7 found their way on to route 28, too. RML2313 (X), a former London Country vehicle, has just passed Kilburn Park Station on 6th February that year.

The unscheduled RTs at Holloway Garage for RM cover also stood in on rare occasions for crew DMs. On 11th November 1976, RT2352 (HT) heads DMS578 (B) along Victoria Street on route 29, which had converted to DM the previous December. When the 29 was diverted and extended north of Wood Green to Enfield Town in March 1977, appearances of these rogue RTs became more frequent on the route, especially as Wood Green Garage had some, too. They remained until the autumn of that year.

When route 29 was extended to Enfield Town in March 1977, Palmers Green Garage joined Holloway and Wood Green in providing a daily RM allocation for it. However, the garage then still had a large RT allocation on routes 34, 102 and 261, and for a few months, they frequently worked the 29 to Enfield, which they never did when the route was officially RT-operated prior to 7th September 1968. On 26th June 1977, RT2068 (AD) leaves a remarkably deserted Enfield Town for Victoria; it is in fact a Sunday, however! Just over six months later, the very last RT based at Palmers Green Garage specially worked the 29 at the end of March 1978.

Long before they took over daily operation of the 29 when it converted to OPO in November 1988, Metrobuses worked the route as crew buses, standing in for RMs. On 12th June 1982, new M641 (WN) does so at Turnpike Lane Station. It wrongly shows the 'pay driver' flap - its conductor may be clearly seen standing next to the driver! Also, an RM following on the same route proves this is when the 29 was still crew-operated.

Route 29 was the last London bus route in the outer North London to be operated by Routemasters. It was RM-operated from September 1968 to November 1988, but had never been scheduled for RML operation. However, from early 1986, RMLs worked its Holloway allocation quite often. One of them, former London Country RML2310 (HT) calls at St. Stephen's Church, Bush Hill Park, on 14th May 1988. Of note is the upper-case via blind, as fitted to several London Northern vehicles at the time, as well as the British Railways Eastern Region sign pointing to Bush Hill Park Station on the lamp post on the right.

Although route 30 converted from RTL to RM operation in the summer of 1964, RTLs still appeared on it until their demise from Hackney Garage on 7th September 1968. RTWs also made odd appearances prior to their withdrawal there early in 1966. On Friday, February 3rd 1967, RTL526 (H) calls at St. Mary's Church, Upper Street, Islington on its long journey from Hackney Wick to Roehampton. This is one of the additional RTLs overhauled in 1965; the only members of their class to have grey waistbands in public service. It also sports a nice set of upper-case via blinds.

On 7th September 1968, Hackney Garage's surviving RTLs on route 106 were replaced by RTs which had themselves been ousted from route 171 at Tottenham Garage by RMs. On the following Monday, 9th, Saunders-bodied RT1054 (H) from this allocation puts in a very rare appearance subbing for an RM on route 30 in the evening rush hour. It is in Islington High Street. Just three months later, these RTs were replaced on route 106 by RMs.

A batch of unscheduled RTs sent to Putney Garage in the autumn of 1976 to help out with RM shortages invariably found their way on to route 30. A shabby RT2812 (AF) calls at St. Paul's Church, Canonbury, on 2nd February 1977. They remained until the spring of that year.

Although they were allocated at weekends from both Hackney (later Ash Grove) and Putney Garages, and also stood in for RMs during the week, RMLs never comprised route 30's full allocation. On 26th October 1986, just over three months before the route's OPO conversion, RML890 (AF) is seen at Highbury Corner. It is promoting Capital Radio, and even has a via blind saying so!

31 - 41

In January 1974, Chalk Farm Garage's allocation on route 31, which had gained RMs in June 1972, officially converted to RML at weekends, though their actual use was erratic. On 27th May 1974, RML2702 (CF) is in Adelaide Road, Swiss Cottage. Their use ceased when route 24, from which the RMLs came, converted to DM in October 1975.

As also with route 28, unscheduled RMLs from Westbourne Park Garage's allocation for route 7 appeared on route 31 in its final weeks of crew operation. On 1st April 1989, RML2486 (X) is one of two at its Chelsea, World's End terminus a fortnight before midibuses took over the route. Note the badly dented roof of the second RML. Holloway Garage, which had gained a small allocation on the 31 a few months previously, also fielded RMLs at this time.

Route 34 converted from RT to DMS OPO in September 1977 without ever having a daily RM allocation. However, RMs were allocated to it from Palmers Green Garage on Sundays during its final years of crew operation. They also, very rarely, ventured onto the route on weekdays. On 31st May 1973, RM624 (AD) runs along Leyton High Road on one of the 34's Monday to Friday rush hour journeys to Leyton, Downsell Road. The route ceased to serve this area a fortnight later, too, thus making this a very rare picture indeed!

On 23rd June 1976, RM782 (AD) had also found its way onto route 34 on a weekday and is heading for Leyton Green at Walthamstow Central Bus Station, the furthest the scheduled Sunday RMs got on this route. The inclusion of Whipps Cross on the blind, and the fact this RM is in 'solid bullseye' livery also proves this shot to have been taken after route 34's re-routeing to Leytonstone in June 1973.

On 7th September 1968, route 35 was converted from RT to RM operation, and withdrawn between Shoreditch and Chingford. It was rerouted from Shoreditch to run via Cambridge Heath to Hackney Station, and subsequently was extended from there to Clapton Pond. A handful of RTs from Camberwell Garage supplemented RMs on Saturdays and sometimes subbed for RMs on other days, and could be seen on the new section of route where they had not run when they were its full scheduled allocation. On 27th October 1969, RT3863 (Q) approaches Hackney Station with an RLH running out to take up service on route 178 following.

The last known new allocation of unscheduled RTs to help out with RM shortages to have actually operated was that of RT1152 (Q), one of two sent to Camberwell Garage. During the evening rush hour of 15th September 1977 it approaches the junction of Shoreditch High Street and Great Eastern Street.

Left: Following route 172's conversion from crew DM to RM, a few of the DMs remained at Camberwell Garage as spares. As such they made unscheduled appearances on route 35. On 23rd January 1982, DM1113 (Q) passes RM782 (Q) on the stand at Clapton Pond. This was the first day of the route's extension to that terminus.

Below: Originally allocated on Saturdays working Hackney Garage's minor share of the route from 7th September 1968, RMLs occasionally stood in for RMs on route 35 right up to the time of its OPO conversion in June 1986. RML2692 (AG) leaves Clapham Common, Old Town terminus for Hackney on 29th December 1985.

Although MDs were ousted by RMs when they returned to the 36 group of routes early in 1980, they still made odd appearances on them, still being allocated to Peckham Garage for route 63. On 30th October 1981, MD110 passes Lewisham Odeon on the 36B. Some eighteen months later, the 164-strong MD class was extinct.

Although route 37 had been one of the first to convert from RT or RTL to RM operation, in December 1962, until RTLs Stockwell Garage were replaced by RMs on the 2 group of routes in the summer of 1967, a number of RTLs supplemented the RM (and later, RML) allocation of the 37 when more buses were needed on Saturdays. RTL403 (SW) stands at its Peckham Garage terminus on 27th February 1965, and is unusually curtailed at Ilseworth.

Busy route 38 converted from RT to RM operation in January 1971, soon receiving visits from RMLs based at its main garage, Leyton, and eventually being mainly operated by that type. RTs continued to substitute for RMs until Leyton's route 55 converted to DMS OMO in October 1972. However, in May 1976, two unscheduled RTs were sent to Clapton Garage, which had a minor share of the 38, to cover RM shortages. They frequently appeared on the 38, as RT1161 (CT) shows at Mildmay Park on 2nd August 1976.

During the late 1960s and the 1970s, routes 40 and 40A had a very complicated operating pattern, involving Camberwell, Poplar, Upton Park and West Ham Garages. The latter had a Saturday RM allocation on it, but for some reason, on 6th July 1968, all of this was operated by RTLs based at the garage for route 25: I was turned to stone when I encountered them at London Bridge Station on my way down to Sidcup, and promptly travelled out towards Wanstead on one of them instead! RTL459 (WH) runs out from its garage at Plaistow, Greengate on a very short working to Stepney, Arbour Square. Poplar RTLs had scheduled rush hour turns on route 40, too. Two months after this picture was taken, all RTLs in the East would perish.

In this period, Upton Park Garage worked route 40 on Sundays and Bank Holidays, usually with RMLs derived from its large allocation for the 15. However, on Bank Holiday Monday 2nd September 1968, Saunders-bodied RT3831 (U) works an extra on the route in connection with Wanstead Flats funfair. It is also at Greengate, and its driver has forgotten to change its blind for the southbound journey.

RML2540 (U) has a very long run ahead of it as it leaves Wanstead Station for Norwood Junction on 24th November 1968. The 40 was extended on Sundays beyond Herne Hill in place of the weekday 196. The RT heading in the other direction is also an Upton Park bus, about to terminate on route 101.

After Poplar Garage lost RTLs on 7th September 1968, RTs based there substituted for RMs on route 40, as did those at Camberwell Garage. On 19th March 1972, RT2541 (PR) waits at the traffic lights at Aldgate East Station, as an early DMS on route 67 turns left into Commercial Street. Poplar RTs continued to appear on routes 40 and 40A as late as the autumn of 1977, one example remaining at the garage after their official RT allocation on route 277A there had ceased more than a year beforehand.

Crew-operated DM1103 (Q), one of those retained at Camberwell as back-up for their RMs, puts in an odd appearance on route 40 on the Aldgate one-way system on Easter Monday, 20th April 1981. By now, the 40's extension south of Herne Hill on Sundays and Bank Holidays had ceased, with the 196 serving that section daily.

Route 40A was introduced in January 1965, originally differing from the 40 by continuing to North Woolwich at its eastern end, rather than Wanstead. It was allocated RMs from Camberwell and Poplar Garages, but RTLs and RTs from both substituted for them on it. Here, on 11th April 1968, RTL1585 (PR) passes Elephant & Castle.

Viewed from the footbridge crossing the railway just north of North Woolwich Station, RT1110 (PR) has just left its terminus there for Herne Hill on 1st July 1976. Route 40A was effectively renumbered 40 when the 40 and 40A were merged in April 1978.

Below: RT1096 (AR) substitutes for an RM on route 41 in Turnpike Lane on 31st August 1968. This was from Tottenham Garage's very short-lived RT allocation on route 171, which replaced the last RTLs at the garage and lasted only from 14th June to 6th September that year. The shop behind the milk float behind the RT became one of the early J.D. Wetherspoons' pubs, The Toll Gate, in the late 1980s.

Above: Route 41 was the pioneer RTW route, and in 1960 gained a long extension from Tottenham Hale to Victoria & Albert Docks in conjunction with trolleybus replacement. This added West Ham RMs to the existing Tottenham RTW allocation. The latter converted to RM early in 1964, but after that RTLs from both garages put in appearances substituting for RMs. On 29th September 1967, RTL289 (AR) splashes through heavy rain at Stratford Broadway. This eastern extension was withdrawn upon 'Reshaping' on 7th September 1968.

Three RTs worked at Tottenham Garage from late December 1973 for a few weeks into 1974 as spare cover for RM shortages, and often appeared on route 41. The process was repeated in the spring of 1976, with another three working there through until the late summer of 1977. On 15th December 1976, a crowd of Christmas shoppers seem not to be interested in RT3141 (AR) at Turnpike Lane Station, and are apparently awaiting the 'pleasure' of riding on a DMS on routes 144 and W2!

During the 1970s, RMLs based at Tottenham Garage for route 243 supposedly worked route 41 on Sundays, though their actual appearance was erratic. However, for a few months before the route's OPO conversion in the autumn of 1985, they could also be seen working the route during the week. In doing so, they sometimes found their way onto the 41's new weekday-only extension to the new Ferry Lane Estate, which they were not scheduled to do. On 16th July 1985, RML2446 (AR) is seen there. This RML had had the distinction of being the very last to operate for London Country, on 1st March 1980. It is preserved today.

43 - 58

Route 43 converted from RT to RM operation in the summer of 1963, though RTs occasionally worked it as RM substitutes until their removal from Muswell Hill Garage in January 1971. A very late appearance is that of RT4456 (MH) at Highbury Corner on 5th May 1970, complete with a nice set of upper-case via blinds.

Route 45 had received RTWs in the late 1950s, and these were replaced by RMs in January 1966. However, following that, RTLs from Walworth Garage sometimes appeared until their replacement by RTs later that year. RTs frequently appeared on the route from that garage, though those from Chalk Farm Garage were less common. That garage's allocation ceased when the route was withdrawn north of Kings Cross in January 1972. Here, a well-laden RT4813 (WL) is at Clapham Common Station on Saturday, 22nd February 1975. Its via blinds reflect that the 45 no longer runs north of Kings Cross, and on this particular day most buses on the 45 were RTs, since new crew-operated DMs took over next day and the RMs they replaced at Walworth had already been transferred elsewhere.

Route 45 was extended north of Kings Cross again in October 1978, replacing route 17 (which it paralleled between Camberwell and Kings Cross anyway) as far as Archway Station. It reverted to RM operation in the spring of 1981, and subsequently the Walworth allocation was transferred to Stockwell. A very odd appearance from that garage is OPO B20 DMS2435 (SW) working with a conductor in Rookery Road, Clapham Common on 17th April 1985.

Above: Holloway Garage gained a share of route 45 upon its extension to Archway, and in the route's last two or three years of crew operation, RMLs from that garage occasionally stood in for RMs. On 7th May 1985, a very smart ex-London Country RML2413 (HT) stands at the MacDonald Road, Archway terminus. The 45 converted to OPO in August 1985, when it was cut back north of Kings Cross again and 'new' route 17 introduced to replace its northern section!

Left: Route 46 (Waterloo - Alperton) was RTW-operated from Willesden Garage until RMs took over in the summer of 1965. Prior to that, RTLs occasionally worked it, as illustrated by Metro-Cammell-bodied RTL734 leaving Waterloo on 15th June 1965.

A very rare working indeed saw RT4539 (AC) on route 46 whilst on loan to Willesden Garage. It approaches the Waterloo terminus on 13th December 1967. It is an Alperton bus, so perhaps the scheduled RM had broken down at Alperton Garage and this RT hastily substituted for it. However, that does not explain why the RT has a correct via blind for a route Alperton Garage did not operate!

Left: The long route 47, which ran from Farnborough to Stamford Hill between July 1961 and January 1970, had a large allocation of RTs from Bromley and Catford Garages, as well as a small allocation of RTLs from Dalston until June 1968. On Saturdays, until their demise in February 1966, RTWs from route 11's allocation were used. RTW375 (D) stands at Shoreditch Church on a wintry 4th December 1965.

Below: Dalston's RTLs on route 47 were always in a minority amid its Bromley and Catford-based RTs. On 14th May 1968, a month before their demise, an empty roofbox RTL1427 (D) speeds along Shoreditch High Street approaching its terminus. This RTL has since been preserved.

Below: Until the very end of RT operation in South London in August 1978, RTs from both Bromley and Catford Garages substituted for RMs, which had replaced them in January 1975, on route 47. RT2146 (TB) is one of two seen here together at Bromley Market Place on 17th August 1978, nine days before the final demise of RTs in this part of London.

By late 1966, Battersea RTLs worked a small Saturday allocation of RTLs on route 49 alongside the main RT allocation the route had from Merton and Streatham Garages. On 19th December 1966, RTL1261 (B) calls at Arding & Hobbs' famous department store at Clapham Junction. These odd workings ceased at the end of the year.

For a while during 1982/83, six new Metrobuses (M800-M805) were used at Sidcup Garage for comparative trials with Titans, which were replacing DMSs there. They were used on all OPO routes at the garage, and this view shows M800 (SP) in Woolwich, Powis Street heading for Orpington on route 51 on 29th August 1983. When the trials were completed, the Ms were transferred to Holloway Garage, which was receiving the type to replace DMSs at that time.

A very rare working on 7th June 1967. A newly-overhauled ex-Green Line RT3226 (GM), which has just been allocated to Battersea Garage to replace RTLs is working on loan to Victoria, Gillingham Street Garage and substituting for an RM on route 52 at Victoria Station. The 52 had converted from RTL to RM operation in the autumn of 1966.

Left: RTLs from Willesden Garage continued to appear on route 52 in substitution for RMs until the class' final demise at the end of November 1968. On 28th March that year, RTL1447 (AC) approaches Hyde Park Corner on a very short evening rush hour working to Notting Hill Gate.

Below: Following the demise of route 46 in 1969, Middle Row Garage gained a small share of route 52, using RMs. RT substitutions for them from that garage was very rare indeed, but on 20th October 1972, RT3920 (X) puts in an appearance and is at Kensal Rise Station in the evening rush hour.

RTs from Willesden Garage occasionally substituted for RMs on route 52 following their replacement of the last RTLs on routes 176 and 226. RT2610 (AC) leaves Victoria for its home garage on 26th June 1970.

Following the misbehaviour of participants at the Notting Hill Carnival on a number of occasions, bus crews at Westbourne Park Garage (which inherited Middle Row's share of route 52 in August 1981) demanded doored buses for the duration of the carnival in 1982, to safeguard conductors from being mugged or otherwise assaulted. Crew DM2571 (X) from route 18's allocation is at Kensington Church on 30th August 1982.

For the same reason, route 52A, which was in effect a bifurcation of route 52 taking buses on short workings to the new Westbourne Park Garage, was also worked by crew DMs for the duration of the 1982 Notting Hill Carnival. DM2550 (X) heads along Kensington Church Street. Its conductor may be seen standing next to the driver.

Route 53 converted from RT to RM operation in the autumn of 1967, but RTs from New Cross Garage continued to appear on it as RM subs until as late as the spring of 1976. On 12th April that year, shortly before the last few RTs based there were replaced by RMs on route 192, a shabby RT3257 (NX) has terminated at Camden Town Station and for some reason stands in Kentish Town Road, rather than at the route's usual terminus around the corner in Camden Gardens.

Left: However, the last workings of an RT on route 53 took place when a solitary example was sent to New Cross to help out with RM shortages late in 1976. RT4681 (NX), complete with an RM side blind in the via box as used on most rogue RT workings at this period, calls at Trafalgar Square on 1st December 1976. Shortly after this, the 53 was converted from RM to MD operation, though RMs returned four years later.

Below: During the MDs' tenure of the 53, crew DMs at New Cross Garage for route 141 occasionally worked route 53 instead. On 4th May 1977, a shabby DM1166 (NX) turns from Parliament Street into Bridge Street heading for its home garage. DMs continued to appear on route 53 after its reversion to RM operation, too.

Following route 53's reversion to RM operation, RMLs based at New Cross Garage for their small share of route 37 also appeared on the 53 until the 37's OPO conversion in June 1986. RML2533 (NX) departs from Camden Gardens for home on 15th April 1982.

Although the original route 55, which ran from Hayes (Middx) to Chiswick, had received RMs in October 1965 from Hanwell and Turnham Green Garages, the latter used mostly RTs on Saturdays on the route until its demise in November 1968. On 30th September 1967, RT461 (V) waits at the traffic lights in Church Road, Hanwell to turn into the Uxbridge Road.

'New' route 55 was introduced with RTs in October 1969, replacing the 170 between Bloomsbury and Leyton, Bakers Arms, and the 38 from the latter point to Walthamstow Garage. However, the section between Bloomsbury and Leyton had been operated by tram route 55 until 1939, when that was replaced by trolleybus route 555. The latter was then replaced by an extension of the 170 in April 1959! The 'new' 55 originally ran to Marylebone, but was later re-routed to terminate at Aldwych. RMs were introduced on Sunday in January 1971, but not RMLs. However, on the route's last day of crew operation, 27th October 1972, prior to conversion to DMS, RMs and RMLs worked it to facilitate the transfer away of Leyton Garage's last RTs. Thus RML2403 (T) is heading along Kingsway that day. So disastrous were the DMSs that route 55 was re-converted to crew operation in January 1981, using RMs, and diverted at Bloomsbury to parallel the 38 to Victoria. It later received RMLs before converting to Titan OPO in June 1986. However, RMLs were never scheduled to work its section to Aldwych.

Trolleybus replacement route 58, RM operated from February 1960 until its OPO conversion in 1982, saw odd workings by West Ham RTLs and the RTs that replaced them. On 6th July 1968, RTL1558 (WH) has run into its home garage after a turn on the route. Just two months later, all RTLs in the East perished.

In its last few years of crew operation, RMLs from West Ham Garage also had visitations on route 58. On 25th October 1980, former London Country RML2318 (WH) negotiates one of the tight curves in Markhouse Road, Walthamstow on its way back to the garage. The 58 converted to OPO in April 1983 without ever having scheduled RML workings.

59 - 86

For many years, the long route 59 (West Hampstead to Chipstead Valley, or latterly Old Coulsdon) ran on Sundays and Bank Holidays only. It converted from RT to RM operation, using South Croydon Garage's buses off the 130 group of routes, early in 1964, and then to RML late in 1967. However, on Bank Holidays, additional buses ran and, until their demise from the garage in the early 1970s, these were RTs. This lucky shot in Parliament Street on Easter Monday, 7th April 1969 - exactly ten years before the last RTs were withdrawn - finds standard RT4690 and Saunders-bodied RT1243 (TC) heading for their home garage. It is interesting that the two show different points on their via blinds, though both are incorrect! The route crossed Westminster Bridge, therefore did not serve Millbank - the display is in fact for the weekday 159, which only reached South Croydon on Saturdays.

A small allocation of RTs from Streatham Garage was later added to route 59, but it converted to RM in May 1971. When South Croydon's main allocation converted to crew DM in March 1975, these became oddities. RM833 (AK), with blinds already turned for its southbound trip, arrives at West Hampstead, West End Green terminus on 9th July 1978. A DM on the stand may just be seen on the left. Route 59 in this format was withdrawn at the end of October 1978, but the number has subsequently reappeared along the Brighton Road corridor, running from Streatham Hill to Kings Cross as this book is compiled.

New route 60 was introduced using RMs on 4th September 1982, running from Brixton Garage (Streatham Hill) to Old Coulsdon, replacing parts of routes 130 and 190. It converted to OPO on 23rd April 1983, but a week before that, on 16th April, B20 D2569 (TC) leaves Brixton Garage substituting for an RM, and clearly showing the 'pay conductor' flap.

Route 63 fully converted from RT to RM operation, from Peckham Garage, in the summer of 1963, having already had a small allocation for RMs from Highgate Garage following its extension to Parliament Hill Fields in conjunction with trolleybus replacement in February 1961. RTs seldom appeared on it once it was fully RM-operated, but on 27th January 1971, RT3086 (PM) puts in an appearance at Elephant & Castle. Its in-house LT adverts, and the modernised Green Line RF about to overtake it, prove this picture was taken long after RM conversion.

Back in July 1960, route 64 had the distinction of being the first RT-operated route to convert to RM operation, which it did at Elmers End Garage as a result of its extension to Tooting in part replacement of trolleybus route 630. The allocation was transferred to Thornton Heath Garage in January 1964, but RTs continued to substitute for RMs until the route converted to OMO DMS in December 1971. RT2073 (TH) calls at Mitcham, Fair Green on 25th September 1970. Once again, its in-house LT adverts leave no doubt that this picture was taken longer after RM conversion.

Right: Route 65 converted from RT to RM operation in October 1975, but odd RTs from Kingston Garage's allocation on route 71 continued to work the route until they themselves were replaced by RMs early in 1978. On 18th May 1977, RT2799 (K) escorts an RM through Ham Common. This RT survives today as part of the London Bus Company's heritage fleet.

Below: Late in route 65's career as a crew-operated route, a small Sunday allocation of RML's from Hanwell Garage was added. On 26th October 1985, former London Country RML2454 (HL) sets off from Ealing Broadway, Haven Green on a short working to Surbiton Station. The Leyland National on the right is in Ealing Broadway Station forecourt, working a rail replacement service.

Throughout much of its time as an XA-operated route between 1966 and 1969, the eight Country Area XFs, still in green livery, worked alongside them from Stamford Hill Garage on route 67 for comparative trials. On 23rd May 1967, XF3 (SF) is seen at Gardiner's Corner, Aldgate. This itself is an oddity in that it was experimentally fitted with a Cummins engine, which necessitated a larger rear bustle, as this view clearly shows.

Until RM conversion in February 1970, the long 68 route from Chalk Farm to South Croydon was RT-operated by South Croydon, Norwood and Chalk Farm Garages. The latter's allocation, had, however been RTL-operated until late in 1965. RTL123 (CF) is heading along Waterloo Road on 29th May of that year.

On a foggy 20th December 1970, RML2752 (TC) calls at the Newington Butts stop south of the Elephant & Castle roundabout working South Croydon Garage's weekend share of route 68, worked by RMLs derived from the 130 group of routes. These workings ceased when they converted to DM in March 1975.

Until South Croydon Garage lost RTs at the start of 1974, their RTs frequently substituted for RMs on route 68. On 30th August 1973, RT3960 (TC) calls at West Croydon Bus Station on a short working to Upper Norwood, Upper Beulah Hill.

At various times, RMLs from both Chalk Farm and South Croydon Garages worked route 68, both officially at weekends and as RM substitutes otherwise. On 13th August 1986, two and a half months before the route converted to OPO, RML2506 (CF) stands at its Chalk Farm Station terminus. The building on the right is the Roundhouse, once an engine shed on the L.N.W.R. main line and now a theatre.

Trolleybus replacement route 69, introduced with RMs in February 1960, was another from West Ham Garage upon which RTLs sometimes appeared. On 29th September 1967, RTL414 (WH) sets off from Stratford Broadway for North Woolwich.

Following its extension to Chingford upon route 249's withdrawal on 7th September 1968, route 69 gained a Saturday-only allocation of RTs from Leyton Garage. Nine months later, on 7th June 1969, newly-overhauled RT3678 (T) is pursued by an MB along Station Road, Chingford.

For some reason, on their last Saturday of operation prior to route 5's conversion to OMO DMS, 9th April 1971, a number of RMLs based at West Ham for it appeared instead of on route 69. RML2380 (WH) is in The Grove, Stratford that day. Later on, West Ham RMLs officially worked route 69 on Sundays.

Left: During 1976, RTs based at Walthamstow Garage for route 34 had scheduled duties on route 69. Additional RTs were sent there during the hot summer of that year and these also worked as unscheduled cover for RM shortages. One of these, RT794 (WW) worked there for only a few days, and on 1st July 1976, crests Chingford Mount complete with chalked number blind.

Below: Towards the end of crew operation on route 69, when its main West Ham allocation had become crew Titan, RMLs from Leyton Garage at weekends had the last regular Routemaster operation on the route. On a wet 26th January 1985, recently-overhauled RML2634 (T) calls at the stop in Tramway Avenue, Stratford. This was the last day RM-types officially worked the route; a week later, almost exactly twenty-five years after their introduction on it, the 69 converted fully to Titan operation.

Route 73 was one of the first to convert from RT-type to RM working, in December 1962, following the completion of the Trolleybus Conversion Programme. After that, the appearance of RTLs on the route was quite rare, but on 10th July 1967, RTL428 (AR) did, it passes Essex Road Station. The adverts for the new Weekender tickets immediately show this to be a long time after RM conversion.

Above: The rogue RTs based at Tottenham Garage in the mid/late 1970s for RM spare cover usually worked routes 41, 76 and 171, but rarely the 73. However, on 15th February 1974, RT489 (AR) has found its way onto the route, it is passing Knightsbridge Station.

Left: Before route 74's conversion from RTW to RM operation in the autumn of 1965, RTLs from Putney Garage sometimes worked the route. RTL1447 (AF) is in Park Lane on 7th June of that year. The vehicle still looks smart a few months after its last overhaul.

Putney Garage's rogue RTs usually worked routes 14 and 30 during their time as cover for RM/RML shortages during the winter of 1976/77. However, they occasionally also worked route 74. Shortly before their transfer away from the garage, RT4363 (AF) is very unusually seen on one of the route's few rush hour journeys extended to Kingston Vale terminus in the evening of 13th April 1977.

Route 74B had also converted from RTW to RM, and then to RML, as the 74 had, and the appearance of RTs from its operating garage, Riverside, where they had replaced RTLs on route 72, were rare. This view shows RT1641 (R) on a short working to Earls Court at Hyde Park Corner on 18th September 1975, just under three months before RTs at the garage were replaced by RMs on route 72.

Both during the tenure of RTWs on route 76, and the RMLs and XAs subsequently used to replace them, until their demise in June 1968, RTLs from Tottenham Garage appeared on route 76. On 17th May 1967, RTL1366 (AR) is in York Road, Waterloo on one of the route's odd journeys that bifurcated from its main route to terminate at Liverpool Street.

The lone FRM1 (AR) was the oddest strange visitor of all on route 76, despite being scheduled to work it for a couple of years. It has just entered service when seen loading up in Stamford Street, Waterloo on 3rd July 1967. Of note are its non-opening windows. At this stage, it has forced air ventilation, but after a fire damaged it later, in 1967, was fitted with standard RM/L-type opening windows.

RTs from Tottenham Garage during their short-lived allocation for route 171 also stood in for the troublesome XAs at Tottenham Garage, but this one, RT4070 (AR) departing Victoria Bus Station on 19th April 1968, is on loan to the garage before that happened. It provides an interesting contrast with the XA behind which is working on the same route.

The rogue RTs covering for RM shortages at Tottenham Garage early in 1974 and again in 1976/77 often worked route 76. On 17th August 1977, RT3141 (AR) finds itself on one of the odd rush hour journeys the route worked to Brimsdown Power Station, where it accompanies another RT on route 135. Both routes ceased serving this point three days later. The power station had already closed!

Mainly during the 76's last years of crew operation, RMLs allocated to Tottenham Garage for route 243 found their way onto the route, substituting for the usual RMs. On 15th August 1984, RML2718 (AR) passes over Stamford Hill Broadway.

The 77 group of routes converted from RT to RM operation in December 1973, but RTs from Merton Garage continued to operate it until finally removed from the garage in January 1977. On 6th August 1976, RT4234 (AL) is at Parliament Square. The DM on the left on route 168 proves this picture to have been taken long after the conversion.

Late in the 77's career as a crew-operated route, following the re-appearance of a Stockwell allocation on it, RMLs began to substitute for RMs. On 11th October 1985, RML2516 (SW) has just passed Wandsworth Road Station. This RML later became the famed 'DRM', rebuilt with a doored back from an RMC.

An oddity to appear on the 77 was the unique dual-door Alexander-bodied Volvo-Ailsa V3 (SW), one of three of that type being evaluated as possible replacements for RM-types. It had a doored platform and back which resembled an RMC or RCL somewhat and two staircases - the latter feature being repeated on the so-called 'New Routemasters' in more recent times. Also based at Stockwell, it is working crew-operated at Vauxhall Station on New Year's Day, 1986. V3 is now part of the London Bus Company's heritage fleet.

In common with the 77, route 77A also had RT substitutions for RMs long after its conversion to the type. On 2nd August 1976, RT2843 (AL) accompanies an RM on route 77 in Whitehall.

RMLs from Stockwell Garage found their way on to route 77A too in its last year of RM operation. On 8th May 1985, RML2389 (SW) climbs West Hill towards Clapham Junction. V3 also worked the route at this time.

For several years, Dalston Garage had a small allocation of RTLs on route 78, which ran alongside its main allocation of Peckham RTs. RTL1412 (D) heads a group of RTs at the route's Shoreditch Church terminus on 14th May 1968, a month before being replaced itself by RTs.

Dalston RMs were never scheduled to work their small allocation on route 78, but shortly before this allocation was withdrawn, RM150 (D) is in Minories, Aldgate on 31st December 1971. However, Peckham RMs worked the route at weekends at this time, until its DMS OMO conversion in May 1972.

Route 83 converted from RT to RM operation in the autumn of 1966, but RTs still appeared on the route on odd occasions until February 1975. On 4th June 1974, RT1213 (ON) is working through West Hendon. The DMS following on route 32 proves this picture to have been taken long after the 83's RM conversion.

Routes 85 and 85A at Putney Garage converted from RTL to RM operation in August 1963, leaving just a small allocation of RTLs there for route 93. These were replaced by RTs two years later. Until this allocation ceased in April 1970, very rare appearances by RTs were made on the 85. Perhaps the rarest of all was that of RT3050 (AF) on 29th September 1969 - this Saunders-bodied RT was the only roofbox RT at the garage. It is leaving Putney Bridge Station on one of the 85's many short journeys to Roehampton. Route 85 succumbed to SMS OMO in January 1971.

Above: Whereas route 85 converted to SMS OMO in January 1971, the 85A retained RMs for another two years, falling to DMS OMO in March 1973. On rare occasions, RMLs found their way on to the route, as shown by RML2465 (AF) approaching Roehampton from Putney Heath on 22nd September 1972.

Right: Busy Romford Road route 86 could not convert to RM operation from its main base, Seven Kings, until the antiquated garage was rebuilt to accommodate anything wider than an RT early in 1976. However, for several years beforehand, RMs and RMLs from Upton Park Garage worked it at weekends. On 4th August 1973, RM677 (U) is at Chadwell Heath.

RMLs derived from Upton Park's large allocation of them on route 15 worked the 86 at weekends long after the Seven Kings allocation had converted to RM. On 12th May 1979, RML2534 (U) sets off from Romford for Stratford. In later years, the route gained an RM allocation from West Ham Garage before its inevitable conversion to Titan, initially as crew buses, in 1984.

87 - 109

The last instance of all of RTs working on RM-operated routes was that of route 87 at Barking, where those off the final allocation for route 62 did so until the very end. On 16th March 1979, RT2671 (BK) accompanies RM377 (also a Barking bus) on the stand at Blake's Corner, Barking, three weeks before the RTs' final withdrawal.

Left: Route 88 converted from RTL (Stockwell) and RT (Shepherd's Bush) to RM operation in the spring of 1967. After this, RTs from both garages worked as RM substitutes. In March 1969, Saunders-bodied RT2515 (S) is doing so in Parliament Street.

Below: A very unusual appearance on 10th September 1972 was that of RT4711 (S) working route 88's Sunday extension to Banstead Hospital. It passes a remarkably traffic-free Trafalgar Square. Unscheduled use of RTs on RM-operated routes on Sundays was most unusual by this time.

Left: During the summer of 1982, spare B20 crew-operated DMs at Stockwell Garage put in appearances on route 88, by which time its official allocation was RML. On 28th June 1982, DM2643 (SW) works short to Charing Cross during the evening rush hour whilst following an RML in Marsham Street, Westminster.

Below: Also in Marsham Street, Metrobus M871 (S) escorts RML880, working as a crew bus on route 88 on 2nd May 1984. By the time the 88 converted to M OPO in August 1992, it no longer ran west of Marble Arch, nor south of Clapham Common, thus making this picture, complete with 'pay conductor' flap a rarity.

Following the replacement of RTs at New Cross Garage by RMs early in 1975 on routes 1 and 21, the garage's Saturday-only share of route 89 converted to the type. This brought RMs to Bexleyheath for the first and only time, and this view shows RM703 (NX) in Welling High Road heading there, complete with the name 'Bexley Heath' wrongly spelt as two separate words on its destination blind! It is shown on 15th April 1978, the last time RMs worked this route, which converted to DMS OPO a week later.

RTLs from Putney Garage had a small RTL allocation on route 93 (the majority of whose buses were Sutton RTs) until they were replaced by RTs in the autumn of 1965. However on Derby Day, 29th May 1968, RTL418 (AF) was loaned to the garage for the day from Cricklewood Garage, thereby making a farewell performance of this class on route 93. It is runs along Epsom High Street. Other Derby Day extras on the 93 included a Saunders-bodied RT from Chalk Farm and an RM from Hounslow that day.

Tram replacement route 95 at Brixton Garage was originally RT-worked, but RTWs replaced them in 1963/64. Subsequently, the route had the distinction of operating the last RTW at London's service in May 1966, when the route converted to RM operation. RTs returned to Brixton Garage at this time to replace other RTWs on route 109, and RTLs on the 133. Inevitably, these also worked as RM subs on the 95 until its conversion to DMS OMO at the beginning of January 1971. Four weeks before that, RT274 (BN) heads along Mitcham Lane on 5th December 1970. Its upper-case via blinds must have come from an RT that had been at Brixton in the first place!

In the same way, RTs found their way onto the Sunday 95A. On 27th December 1970, RT2361 (BN) calls at a snowy London Bridge Station. This was the last day the route ran, and also my 23rd birthday!

Left: By the time of irregular route 100's withdrawal, it was worked by RMLs interworked with Upton Park's allocation on route 15. However, on its last day of operation, 6th October 1971, RT4071 (U) worked it. It is seen amid the desolation of Beckton Gas Works prior to running back to East Ham, White Horse, to take up route 15.

Below: Following the appearance of RMs at Muswell Hill Garage for route 43 in the summer of 1963, these worked their weekend allocation on route 102 alongside Palmers Green RTs. They were subsequently supplanted by RMLs, as depicted here by RML897 (MH) passing Pymmes Park, Edmonton on 8th October 1978, by which time the main Palmers Green share of the route was RM operated. Three weeks later, the RML allocation (by now derived from Muswell Hill's small weekday share of route 13) was withdrawn on Saturdays and replaced by crew DMs from routes 43 and 134 on Sundays.

Muswell Hill's RMLs were in turn replaced by crew DMs on their weekend allocation at the end of October 1978, and reduced to Sundays only, as related above. On 30th May 1982, my two little girls, Felicity (left) and Margaret (right) have 'enjoyed' a bumpy ride on DM2599 (MH) which has just arrived at Chingford Station. This allocation on the 102 ceased when the 102 converted to Metrobus OPO just over three months later.

A real oddity on route 102 in the summer of 1986 was Alexander-bodied Volvo Citybus C1 (AD), used in trial service. This was a diesel/battery electric hybrid, in some ways the predecessor of such vehicles in use in London in 2017. It is passing Pymmes Park, Edmonton on 11th July 1986. The bus was only on loan to London Buses, and subsequently returned to its manufacturer.

Route 104 replaced trolleybus route 609 (Moorgate to Barnet Church) on 8th November 1961, using most of the 24 pilot-batch RMLs at Finchley Depot. Subsequently, RMs from this garage (as it became in January 1962 when its last trolleybuses left) as well as those from Highgate (later named Holloway) worked it too, but never a batch of RTs allocated to Finchley for route 125, despite their having blind displays for the 104....until 8th May 1970, when I was turned to stone to be confronted with RT965 (FY) bearing down on me on the 104 as I emerged from Highbury & Islington Station on my way home from 'work'! Fortunately, I always carried a camera and the bus was caught in homegoing rush hour traffic! It even has a correct blind display. In all probability, a fitter at the garage had mistaken it for RM965, also based at Finchley, when dressing it for service. RTs left Finchley in January 1971, as did the main allocation of RMLs for route 104.

I've heard it said that 'lightning never strikes twice', but it DID on 28th September 1977 when I was again gorgonised by the sight of an RT on route 104! At this period, the route was operated by the unreliable DM types, thus one of the rogue RTs allocated to Holloway has come to help out with vehicle shortages. RT2352 (HT), has come to the rescue rather than an RM or RML! It is in City Road approaching the junction with East Road in the evening rush hour. The 104 reverted to RM-type working for a few months in the early 1980s, but converted to OPO in September 1982 and was withdrawn just under three years later.

Above: Route 105 did not convert from RT to RM operation until the end of April 1978, but before that RM's based at Shepherd's Bush substituted for RTs on rare occasions. One of these, RM1186 (S) calls at Greenford, Red Lion on 14th February 1976.

Below: Although route 105 converted from RM to crew Metrobus operation on 4th September 1982, a handful of RMs returned to Southall Garage a few days later owing to a shortage of them. Two of these, Leyland-engined RM1365 and 1377 (HW) both of which had been withdrawn from Tottenham Garage whose codes they still show, are seen at Perivale, Bideford Avenue on 23rd September. They were finally withdrawn shortly afterwards.

Running from Finsbury Park to Becontree, route 106 had been an RTW stronghold, worked principally from Hackney Garage, but following their demise there in the winter of 1965/66, RTLs then worked the route from Hackney as well as Clapton Garages. The latter then received RTs in the late summer and autumn of 1967, but these were always in a minority on the route. On 30th December 1967, RT558 (CT) has just passed Stoke Newington Station on its way to Becontree, Chitty's Lane.

Because Hackney Garage did not have enough RTLs to operate route 106 on Saturdays, and at the time crews did not apparently agree to RMs being used, RTs had to be borrowed from nearby Dalston Garage to make the number up. On 17th August 1968, Saunders-bodied RT220 (H) waits on the stand at Blackwall Tunnel. Three weeks later, Hackney and Poplar's RTLs on this route were replaced by RTs, but these lasted only until December when they were in turn replaced by RMs.

RTs from Poplar Garage put in occasional appearances on route 106 substituting for RMs until their allocation was withdrawn when the route was cut back to Blackwall Tunnel in January 1971. Their RT2343 (PR), another Saunders, heads along Grove Road, Mile End on 14th April 1970. By now, buses of this type were rapidly becoming extinct.

Route 106 converted to DMS OMO in July 1972, but made history at the end of March 1979 when it converted back to RM crew operation as a result of the atrocious performance of the DMSs and their consequent unpopularity with all and sundry. Before the DMS conversion, RMLs from Hackney Garage made very rare appearances on it, but after RMs had been re-established on the route, they worked fairly often. On 20th May 1979, RML2482 (H) shows off its length at the then-new 'mini-bus station' at Limehouse, where it has turned short. This facility was primarily used as the terminus for route 86, at the junction of Burdett Road and Commercial Road, but was closed after only a few years and later built upon. Meanwhile, route 106 converted to OPO for good, using T-class Titans, on 4th September 1982 amid the infamous 'Law Lords' cuts.

For a few months in 1967/68, Clapton RTs worked Sunday route 106A alongside Dalston RMs, once all their RTLs had been replaced by them. On 10th December 1967, RT378 (CT) is seen in Brownswood Road, Finsbury Park, three months before RMs took over this allocation too. In September 1968, that allocation was transferred to Hackney. This route differed from the weekday 106 in terminating at Dagenham New Road rather than Becontree and was withdrawn in January 1971.

Tram replacement route 109 had been operated by RT-types from its inception in 1951 to the summer of 1976, when a very gradual conversion to RM took place. These lasted only a couple of years before being replaced by crew DMs. As so often, the DMs were a disaster so the 109 was reverted to RM working during 1981/82. In its last month or so of crew operation prior to OPO conversion in February 1987, RMLs allocated to Brixton Garage for routes 137 and 159 began to appear on it. On a wintry 24th January 1987, two weeks before the conversion, RML2589 (BN) climbs Brixton Hill. In contrast to its shabby state here, this RML is now smartly preserved.

117 - 140

Route 117 was extended eastwards from Hounslow to Shepherds Bush and converted to RM operation to replace trolleybus route 657 in May 1962. Odd RTs from Hounslow Garage continued to substitute for RMs, and subsequently a small allocation of RTs was added from Turnham Green Garage on Saturdays. One of these, RT2281 (V) passes Feltham Parish Church on 8th November 1969.

Route 119 converted from RT to RM operation in May 1976, but RTs from Bromley Garage continued to put in appearances on it until the very end of RT operation in South London in August 1978. RT1790 (TB) is passing Bromley South Station on 17th of that month, just over a week before the end. This RT is another to survive in preservation today.

Above: Following route 29's conversion from RM to crew DM operation in December 1975, Wood Green Garage's small Saturday allocation on route 123 converted to this type. On 13th March 1976, DM1192 (WN) still looks quite smart when passing Southgate Town Hall. The route converted to DMS OPO just over a year later, whereupon these workings ceased.

Left: The 130 group of routes converted from RT to RM operation early in 1964, then to RML late in 1967. Croydon RTs continued to substitute for RMs occasionally, with RTs from Thornton Heath Garage providing the Sunday service on the 130 and 130A for a while in the early 1970s. In March 1975, the routes converted to crew DM, but RMs still made occasional appearances. Thus RM911 (TC) heads through New Addington on 1st June 1977 on the 130B.

In similar fashion to the 117 seen earlier, route 131 was converted from RT to RM operation and extended from Kingston to Wimbledon in May 1962, replacing trolleybus route 604. Subsequently, Merton Garage had a small RT allocation on this route on Saturdays. RT4721 (AL) works this through Kingston town centre on 15th April 1971.

Owing to the disastrous failure of DMSs at Bexleyheath Garage, redundant SMs were sent there to prop up route 132 in the autumn/winter of 1979/80. On 1st October 1979, SM5 and SM45 (BX) are at Bexleyheath Clock Tower. The temporary blind arrangements are of note! These vehicles had originally replaced RTs on the route in April 1970, but had been replaced by DMSs early in 1974.

Crew DMs replaced RTs on route 133 in March 1975, but on 21st May 1976, RM140 (BN) is seen on the route at its Elephant & Castle terminus. RMs were just being introduced to the 109 to replace RTs at Brixton Garage at this time.

Left: On the same day as the previous picture, RT3408 (BN) also works the 133 as it passes Kennington Park. A number of other RTs reappeared on the route at this time, implying the DMs were in trouble, though they staggered on until being replaced by RMs in August 1981.

Below: Long after their dubious tenure on route 133, DMS2463 (BN) has been pressed into service on a wet 24th June 1984, when RMs were its regular allocation. It is leaving London Bridge Station for Norbury. These contraptions returned to the route for another year in November 1984. The all-red livery does not enhance the angular body design of the B20 DMSs, which at least looked better as originally delivered with white upper-deck window surrounds.

The long route 134, running from Potters Bar to Pimlico, converted from RT to RM operation in the summer of 1964, when it was operated by Potters Bar, Muswell Hill and the old Holloway Garage. Following their appearance at the latter for route 14 in early 1967, RMLs made occasional forays onto the route, as illustrated by RML2620 (J) running along Victoria Street in April 1968. This RML is also smartly preserved today.

In contrast, at the same period, Muswell Hill RTs had regular Saturday workings on the 134. On 20th April 1968, RT2141 (MH) collects passengers in Colney Hatch Lane. The RM on the other side of the road is having a crew change - Muswell Hill Garage was just around the corner in Sydney Road, and would be closed in 1990. RTs from Muswell Hill and Potters Bar Garages could still be seen on the 134 as late as 1971, though by then its Holloway allocation had ceased.

Holloway RT1851 (J) puts in an appearance on the 134 at Parliament Square in October 1968. The RM following on the 29, and that on the right advertising the new Victoria Line, prove that this picture is taken more than four years after the 134 converted to RM.

In the autumn of 1973, crew-operated DMSs were foisted onto the 134, which by now had been cut back to Warren Street Station, but RMs still made occasional appearances. On 23rd January 1975, RM2017 (MH) approaches Kentish Town Station. Two days later, route 43 also converted to DM making such appearances very rare until the 134 reverted to RM operation in September 1982.

Such was the abysmal failure of the DMs working route 134 that a few unscheduled RMs were allocated to Potters Bar Garage in the winter of 1978/79 to help them out. RM1011 (PB) is loading up in Archway Road on 8th November 1978.

In early 1982, new Metrobuses replaced DMSs on OPO routes at Potters Bar and inevitably had to assist the struggling DMs on the 134. In February 1982, M684 (PB) is at Muswell Hill Broadway working short to Camden Town. It wrongly shows the 'pay driver' flap; its conductor may be clearly seen standing by the driver! These Ms were never allocated to the 134, since it was cut back to Barnet Church in May 1983 and the Potters Bar allocation consequently ceased.

During its last few months of crew operation, RMLs from route 43 (which had regained them after several unhappy years of crew DM operation) often found their way onto the 134. In October 1985, RML2353 (MH) passes beneath the famous Highgate Archway Bridge, which once formed the border between the counties of London and Middlesex, and now is the border between the boroughs of Haringey and Islington. The latter's placards as seen on the lamp-post on the right are having a dig at the Thatcher regime; in such respects, things seem to have gone 'full circle' as this book is being compiled in the summer of 2017! Meanwhile, route 134 converted to OPO in May 1986, and RML2353 is now preserved in Australia!

Route 134A was a Sunday variation of the 134, replacing the 34 that day in the Barnet and New Barnet area, and then continuing to Victoria. Operated by Muswell Hill and Holloway Garages, it too converted to RM in the summer of 1964, and Holloway RMLs from route 14 made occasional appearances. On 3rd December 1967, Archway Road is being widened as RML2606 (J) heads south towards Archway Station.

Route 135 was a local route in the Enfield area, which had rush hour extensions across the Lea Valley railway line at Brimsdown to serve the nearby Power Station. On 24th May 1977, RM2210 (E) substitutes for an RT working one of these journeys after arrival at Carterhatch terminus. The section of route to Brimsdown Power Station was withdrawn three months later, and the route did not convert to RM until January 1978. Therefore RMs were never scheduled to work there on route 135, as this one is doing.

The 135A differed from the 135 in that its rush hour journeys were extended from Brimsdown Station to Lockfield Avenue, rather than the Power Station. Also on 24th May 1977, RM673 (E) has substituted for an RT on this route too, and is seen turning from Baker Street into Lancaster Road, by 'The Hop Poles' pub. Route 135A was withdrawn three months later, or rather simply renumbered 135, since that route was diverted to terminate at Lockfield Avenue. Therefore RMs were never scheduled to operate route 135A.

The Enfield route 135 was withdrawn in September 1980, and a new 135 was introduced in November 1987, replacing the northern end of route 137. Initially worked by RMs from Holloway Garage, it often had visits from their RMLs. On 22nd October 1988 a very shabby RML2284 (HT), whose registration plate has fallen off and been placed in the driver's cab window, passes the former Kentish Town West Station. The station closed during the 1926 General Strike and never reopened. The 135 was converted to OPO a fortnight later and has subsequently been withdrawn.

Route 137 converted from RTL (Victoria Gillingham Street) and RT (Norwood) to RM operation in the autumn of 1964, after which appearances by RT-types on it were initially rare. However, in the early part of 1967, shortages of RMs at Gillingham Street Garage caused one or two RTLs to be retained at the garage as spares for it, and also on occasion RTLs were borrowed from Battersea Garage. In April 1967, RTL1031 (GM) is one of two of these that was outside the garage. A close look will show it has an RM side-blind above its platform - a taste of things to come!

As a result of the severe shortage of spare parts that afflicted London Transport in the mid-1970s, a batch of RTs was sent to Gillingham Street Garage at the beginning of August 1976, and used to cover for RM shortages on route 137, where they would remain well into 1977. On 29th March 1977, RT3799 (GM) loads up in Hampstead Road by Euston Tower as an RM overtakes it. Masked RM side-blinds were used in the front, side and rear on these RTs and this one's front number blind appears to have been torn!

Above: An oddity to be seen on route 137 at Hyde Park Corner on 2nd August 1987 is Dennis Dominator H3, one of three of this type bought by London Transport for evaluatory trials as possible replacements for Routemasters. Although no more of these were purchased, many Dennis low-floor double-deck buses would work in London from the late 1990s onwards. Meanwhile, Routemasters survived on route 137 until July 2004!

Right: Route 139 ran between Ilford and Dagenham and was never RM operated. However, following transfer of its RT operation to Romford (North Street) Garage, RMs based there for other routes made rare appearances on it. This has happened in February 1976, when RM139 (NS) is in Whalebone Lane, Chadwell Heath awaiting a crew change on 27th of that month. As is apparent, the RMs did not have via blinds for the 139, so they were left blank.

On the last evening of route 139's crew operation, Friday 22nd July 1977, RM1869 (NS) has been pressed into service on its rush hour extension to Dagenham Dock, a point where RMs never normally reached. Again with blank via blinds, it passes Gants Hill Station.

A new route 139 was introduced in February 1992, replacing the 159's northern section to West Hampstead and continuing to Golders Green. Originally operated by RMs from Chalk Farm Garage, it was transferred to Holloway when that garage closed in 1993. After that, RMLs began to appear on rare occasions. Recently refurbished RML2479 (HT) heads down Oxford Street on the evening of 28th January 1994.. Following privatisation, RMLs became more common on this route, until its OPO conversion in March 1998.

Route 140 should have converted to RM operation in the summer of 1962, but its operative crews refused to accept them on a 'nine for ten' basis replacing RTs. As if for Schadenfreude at the busmens' union's expense, the route, by then daily operated by Harrow Weald Garage, did not finally receive them until July 1978! However, in the early 1970s, Hendon Garage had a small Sunday allocation of RMs on the route, which was converted to RML when they took over routes 13 and 113 early in 1976. On 11th July that year, RML2383 (AE) sets off from Mill Hill Broadway Station for Heathrow Airport. Its offside advert is particularly appropriate for this long Routemaster!

141 - 175

Route 141 was introduced in November 1961 to replace the 641 trolleybus route between Winchmore Hill and Moorgate. However, initially on weekdays only, the route was extended along London Wall to St. Paul's from where it replaced tram replacement route 179 all the way down to Grove Park. Although it was allocated RMs, RTs from New Cross Garage frequently worked it, as occasionally did those from Wood Green. A late appearance of one of the former was on 5th November 1974. RT3395 (NX) overheats as it calls at a stop in Downham Way, and has a long way to go to reach Newington Green - will it make it? Perhaps it will be given a 'drink' at Catford and New Cross Garages which it has to pass on the way!

Two unscheduled RTs were allocated to Wood Green Garage in November 1976 to cover for RM shortages on route 141. On 11th March 1977, RT2506 (WN) is actually at Newington Green in the morning rush hour. A week later, these two RTs were switched to work route 29, which exchanged its crew DMs with the 141s RMs.

A very rare working on the 141 on 21st June 1977 is that of MD139 (NX), one of the Metro-Scanias allocated to New Cross Garage for route 53 at the time. It is substituting for a DM, and about to pass The Mildmay Tavern, where the eccentric record producer Joe Meek discovered The Honeycombs pop group in 1964, with whom he recorded his last No.1 hit - 'Have I The Right'.

In its last two or three years of crew operation, RMLs from New Cross's small allocation of them on route 37 found their way onto the 141. On 3rd September 1982, RML2461 (NX) leaves the garage on a short working to Moorgate from the south. Its via blind is incorrect, as it relates to the northern Wood Green to Moorgate section!

Upon the introduction of the 141 in November 1961, the 179 which it largely replaced continued to run from Grove Park to Finsbury Park at weekends, but was renumbered 141A. It was operated by Holloway RTs and New Cross RMs. RT substitutions for the latter were very rare indeed, but on its last day of operation, 28th August 1971, RT2304 (NX) has found its way onto the route, and is at Highbury Corner.

Following the rebuilding of Seven Kings Garage to allow RMs to operate there, those allocated to route 86 in February 1976 worked that garage's share of route 148 on Sundays until its conversion to OPO in July 1977. The route retained RTs from Barking that day, thus making the RMs working it, oddities. On 4th July 1976, RM2006 (AP) turns from Longbridge Road into Bennett's Castle Lane, Dagenham.

Route 149 was introduced with RMs in July 1961 to replace trolleybus route 649, being extended during the week from the latter's Liverpool Street terminus to Victoria. Crew DMs replaced the RMs in February 1974 but, as usual, were so unreliable that RMs from both Edmonton and Stamford Hill Garages had to cover for them. In addition, RMLs at the latter garage on route 243 also appeared on the route, as illustrated by RML2273 (SF) heading for home in Bressenden Place, Victoria on 25th July 1977.

The atrocious service record of the DMs working route 149 meant that forty former Green Line coach RCL-class Routemasters retrieved from London Country were overhauled to replace them on route 149 in the latter part of 1980. Sadly, their reign was short-lived, and RMs had supplanted them by the end of 1984. Shortly after this, Tottenham Garage gained a small Saturday RM allocation on the route, on which freshly repainted RML2626 (AR) has substituted for one when seen at Snell's Park, on the Edmonton/Tottenham border on 25th June 1986. It incorrectly shows the via blind (most of it, anyway!) for the section of route beyond Liverpool Street to Victoria. The 149 converted to OPO in February 1987.

As with route 148, RMs from Seven Kings Garage worked the 150 on Sundays until its OPO conversion. However, they never usually did so during the week, meaning they did not reach its weekday-only termini of Becontree Heath and Lambourne End. On Friday, 14th October 1977, its last day of crew operation, Seven Kings Garage's last RTs were being moved away, so RMs did reach those termini then. Thus newly overhauled RM802 (AP) is at 'The Beehive' pub, terminus of the 150's odd journeys to Lambourne End, awaiting departure for Becontree Heath.

Route 155 was extended westwards on Sundays to replace route 131, which had in turn been extended eastwards to replace the 604 trolleybus, the two routes usually meeting end-on at Wimbledon during the week. It was RM-operated on Sundays and Bank Holidays, but RTs from Norbiton Garage worked Bank Holiday extras. On Easter Monday, 23rd April 1973, RT344 (NB) passes a broken-down RM on the same route as it approaches Hampton Court. Three weeks later, this extension ceased upon the 131's conversion to OMO and introduction on Sundays, and the consequent discontinuation of the 155's Sunday extension. Route 155's daily allocation did not convert to RM until January 1977.

Although the 168A was never scheduled to be Routemaster operated, RMLs also appeared later on in its life. With a via blind meant for night route N93, as it did not have a correct display for the 168A, and driver Jon Batchelor at the wheel, RML2696 (HT) calls at the site of the old Hornsey Road Station on 6th October 1980. The route had by now been diverted at Holloway, Nag's Head to run to Hornsey Rise, but was withdrawn on 4th September 1982.

Despite being a trolleybus replacement route for the 691, route 169 was not allocated RMs, since surplus RTs were used instead, working from Barking and later Seven Kings Garages, Ilford Trolleybus Depot having been closed. However, when Barking Garage received RMs for route 23 in March 1964, these were used on the 169 alongside Seven Kings RTs at weekends, though latterly only on Saturdays. RM1794 (BK) is seen in Fanshawe Avenue, Barking on 20th June 1970, shortly before Barking lost its RMs. The 169 converted to DMS OMO in September 1973, without ever having a daily RM allocation.

Tram replacement route 171 was operated by RTLs and RTs until its conversion to RM operation in September 1968. At that time, its southern allocation was transferred from Walworth to New Cross Garage, who often put out RTs on the route, which had never been scheduled to work it from that garage before. On 5th July 1973, RT4782 (NX) has been brought out to replace an RM that has broken down in the morning rush hour, which fitters attend to in the background. The buses are standing beside Finsbury Town Hall rather than on the usual Rosebery Avenue stand nearby owing to building works obstructing the latter.

The first of many unscheduled RTs allocated to various garages to cover RM shortages in the mid-1970s was RT3925 (AR), which appeared at Tottenham Garage in late December 1973 working route 171. On 29/12/73, it calls at Queens Road, Peckham Station on its way back from Forest Hill to Tottenham. Seeing it unexpectedly the previous evening nearly gave me a seizure; fortunately I lived just by the route in Canonbury at the time, so was able to hunt it down next day - successfully, as I used an entire reel of film on it! In this view, it is remarkable how little traffic there is on this main road between New Cross and Peckham!

RMLs were supposedly allocated to the 171 for a while on Sundays in the mid-1970s from New Cross Garage, though their actual appearance was erratic. However, in later years, those from the small allocation for the 37 occasionally appeared on all days of the week. On 20th April 1984, RML2580 (NX) approaches Essex Road Station on its way south.

The 172 was also a tram replacement route, initially also worked by RTs and RTLs. RMs worked it at weekends in the early 1970s, but the route converted to crew DM in August 1975. These were, as usual, unsuccessful so were replaced daily by RMs early in 1981. From then until the withdrawal of the 172's Holloway allocation in September 1982 when the route was diverted to terminate at Kings Cross, RMLs from that garage made rare appearances. On 20th April 1982, RML2391 (HT) is in Offord Road, Barnsbury on the route's 'backstreet' section around which it had been diverted in March 1971.

Although it was converted from RT to RM operation in 1966, route 174 often had RT's working it from Romford, North Street Garage until their removal from there in 1977. On 22nd August 1973, RT4430 (NS) is one of two RTs working Express journeys on the route at its Harold Hill, Gooshays Drive terminus.

Just a week before RTs were finally withdrawn from Romford, North Street Garage, RT2127 (NS) is at the Dagenham New Road terminus of route 174, setting off for Noak Hill, Tees Drive on 16th July 1977.

Route 175 had the unusual distinction of converting from RT to RMA operation, using thirteen former British Airways forward-entrance RMs, in October 1975. They were replaced by RTs eleven months later because their internal layout was unpopular with crews, especially since there was no conductor's 'cubby hole' under the stairs as on RTs and RM/RMLs. Standard RMs in turn took the route over in March 1977. However, during the RMAs' tenure of the 175, not only did RTs still appear on occasion, so also did standard RMs, albeit rarely. On 9th April 1976, RM468 (NS) calls at Collier Row on its way from Chase Cross to Dagenham. It wrongly shows a via blind for route 175A, perhaps because no such blind was then available for the 175. RMs eventually worked route 175 for five years between March 1977 and the spring of 1982.

176 - 234

Route 176 had the distinction of not only being the one of the last two routes to operate RTLs, but also of being the last to have a full RT allocation running through the West End. After the RTs were replaced by RMs in the spring of 1976, a number of RTs were kept at Walworth Garage as spares for RM shortages until late in 1977. On 11th October 1976, RT2794 (WL) is at Charing Cross. The work going on in the background is in connection with the new Jubilee Line and the combination of Strand and Trafalgar Square tube stations, which were renamed Charing Cross.

Oddly, route 180 was converted from RT to RM operation in the autumn of 1968, being the first outer suburban route in South East London to receive them. However, RTs continued to appear on it, those from its Catford allocation doing so until the very end. On 26th July 1978, RT4210 (TL), complete with a crumpled via blind, arrives at Lower Sydenham Station a month before RTs finally left this part of London.

Sunday-only route 180A was introduced in January 1972, replacing routes 171A and 180 that day, and withdrawn in April 1978. It was always RM-operated, but on Good Friday 28th March 1975, RT2296 (AW) is working the route, and is in Woolwich Road, Charlton, very close to where London's last trams were scrapped in Penhall Road.

The vehicle spares shortage in the mid-1970s caused route 190 at South Croydon Garage to be turned over almost entirely to Leyland Titan PD3's loaned from Southend Corporation for a few months during the autumn and winter of 1975/76! The majority were Massey-bodied examples dating from 1965. On 9th October 1975, one of these, Southend No.338, is at the Old Coulsdon, Tudor Rose terminus.

A handful of newer, 1967-built East Lancs-bodied PD3s also worked route 190. No.346 is working short to Coulsdon, Red Lion as it runs along London Road, Coulsdon on the same day.

Not long after the Southend PD3's had left route 190, and in fact been loaned to London Country at Harlow Garage for route 339, a sole RT was sent to South Croydon Garage to help out the RMs which had returned to the route. RT4154 (TC) is approaching Thornton Heath Pond on 21st May 1976. Unlike the other rogue RTs at this period which carried RM side blinds in their front via boxes, this one had a masked RM front via blind.

For many years, route 191 was RT-operated from Ponders End Garage prior to its DMS conversion in the spring of 1974. However, for a few years, RTLs from Tottenham Garage had a small Saturday allocation on the route. Their RTL1558 (AR) approaches Edmonton, Regal on 16th December 1967. RTs replaced these RTLs upon the 171's conversion to that type six months later, but lasted only until 6th September 1968.

Upon the conversion of route 171 to RM operation in September 1968, these also worked Tottenham's Saturday share of route 191. On 19th December 1970, RM1527 (AR) crosses the level crossing on the former Great Eastern Chingford Branch at Chingford Hatch, four weeks before this allocation was withdrawn. A couple of years later, the level crossing was replaced by a bridge, taking route 191 beneath the railway instead.

Though route 191 would never have a full, daily allocation of RMs, on rare occasions those from Ponders End Garage's small allocation for routes 127 and 279 substituted for RTs on the route. RM268 (E) calls at Chingford Mount, Prince Albert, on 15th December 1973 to illustrate this.

Trolleybus replacement route 193 had to wait until October 1976 before it gained Routemasters! This was because, upon the closure of Ilford Depot, the route was allocated to Seven Kings Garage with RTs in August 1959, and that garage could not accommodate anything larger than them until it was rebuilt in the mid-1970s. By the time of its RM conversion, the 193 ran from Barking, Blake's Corner to Hornchurch, White Hart, but in March 1977 was extended through 'new territory' to Hacton Lane Estate. RTs were never scheduled to work to that point, but on 22nd July 1977, RT4625 (AP) stands in for an RM at the new terminus. That same day was the last day of RT operation on route 148 from Seven Kings Garage, and just under three months later, the last examples there were replaced by DMSs on route 150.

Following tram replacement route 196's withdrawal north of Euston, and conversion to RM operation in March 1971, RT's appeared on it very rarely indeed. However, on 14th November 1973, RT4702 (Q) leaves Waterloo in the morning rush hour on a short journey to Bloomsbury. It sports a rarely seen RT via blind reflecting its 1971 cutback. In January 1974, the route was rerouted north of Herne Hill to terminate at Brixton, and also reallocated to Norwood Garage, meaning that RTs never worked it again.

Traditionally operated by Hanwell Garage, trolleybus replacement route 207 also had a small allocation of RMs (and later RMLs) at Uxbridge Garage. For a while, RTs from there were used on Sundays, but by the time RT2760 (UX) was seen on 29th February 1972 leaving Shepherd's Bush Green for the long trek along the Uxbridge Road, they only appeared as substitutes for Routemasters. Uxbridge lost its last RTs in the summer of that year.

Although they comprised virtually the entire weekday allocation of route 210 from Muswell Hill Garage until they were gradually repainted red, a batch of green Country Area RFs allocated there at the beginning of 1969 were in effect 'Strange Visitors', especially as the route ran so close to Central London. This also recalled the green Q-type single-deckers that I remember on the route shortly before it gained the first red RFs in 1952. On 4th January 1969, RF598 (MH) approaches the famous Jack Straw's Castle pub in the middle of Hampstead Heath. Alas, this London landmark has in recent years been converted into apartments! The green RFs were surplus to requirements in the Country Area and, in effect, provided a float for red ones used to convert a number of routes from RT to RF operation in 1970/71; most of these RFs also had to be converted to OMO.

Moving ahead to 28th September 1986, route 210 by now was operated by Leyland Nationals, but for some reason at least three BL-class E.C.W.-bodied Bristol LHs were out on the route! These usually operated route C11 from Holloway Garage. BL84 (HT) descends the steep Highgate Hill.

Although RMs were scheduled to work route 213 from Norbiton Garage on Sundays between 1966 and 1970, they were not when RM1238 (NB) was seen passing through Sutton on 16th June 1972 - which in any case was a Friday! The RM is therefore obviously substituting for an RT. A Country RT on route 408 follows.

In common with the 213, route 213A was shared between Sutton and Norbiton Garages, the latter providing RMs on Sundays only between March 1969 and April 1970. On 5th of the latter month, RM1219 (NB) passes beneath the bridge at Worcester Park Station which once enforced RFs on the 213 group of routes and RLHs on the 127. A 4SUB Southern Region EMU completes the picture, whilst this RM allocation ceased a fortnight later.

Route 214 was introduced with RMs in February 1961 as a replacement for trolleybus route 615. Later in the 1960s, a small allocation of RTLs, supplanted after two years by RTs, was given to Highgate Garage on route 196, and these made odd appearances subbing for RMs on the 214. On 10th June 1970, RT1756 (HT) passes the famous Dawson Brothers department store in City Road on its way to Moorgate. The garage lost its RTs in March 1971, but gained many more six months later when it inherited the allocation from the old Holloway (J) Garage. However, these did not have blind displays for route 214.

Along with several other routes at Holloway, as the former Highgate Garage was renamed upon the closure of the old Holloway (J) Garage, the 214 converted from RM to crew DM operation in 1975, after which use of RMs on it was rare. Unlike most of the others, it did not convert back to RM in 1981, instead going DMS OPO. However, on New Year's Day 1978, RM979 (HT) also passes Dawson's in City Road, but in the opposite direction. The in-house LT adverts either side of its front blinds clearly date this picture to that period.

Route 216 converted from RF to BL operation in the summer of 1976, but after that, following the recertification and repaint of RFs at Kingston Garage to keep routes 218 and 219 going in 1977/78, they sometimes substituted for BLs on the 216. On 30th March 1979, the very last day of red RF operation, RF522 (K) accompanies BL34 at Staines, Moor Lane terminus. The fact that this RF has the 'solid roundel' livery proves this view to date from after RFs had been replaced by BLs on the route.

Route 217 was RT-operated by Ponders End Garage for many years, and when its variant the 217B was introduced in April 1976, all RTs and RMs at the garage were given new blinds which allowed them to work one another's routes. This led to RMs appearing on odd occasions as substitutes for RTs; never having been scheduled to work the 217. Whether it is a coincidence that newly-overhauled RM217 (E) worked the route with the same number on 16th August 1977, four days before its OPO conversion, I cannot say! It is on the Great Cambridge Road, just north of the junction with Southbury Road. A close look shows it has no canopy number blind, and has a rear number-only blind above the platform.

On the same day as the previous picture, 16th August 1977, RM2210 (E) approaches the junction of Bullsmoor Lane and the Great Cambridge Road on the new 217B, which also converted to OPO four days later. This one is fully-blinded for the route.

The 221 was another trolleybus replacement route in North London, covering the 521 and 621 in November 1961. Introduced with RMs from Finchley Garage, it theoretically had a solitary RT allocated on Sundays during 1967/68, though more often than not, RMs were used anyway. However, on Sunday 1st September 1968, RT965 (FY) was fielded, and is seen leaving the Finsbury Park, Astoria stand. This RT allocation was withdrawn the following Saturday.

Another theoretical Sunday part-allocation on route 221 was that of Wood Green RMLs, which worked a small share of route 243 during the week between June 1970 and January 1971. Once again, RMs were usually used anyway, but on 18th October 1970, RML2503 (WN) stands at a foggy North Finchley, Tally Ho! Bus Station. Curiously, RMLs from Finchley Garage did not work the route, probably owing to stringent Union rules that restricted them to routes 104 and 104A only.

An odd case of a solitary green Country Area bus working a red Central Area route was that of RLH27 being sent to Harrow Weald Garage in the summer of 1968 for route 230. It is seen shortly after its arrival there at Rayners Lane Station terminus on 5th July 1968. Interestingly, it still carries adverts for new Green Line route 727 which had been introduced over a year previously - the bus had been in store at Garston most of that year. It had been kept as a spare - just as well, since a red RLH at Dalston Garage where they were needed for route 178 was deroofed! Curiously, instead of sending the green RLH to Dalston, one of Harrow Weald's red ones went, leaving that garage short! In the end, this green RLH remained there until the 230 was replaced by MBSs on route H1 in June 1969.

New route 230, introduced with RMs in June 1973, had never been RT-operated, but at the end of November 1976, at the height of the vehicle spares shortage, a batch of rogue RTs was sent to Leyton Garage, in effect taking over the entire route. Instead of having RM side-blinds in their front via blind boxes, as most rogue RTs did, they had carefully-masked RM front blinds; their side-blinds were masked, too as seen here on RT2554 (T) at Bakers Arms, Leyton on 23rd August 1977, a couple of weeks before RMs returned to the route.

During its period of operation by OMO XAs, between April 1970 and early 1973, route 234 was also occasionally worked by the lone FRM1 (TC), which is seen here at Purley Cross on 15th July 1972. This outer suburban route was a far cry from when it worked route 76 in the City and inner north London!

237 - 266

Route 237 had an odd history in the late 1970s and 1980s, in that it had run from Hounslow to Chertsey with RFs, which were replaced by BLs in April 1977. However, in January 1978, it was withdrawn beyond Sunbury Village, but extended all the way from Hounslow to Shepherd's Bush, replacing route 117, and converted to RM operation! Subsequently, the route had visits from RMLs based at Hounslow for route 37. On 9th May 1986, one of these, RML2742 (AV) escorts a Leyland National on route 117 at Ilseworth Station, as a Class 37 diesel noses its way in on the bridge above.

'New' route 239 was introduced in March 1971 replacing the northern end of the 196, using SMSs. The original route 239 had in fact itself been replaced by the 196 in 1951! It graduated to DMS and then LS operation in the 1970s, officially using DMSs again in its last year or so before withdrawal in September 1982. In practice, however, DMSs were still used when the route was officially LS-operated, as shown by DMS837 (HT) at St. Pancras on 6th May 1981.

Route 241 was introduced with RMs on 7th September 1968, replacing parts of route 123 and 41, running from Manor House to Victoria & Albert Docks. On rare occasions, RTs from West Ham Garage appeared as substitutions on it until their removal from the garage in January 1972. Just six weeks after its introduction, RT358 (WH) is at the Manor House terminus on 15th October 1968.

The 241 converted to DMS operation in June 1973, however due to the dreadful unreliability of these vehicles, SMSs based there for route S3 occasionally had to stand in for them. SMS603 (WH) is seen at Plaistow Broadway on 14th July 1979.

Another North London trolleybus replacement route was the 243, which replaced the 543 and 643 in July 1961 using RMs. It was upgraded to RML in June 1970. Following their appearance at Stamford Hill Garage in August 1980 replacing crew DMs on route 149, resurrected ex-Green Line and London Country RCLs often appeared on the 243 until their all-too-hasty withdrawal in 1984. A gleaming RCL2240 (SF) arrives at Wood Green, Spouters Corner on 11th August 1980, the first day they did so.

Similarly, though more rarely, Sunday-only route 243A also had guest appearances by RCLs. On 17th August 1980, the second Sunday this could happen, RCL2246 (SF) passes beneath what was then the Kentish Town to Barking Line at South Tottenham Station. It was in fact one of two RCLs working the route that day! Also it is unusually turning short at Tottenham, Bruce Grove rather than going through to Wood Green.

Prior to its final demise in the spring of 1982, when LSs from Muswell Hill Garage were allocated to route 244 on Mondays to Fridays, and Metrobuses on Saturdays, DMs based there for route crew-operated route 134 were used instead. On 5th March 1982, a few weeks before the end, DM1726 (MH) sets off from Muswell Hill Broadway for the short journey to Southgate.

Although route 248 needed low-height RLHs to negotiate the low bridge east of Upminster Station, the one-bus 248A did not have such a restriction despite being allocated them. On the routes' last day of crew operation, RT3261 (RD) worked the route and is seen leaving Corbets Tey for the short run to Upminster. Quite why its driver has wound up the destination of Blackwall Tunnel (served by RTs at the same garage working route 175) is anyone's guess - to annoy the enthusiasts, maybe?

Nicknamed 'The Yiddisher Flyer' by crews and passengers alike due to its linking the Jewish communities of the East End with those in Stamford Hill, route 253 replaced the 653 trolleybus in February 1961. From 1975 onwards, DMs from Holloway Garage and RMLs from Stamford Hill were scheduled to work it on Sundays, though their appearance was sporadic. Conversely, RMLs from both Stamford Hill and, after April 1981, Ash Grove Garage appeared during the week. On 24th August 1975, RML2700 (SF) calls at the then-closed Hackney Station.

Loading outside Clapton Garage in Mare Street on 13th May 1979, DM1732 (SF) is actually a Stamford Hill bus not scheduled at all on route 253! Appearances of these when they were supposed to be working route 149 were in fact quite common, and even after its conversion to RCL they were kept at the garage as spares for a year or two.

Inevitably, the RCLs allocated to Stamford Hill Garage for route 149 often found their way onto the 253 in place of RMs. However, on 22nd April 1983, RCL2246 (SF) has let the side down by breaking down in Camden Road, Holloway, where it awaits rescue by fitters. In the event, it never ran again, as by now there were sufficient spare RMs to replace the non-standard RCLs.

Until the very end of crew operation on the 253, RMLs from Ash Grove Garage subbed for RMs, and a good few RMLs had been allocated to Stamford Hill already for it, too. On 14th November 1987, a week before its OPO conversion, RML2301 (AG) passes over Clapton Common heading for its home garage.

Following the allocation of RTs for routes 34 and 144 to Walthamstow Garage in 1964 when Leyton Garage was being enlarged, these could appear on trolleybus replacement routes there hitherto only worked by RMs and in fact would remain there until as late as September 1977. Route 257 was one, and RT2707 (WW) is seen working it at Clapton Pond on 11th July 1967. The 257, however, was withdrawn upon 'Reshaping' on 7th September 1968.

Above: Trolleybus replacement route 259 was introduced with RMs in April 1961, and following a part allocation on the route being given to Tottenham Garage, RTWs and RTLs occasionally appeared as substitutes. It was converted to DMS OMO in February 1973, and immediately prior to their replacement by new Metrobuses in the spring of 1982, converted former crew-operated DMs, now redefined Ds, could be found on the route. D1809 (AR) is on the southbound side of the Holloway, Nag's Head one-way system on 13th May 1982. By this time, the Edmonton share of the route was already operated by Metrobuses, which soon replaced the Ds at Tottenham.

Right: The second route to be numbered 260 was yet another trolleybus replacement route, introduced with RMs in January 1962 to replace the 660, and at first both Hendon and Cricklewood RTs had small allocations on Sundays during that year. Subsequently, RMLs from Stonebridge Park Garage's allocation on route 18 worked on Sundays for a few months in 1978/79. However, these also found their way onto the route during the week, as depicted by RML2743 (SE) in a wet Willesden High Road on 16th September 1975.

In its final years of crew operation, route 260 had unscheduled visits from Willesden RMLs, too. On 25th April 1984, RML2681 (AC) speeds along Finchley Road, Golders Green.

Route 262 was introduced with RMLs at the end of 1966, replacing parts of routes 26 and 249A, and operated by Leyton and West Ham Garages. Initially, Leyton did not have any RMs, so had to use RTs as RML substitutes. RT1760 (T) is in Greengate Street, Plaistow on 18th May 1968.

Trolleybus replacement route 266, covering the former 666, was another operated by Stonebridge Park Garage, whose small allocation of RTLs on route 112 strayed onto it on rare occasions. On 11th July 1967, RTL1599 (SE) storms along Edwgare Road through Cricklewood, working short to West Hendon.

Subsequently, Riverside Garage had a hand in operating route 266, and on rare occasions until RTs left that garage in December 1975, they found their way onto it. On 30th December 1974, RT4411 (R) keeps a DMS company at Hammersmith Grove terminus.

Between 1975 and 1979, both Willesden and Stonebridge Park RMLs were allocated to route 266 on Sundays. On 16th April 1979, a week before they were replaced on route 18 by crew DMs, RML2357 (SE) is heading along a very deserted Kings Street, Hammersmith.

The crew DMs on route 18 also worked route 266 on Sundays. On 5th August 1979, DM1150 (SE) is at Starch Green. This type of bus never worked the route daily - when it went OPO in the mid-1980s, it converted to Metrobus operation.

For the last few months before the 266's OPO conversion, RMLs allocated to Cricklewood Garage for routes 16 and 16A were sometimes used on the route instead of RMs during the week. On 28th April 1984, RML2463 (W) leaves Brent Cross Shopping Centre. This RML is preserved today.

271 - W8

Replacing trolleybus route 611 in July 1960, RM-operated route 271 passed my original home in Canonbury Avenue. On New Year's Day 1966, the experimental XA-class Leyland Atlanteans replaced them, then for a few weeks in the spring of that year, the eight Country Area XFs, still in green livery, were also allocated for comparative trials. On 10th May 1966, XF6 (HT) waits at the traffic lights by Canonbury Square, just a few seconds from my home. RMLs took the 271 over in June 1966, remaining on it until replaced by OMO DMSs in January 1971.

Less exotic were strange visitors on route 271 in the 1980s. Such was the absymal performance of the DMSs that replaced Routemasters on it in January 1971 that Leyland Nationals from route 210 had to stand in for them. In April 1981, LS136 (HT) is seen doing so at Essex Road Station. Even when Metrobuses had replaced DMS, LSs still occasionally stood in for them as late as the early 1990s.

Route 273 was introduced with RTs in November 1968, running between Ealing Broadway and Ruislip, and later received an extension to Hayes Station, replacing route 98A. In January 1973, it converted to SMS OMO. By late 1979, these troublesome, under-powered single-deckers were on their last legs, so some single-door SMs which had in the meantime been overhauled were sent to Southall Garage to help them out. One of these, SM14 (HW) is clearly having difficulty with crowds wishing to board at Greenford, Red Lion in August 1980. Soon afterwards, the route was double-decked with new Metrobuses, but has since disappeared.

An oddity to be seen at the Stoke Newington Common terminus of route 276, which by now had replaced route S3 through the Hackney Wick area where low bridges enforced single-deck operation, is LS53 (WH), carrying the special livery for Docklands Clipper route D1. The date is 8th February 1986, thirty years to the day after the first Routemaster entered service. Those in the background are on route 73, which will retain them until July 2004!

Route 279 replaced trolleybus 679 in April 1961, and remained staunchly RM-operated for the next twenty-five years. However upon the introduction of new RT-operated route 217B at Ponders End Garage in April 1976, all RMs and RTs there were given compatible blinds meaning that RTs could be used on their small share of the 279, the majority of which was operated by Edmonton Garage. On a very hot 26th June 1976, RT739 (E) heads south from Waltham Cross. Its shabby appearance is typical of those at Ponders End at the time, as their RTs were usually left outside in the yard, whilst their OPO buses were kept under cover! The M25 passes beneath this location today.

Crew-operated DMs from Edmonton Garage for route 149 occasionally stood in for RMs on route 279, but a decidedly odd appearance in July 1980 is that of DMS780 (EM), working as a crew-operated bus, at Scotland Green, Tottenham. By the time the 279 converted to OPO in September 1987, the DM-types had long gone from this area, and it gained Metrobuses instead.

Upon the appearance of RCLs at Edmonton Garage on route 149 in the autumn of 1980, they inevitably found their way onto the 279, substituting for RMs. On 15th February 1984, RCL2260 (EM) is in Islington High Street, after being adopted as the garage's showbus and adorned in a 'red' version of its original Green Line livery, also regaining its original twin headlamps. Fortunately, this RCL survives today in the heritage fleet of The London Bus Company.

RCLs found their way onto the Sunday 279A, too. This chance shot finds two of them, RCL2231 and RCL2226 (EM) passing outside Islington Town Hall on 21st December 1980. This view clearly shows the rear-end arrangement of the RCLs with emergency door. However their folding platform doors were removed when they were overhauled as red buses.

Crew-operated DMs rarely found their way onto the 279A, too. On the same occasion also at Islington Town Hall is DM983 (EM). The route had originally been operated by OMO DMSs, but was converted to crew operation using RMs in October 1978. These remained its official allocation until OPO Metrobuses ousted them on this and the weekday 279 in September 1987.

Route 298 was introduced using RTs on 7th September 1968, replacing the northern section of route 29 (and the localised journeys numbered 29B) between Turnpike Lane and South Mimms. It was converted to RM operation in January 1970, and withdrawn north of Cockfosters, but at the same time extended south to Finsbury Park Station. Until their final removal from Palmers Green Garage in the spring of 1978, RTs often substituted for RMs. On 21st March 1973, RT4295 (AD) arrives at Finsbury Park Station, on the section of route 298 RTs were never scheduled to work. It contrasts with a new DMS on route 259 going in the other direction.

Similarly, route 29A which ran from Turnpike Lane to Oakwood Station was in effect renumbered 298A on 7th September 1968. It too converted to RM in January 1970, but Palmers Green also fielded RTs on it until their demise there. On 12th November 1977, RT921 (AD) is in Wood Green High Road. This RT had just been transferred from Holloway where it had been for the past year as an unscheduled RM spare, hence the masks still in its front via box. This, along with its LT in-house front adverts and the DMS following, leave no doubt that this picture was taken long after RTs officially worked the 298A!

Left: In the last year or so of the MBAs' disastrous reign on Red Arrow services, and despite most of the survivors having had an Aldenham overhaul, SMSs had to help them out. On 22nd April 1981, SMS739 (GM) passes an RM on their short-lived allocation on route 168 in Parliament Street.

Below: A disastrous fire on the Victoria Line at Oxford Circus Station late in 1984 caused extra buses to be allocated to pioneer Red Arrow route 500 to cover for the line's closure. On 14th December 1984, Metrobus M401 (GM) has been working it complete with a conductor, and passes Hyde Park Corner on its way back to Gillingham Street Garage.

The Forester was a limited-stop route operated by London Buses' Forest District, numbered X99 running between Romford and Basildon, beyond the former London Transport area. It was not a success and Leyland Nationals carrying its dedicated livery worked other services. On 23rd October 1988, LS71 (T) is at Epping Station on London Buses' Sunday working of route 500, which ran via the former Country Area route 339 between Harlow and Epping, and then to Romford via the former 250 (later 247) route.

As with route 500, SMSs also helped out ailing MBAs on route 501. SMS720 (WL) passes an RML on route 8 at Holborn Circus on 17th September 1979.

The 502 had SMSs as well. SMS700 (H) sets off from Waterloo for Liverpool Street on 18th August 1980. This SMS has had a repaint into 'solid roundel' livery, whereas the two previous examples were still in original livery and 'working their ticket' until their original seven year certificates of fitness expired, and they were withdrawn.

On the same day as the previous picture was taken, 18th August 1980, a very shabby SMS744 (GM) is on its last legs when seen standing in for defunct MBAs on route 503 in York Road, Waterloo.

Left: The horribly dented front of SMS708 (GM), shown working route 507 at Lambeth Palace on 2nd April 1981, says it all about this unfortunate class!

Below: Yet another Red Arrow route on which SMSs helped out was the 513. On 14th April 1981, SMS327 (WL), also complete with dents despite having received the later livery, passes the Law Courts in Fleet Street.

Route C11 had originally been one of the first four London Transport minibus routes, introduced in September 1971 and running from Archway Station to Cricklewood along previously unserved roads. FS-class Ford Transits were used initially, and on 25th January 1975, Bedford Dormobile demonstrator SGC385M arrives at Archway Station while being tried out on the route. Nothing came of this, though further FSs with similar bodywork were delivered.

In the event, the FSs on route C11 were replaced by BS-class ECW-bodied Bristol LHSs in the summer of 1975. These in turn gave way to larger BL-class full-size single-deckers of the same make in the spring of 1981. However, on 2nd July 1981, BS2 (HT) is still helping out on the route, and makes an interesting comparison with BL67 (HT) behind it at Archway Station.

In September 1982, route C11 was extended from Archway to Kings Cross to replace route 239. On 23rd May 1985, another case of a route-branded vehicle working an inappropriate route is that of BL95 (HT), fully adorned in the special livery of the Hillingdon Local Service 128, it approaches Archway Station heading for Kings Cross.

By July 1990, the former minibus route C11 had become so busy that Metrobuses sometimes worked it - a far cry from the Ford Transits that inaugurated it. On 20th July 1990, M779 (HT) passes BL86 (HT) as it runs around the block at Kings Cross. Later on the C11 would be cut back again to Archway and replaced by Routemaster-operated route 10 on that section, which is today covered by the so-called 'New Routemasters' on route 390.

An oddity to be found on OPO DMS-operated flat fare route W7 during 1981/82 was DM2582 (MH), which was actually a crew-operated bus adapted for operation in either crew or OPO mode. This view outside Hornsey Town Hall in Crouch End Broadway on 6th January 1981 clearly shows it has been fitted with the flap to instruct passengers to pay either the driver or conductor, but still has 'pay as you enter - exact fare please' painted on the front. It later was used on Muswell Hill Garage's OPO routes 244 and W7, and crew routes 43 and 134, and many DM and DMS-type vehicles were subsequently converted for dual operation in the same way, usually being reclassified as Ds.

Illustrating the cheapskate policies of London Regional Transport during the Thatcher years, route W8 at Potters Bar Garage was operated by a variety of secondhand Volvo Ailsas in the late 1980s and early 1990s. However an oddity to appear on the route on 2nd November 1988 is LS175 (PB). It carries the livery of semi-autonomous LRT susbsidiary Harrow Buses, with names painted out and is shown at Bush Hill Park Library.